Environmental Communication
and the *Public Sphere*

4
EDITION

For Niko, Cruz, Michelle, Harrison, Daniel, Keenan, and Cam.
May the work of those reported here and those reading this book leave
a more sustainable and just world for you.

SAGE was founded in 1965 by Sara Miller McCune to support the dissemination of usable knowledge by publishing innovative and high-quality research and teaching content. Today, we publish more than 750 journals, including those of more than 300 learned societies, more than 800 new books per year, and a growing range of library products including archives, data, case studies, reports, conference highlights, and video. SAGE remains majority-owned by our founder, and after Sara's lifetime will become owned by a charitable trust that secures our continued independence.

Los Angeles | London | Washington DC | New Delhi | Singapore | Boston

Robert Cox

*University of North
Carolina at Chapel Hill*

Phaedra C. Pezzullo

University of Colorado at Boulder

Environmental Communication
and the *Public Sphere*

EDITION

Los Angeles | London | New Delhi
Singapore | Washington DC | Boston

Los Angeles | London | New Delhi
Singapore | Washington DC | Boston

FOR INFORMATION:

SAGE Publications, Inc.
2455 Teller Road
Thousand Oaks, California 91320
E-mail: order@sagepub.com

SAGE Publications Ltd.
1 Oliver's Yard
55 City Road
London EC1Y 1SP
United Kingdom

SAGE Publications India Pvt. Ltd.
B 1/I 1 Mohan Cooperative Industrial Area
Mathura Road, New Delhi 110 044
India

SAGE Publications Asia-Pacific Pte. Ltd.
3 Church Street
#10-04 Samsung Hub
Singapore 049483

Acquisitions Editor: Matthew Byrnie
Editorial Assistant: Janae Masnovi
Production Editor: Jane Haenel
Copy Editor: Katharine Blankenship
Typesetter: C&M Digitals (P) Ltd.
Proofreader: Kristin Bergstad
Indexer: Karen Wiley
Cover Designer: Candice Harman
Marketing Manager: Ashlee Blunk

Printed in the United States of America

Library of Congress Cataloging-in-Publication Data

Cox, J. Robert.
Environmental communication and the public sphere / Robert Cox, Phaedra C. Pezzullo. — 4th edition.

pages cm
Includes bibliographical references and index.

ISBN 978-1-4833-4433-1 (pbk. : alk. paper)
1. Communication in the environmental sciences—Textbooks. 2. Mass media and the environment—Textbooks. I. Pezzullo, Phaedra C. II. Title.

GE25.C69 2016
333.7201′4—dc23   2015006459

This book is printed on acid-free paper.

16 17 18 19 10 9 8 7 6 5 4 3 2

Brief Contents

PART V: CITIZEN VOICES AND ENVIRONMENTAL FORUMS

Detailed Contents

PART IV: ENVIRONMENTAL CAMPAIGNS AND MOVEMENTS

Preface for the Fourth Edition

C oncerns about the environment, and the ways we communicate those concerns, continue to change since earlier editions of this book: For example, even as more traditional media—newspapers and broadcast TV—decline, environmental news proliferates online at sites such as Grist (http://grist.org/) and Environmental Health News (www.environmentalhealthnews.org/). Social media are enabling users to report, tag, and distribute environmental content widely, such as through the #ecoMonday trending topic on Twitter, and activists are mobilizing citizens globally through social networking sites like Greenpeace International's multimedia resources (www.greenpeace.org/international/en/multimedia/), as well as 350. org and Avaaz.org, to call attention to the worsening impacts of a range of topics, including fossil fuels, e-waste, and climate change.

Our knowledge of the many forms of environmental communication also continues to grow. The fourth edition of *Environmental Communication and the Public Sphere* gives us the opportunity to share these new developments—digital media and environmental activism, visual communication and popular culture, climate change communication, sustainability and the "greening" of campuses and corporations, and more. This edition also explores recent controversies—North American conflicts over "fracking" for natural gas, World Bank lending for fossil fuel projects, global struggles for public access to clean water, and more—to illustrate principles of environmental communication.

A book attempting to introduce such a wide range of communication about the environment could not have been conceived, initially, or revised for this fourth edition without the help of many of our colleagues, students, and friends, nor without the many helpful suggestions from colleagues with the International Environmental Communication Association and various environmental and civil society groups whose work we admire. And always, we thank our students who have inspired us every day with their intelligence, dedication, and passion for a better world.

The following reviewers are gratefully acknowledged:

Alexa Dare, *University of Portland*

Larry A. Erbert, *University of Colorado Denver*

Jennifer Follis, *University of Illinois at Urbana-Champaign*

Paul Hanna, *University of Brighton*

Anders Hansen, *University of Leicester, UK*

Joy L. Hart, *University of Louisville*

Deborah A. Wertanen, *Minneapolis Community and Technical College*

At SAGE, our thanks go to our editor Matt Byrnie and assistant editors Gabrielle Piccininni and Janae Masnovi for their help in this edition, to Jane Haenel for her skillful work as production director, and also to Katharine Blankenship for her careful eye in saving us from many embarrassing errors in the text itself. Although we have benefited from the suggestions of many, we are clearly responsible for any mistakes that have found their way into the text.

Finally, none of this would be possible without our wonderful partners, Julia Wood and Ted Striphas. Thank you for your support, patience, insights, humor, and judicious criticisms of earlier drafts of this book.

Is the environment silent? Who speaks for (or about) the environment? What constitutes an *environmental problem?* How does seeing, listening to, and hanging out in different environments move you?

Introduction: Speaking for/ About the Environment

Communication about the environment is occurring all around us and seeming more urgent than ever—from news of toxic chemicals in children's toys to warnings that, due to human influence on Earth's atmosphere, some are calling our geological era the "Anthropocene" or new "human" epoch (Stromberg, 2013). And, as we write, climate scientists are warning that by 2047, "the coldest year in the future will be warmer than the hottest year in the past" (Gillis, 2013, p. A9). Such communication cuts across all media—cable TV news, social media, YouTube, newspapers, public rallies, and in classrooms. Showtime TV episodes like *Years of Living Dangerously* and documentary films like *Gasland 2* portray threats to the environment, as well as its wonders.

The ways we learn and talk about the environment are also changing. Online sites and popular blogs like *Treehugger* (treehugger.com) and *HuffPost Green* (huffington post.com/green) showcase breaking environmental news and video daily. And while individuals still speak at public hearings about pollution in their communities, others are organizing through social networking sites like 350.org to address the harmful consequences of climate change.

The importance and sense of urgency about environmental problems have also invited interest in the field of environmental communication, the subject of this book. But why *communication*? As we'll explain in the following pages, our understanding of the environment and our ability to work together can't be separated from *the need to communicate with others*. Indeed, our language, visual images, and ways of interacting with others influence our most basic perceptions of the world and what we understand to be an environmental "problem" itself.

But we do not speak alone. As we'll see throughout this book, many different voices claim to speak for or about the environment. The public sphere is filled with competing visions, agendas, and styles of speaking. It is these different voices, media, and forums that influence our understanding and relationships with the environment that we'll explore in this book.

Communication and the Environment's Meaning

Not everyone sees herself or himself as an "environmentalist" or envisions being an environmental communication professional such as a journalist, science educator, or filmmaker. Some of you may be reading this book simply to learn more about environmental issues. Yet it is impossible to separate our knowledge about environmental issues from the ways we communicate about these issues. As environmental communication scholars James Cantrill and Christine Oravec (1996) observed, the "environment we experience and affect is largely a product of how we come to talk about the world" (p. 2). That is, the way we communicate with one another about the environment powerfully affects how we perceive both it and ourselves and, therefore, how we define our relationship with the natural world. For example, scientist E. O. Wilson (2002) used the language of biology to describe the environment as "a membrane of organisms wrapped around Earth so thin it cannot be seen edgewise from a space shuttle, yet so internally complex that most species composing it remain undiscovered" (p. 3).

Furthermore, the images of the planet and information we receive from friends, blogs, news media, teachers, or popular films play a powerful role in influencing not only how we perceive the environment but also what actions we take. Can the world, for example, meet its energy needs through renewable sources like wind and solar power? Is "fracking" for natural gas in deep shale deposits safe? Is it safe for the U.S. Army to burn stores of chemical weapons near schools and residential neighborhoods? Can the rainforests of Southeast Asia provide sustainable wood products without endangering wildlife or traditional communities? In engaging such questions, particularly in public forums, we rely on

©iStockphoto.com/jordanchez

In 2007, the African country of Rwanda became the first to ban plastic bags. This landmark was achieved through people and leaders talking about the ban during community cleanups, farmers providing testimony to the negative impact, scientists conducting an environmental assessment of plastic bags to establish the problems associated overall, TV and radio campaigns to publicize the ban, videos played on buses and airlines for further education, and more (Nicodemus, 2012).

conversation with friends, news reports, and public debates to imagine, describe, debate, and celebrate our multiple relations with the natural world.

That's one reason we wrote this book: We believe that communication about the environment matters. It matters in the ways we interact with others and in naming certain conditions as "an environment problem." And it matters ultimately in the choices we make in response to these problems. This book, therefore, focuses on the role of communication in helping us negotiate the relationship between ourselves and the thin "membrane of organisms" that envelops our natural and human environments.

The purpose of *Environmental Communication and the Public Sphere* is threefold: (1) to deepen your insight into how communication shapes our perceptions of environmental issues; (2) to acquaint you with some of the media and public forums that are used for environmental communication, along with the communication of scientists, corporate lobbyists, ordinary citizens, and others who seek to influence decisions about the environment; and (3) to enable you to join in conversations and debates that are already taking place locally and globally that may affect the environments where you yourself live, study, work, meditate, and recreate.

Why Do We Need to Speak for the Environment?

At first glance, there appears to be little need for persuasion or debate about environmental issues. Who supports dirty air or polluted water? Although public opinion about environmental issues varies—depending on worries about the economy or war—the U.S. public has generally shown strong support for environmental values. Even so, differences exist among the public about how society should solve specific environmental problems. As we write, a controversy over the uses of "hydraulic fracking," a technology for drilling for natural gas, has embroiled local communities in the United States and some European countries, and some scientists are proposing, while others disagree, that we should be "geo-engineering" projects to address climate change. The complexity of issues such as these makes a consensus difficult. And as the different voices of scientists, the oil and gas industry, and environmental groups enter the public debate, widely divergent viewpoints compete for our support.

There exists, therefore, a dilemma. Although the natural world is alive with sound, the wild species, streams, and forests have little voice in the public sphere. Instead, politicians, media commentators, environmentalists, business leaders, and others claim the right to speak for nature or for their own interests in the use of natural resources. Here then is the dilemma: If the environment cannot speak (at least not in public forums), who has the right to speak on its behalf? Who should define the interests of society in relation to the natural world? Is it appropriate, for example, to drill for oil along fragile coastlines? Who should bear the cost of cleaning up abandoned toxic waste sites—taxpayers or the businesses responsible for the contamination? Where do we define our limits in terms of contamination and growth?

These questions illustrate the inescapable role of communication in environmental controversies. Only in a society that allows debate can the public choose among the differing voices and ways of relating to the natural world. That is one of our purposes in writing this book: We believe that you, we, and everyone in a democratic society have a pivotal role in speaking about these larger environmental issues.

Background and Perspectives of the Authors

After inviting you to join in conversations about the environment, it's time we described our own involvement in this challenging field.

Robert: For a number of years, I've been a professor of communication studies and also the curriculum in the environment at the University of North Carolina at Chapel Hill. In my research, I have focused on the challenges of communicating to the public about climate change and other environmental problems. I have also worked actively in the U.S. environmental movement, as president of the Sierra Club in San Francisco, California, and with Earth Echo International in Washington, DC. And I often advise other environmental organizations about their communication programs.

My interest in the environment, however, arose long before I heard the word *environment*. As a boy growing up in the Appalachian region of southern West Virginia, I fell in love with the wild beauty of the mountains near my home and the graceful flow of the Greenbrier River. As I grew older, I saw coal mining's devastating effects on both miners and the natural landscape, including the streams and water supplies of local communities. In graduate school, I saw the health effects of air pollution from steel mills in Pittsburgh and later from chemical plants in poor neighborhoods in Mississippi. I began to realize how intimately people and their environments are bound together and have come to respect the diverse voices that have spoken about both their communities and their awe of the natural world.

Phaedra: I started my undergraduate training in natural resources but then realized that scientists knew a good deal about what we needed to do to make the world more sustainable; they just hadn't figured out how to communicate those ideas in compelling ways. To learn more about the systems that shape cultural attitudes, I studied political economy and social theory. Then, I met Robbie when I was 20, and he was president of the Sierra Club; I joined him for graduate school and, with his mentorship, I became his first PhD student. I became a professor in 2002 in communication studies, cultural studies, and American studies at Indiana University in Bloomington. My scholarship and advocacy primarily is motivated by how everyday people mobilize resistance to environmental injustices and the barriers constraining these efforts.

Since as long I can remember, I have cared about nonhuman life and social justice. Growing up in the sprawl of Philadelphia, I became a vegetarian at 9 and quickly identified with feminist, labor, and civil rights advocacy. In North Carolina, I advocated with residents of Warren County, North Carolina, to help clean up a toxic dump and with migrant farmworkers for better working conditions, and I participated in many toxic tours throughout North America as a volunteer for the Sierra Club's Environmental Justice Committee. In Indiana, I have served on the City of Bloomington's Environmental

Commission and have been the adviser to inspiring student-led groups. In Colorado, it is exciting and unsurprising to me that the environmental humanities and sciences are growing fields of research and employment; environmental challenges could not be more pressing and they impact every facet of our lives.

As a result of these experiences and also from our own research and teaching in environmental communication, we've become more firmly persuaded of several things:

1. Individuals and communities have stronger chances to safeguard the environmental health and quality of their local environments if they understand some of the dynamics and opportunities for communication about their concerns.

2. Environmental issues and public agencies do not need to remain remote, mysterious, or impenetrable. The environmental movement, legal action, and both new and "old" media have helped to demystify governmental procedures and open the doors and computer files of government bureaucracies to greater public access and participation in environmental decisions.

3. As a consequence, individuals have many opportunities to participate in meaningful ways in public debates about our environment, and indeed, there is more urgency than ever in doing so. That is why we wrote *Environmental Communication and the Public Sphere*.

One other thing: Because of our experiences, we cannot avoid personal perspectives on some of the issues discussed in this book, nor do we wish to. In this sense, we bring certain values and insights to our writing. We do two things, however, to balance this as we cover the topics in this book. First, when we introduce views or positions, we explain how we arrived at them, based on our experience or research. Second, we include "Another Viewpoint" in each chapter to alert you to important disagreements about a topic. Our aim is not to set up false dichotomies but to introduce a diversity of perspectives. We also refer you to suggested resources that allow you to learn about the issues in each chapter.

Distinctive Features of the Book

As its title suggests, the framework for *Environmental Communication and the Public Sphere* is organized around two core concepts:

1. The importance of *human communication* in influencing our perceptions of the natural world and our relations with the environment

2. The role of the *public sphere* in providing opportunities for different voices seeking to influence decisions about the environment

Communication is the symbolic mode of interaction that we use in constructing environmental problems and in negotiating society's different responses to them. We use the idea of the **public sphere** throughout this book to refer to the forums and interactions in which different individuals engage each other in conversation, argument, debate, and questions about subjects of shared concern or that affect a wider community. (We describe the ideas of communication and the public sphere more in Chapter 1.) As this book emphasizes throughout, communication is not limited to words: visual images and nonverbal symbolic actions such as photographs, videos, marches, sit-ins, banners, and documentary films have prompted discussion, debate, and questioning of environmental policy as readily as editorials, speeches, and TV newscasts.

AP Photo/J. Scott Applewhite

Visual images and nonverbal actions such as documentary films, protests, or sit-ins have prompted discussion, debate, and questioning of environmental policy in the public sphere as readily as editorials, speeches, and TV newscasts.

Along with the focus on human communication and the public sphere, this fourth edition includes a number of distinctive features:

1. A comprehensive introduction to the study of environmental communication, including different meanings of *environment*; visual and popular culture portrayals of the environment; managing environmental conflicts; and digital media and environmental activism to scientists, technology, and environmental controversies

2. *New chapters*: "Visual and Popular Culture Portrayals of the Environment," "Sustainability and the 'Greening' of Campuses," "Digital Media and Environmental Activism," and "Citizens' 'Standing': Environmental Protection and the Law"

3. Updated emphasis on communicating about climate change, the movement for climate justice, and challenges to the credibility of climate scientists

4. Use of case studies from recent environmental controversies to illustrate key points

5. A Suggested Resources section (films, websites, and readings) in each chapter

6. Opportunities to apply the principles of environmental communication in Act Locally! exercises

7. A comprehensive Glossary of key terms at the end of the book

Interest in environmental communication is growing worldwide. Although this book highlights the U.S. context, the fourth edition (1) illustrates the application of many concepts globally and (2) recognizes recent developments in environmental communication in many non-U.S. contexts. For example, European nations are making great strides in implementing the Aarhus Convention—an agreement ensuring access to environmental information and public participation in environmental decisions in Europe (see Chapter 12).

New Terrain and New Questions

We recognize you may have questions about a class linking communication and environment. Many of you—perhaps the majority—may not label yourself an environmentalist. Or you may question your ability to affect any of the big problems, such as climate change or the destruction of rainforests. So this is new terrain; that's OK.

In this book, we do not assume any special knowledge on your part about environmental science or politics. Nor do we assume that you know about particular theories of communication. We will take this step-by-step. For example, we use **boldface** type when we introduce an important term. We include a list of these key terms at the end of each chapter and a glossary of terms in an appendix. In some cases, an FYI feature provides background information to help you become familiar with theories or issues raised in a chapter.

In turn, we hope you'll be open to exploring the distinct perspective of this book—the ways that language, symbols, and visual images shape our perceptions of the environment and our own relationships with the environment. By becoming aware of the ways people speak about the environment or what is called an environmental "problem," we hope you'll be interested in joining public conversations, not only about the big questions but also about the fate of the places where you live, study, and enjoy everyday life.

KEY TERMS

Communication   5   **Public sphere**   5

PART I

Communicating for/About the Environment

The natural world affects us, but our language and other symbolic action also have the capacity to affect or construct our perceptions of nature itself.

CHAPTER 1

Studying/Practicing Environmental Communication

This chapter describes environmental communication as a subject of study and an activity that occurs in everyday life. As a study, this chapter points out that our understanding of nature and our actions toward the environment depend not only on information but on the ways our views of the environment are shaped by news media, films, social networks, public debate, popular culture, everyday conversations, and more. As an activity, this and other chapters trace the many ways and settings in which individuals, journalists, scientists, public officials, environmentalists, corporations, and others raise concerns and attempt to influence the decisions affecting our communities and the planet.

Chapter Preview

- The first section of this chapter describes environmental communication, defines the term, and identifies seven principal areas of study and practice in this field.
- The second section introduces three themes that constitute the framework for this book:

  1. Human communication is a form of *symbolic action*, that is, our language and other ways of conveying purpose and meaning affect our consciousness itself, shaping our perceptions and motivating our actions

  2. As a result, our beliefs and behaviors about nature and environmental problems are mediated or influenced by such communication

  3. The public sphere (or spheres) emerge as a discursive space in which competing voices engage us about environmental concerns

- The final section describes some of these diverse voices, whose communication practices we'll study in this book.

After reading this chapter, you should have an understanding of environmental communication as an area of study and an important practice in public life. You should also be able to recognize the range of voices and practices through which environmental groups, ordinary citizens, businesses, and others discuss important environmental problems—from management of public lands to global climate change. As a result, we hope you'll not only become a more critical consumer of such communication but also discover opportunities to add your own voice to the vibrant conversations about the environment that are already in progress.

The Study of Environmental Communication

Along with the growth of environmental studies on college campuses, classes that focus on the role of human communication in environmental affairs have also emerged. On many campuses, environmental communication courses include a wide range, from environmental rhetoric, climate change communication, environmental journalism, risk communication, and environmental advocacy campaigns to "green" marketing and popular culture images of nature. Along with such courses, scholars in communication, journalism, literature, and science communication are pioneering research in the role and influence of communication in the many public settings where the environment is a concern.

Frankly, the wide range of subjects included in environmental communication makes a definition of the field challenging. So first, let's look at some of the areas that you might study.

Areas of Study

Although the study of environmental communication covers many topics, most research and the practice of communication fall into one of seven areas. We explore many of these areas in later chapters. For now, we'll briefly identify these seven areas.

1. *Environmental rhetoric and the social–symbolic "construction" of nature.* Studies of the rhetoric of environmental writers and campaigns emerged as an early focus of the field. Along with the related interest in how our language helps to construct or represent nature to us, this is one of the broadest areas of study. For example, Marafiote (2008) has described the ways in which U.S. environmentalists' rhetoric reshaped the idea of *wilderness* to win passage of the 1964 Wilderness Act.

Relatedly, studies of language and other symbolic forms help us understand the constitutive power of communication to generate ideas and meanings about the environment. For example, Jennifer Peeples (2013) has documented the constitutive power of visual images to convey the effects of environmental toxins on children in ways that mere words cannot. She quotes a photographer describing his use of the camera: "I kept my camera's eye fixed on the haunting faces of children. . . . Their expressions and circumstances bespoke the consequences of the environmental tragedies in ways that any retelling of the experts' verbal arguments never could" (p. 206). (We'll explore this area more in Chapters 2 and 3.)

2. *Public participation in environmental decision making.* When citizens are given a voice in environmental affairs, scholars report that their "participation improves the quality and legitimacy of a decision and . . . can lead to better results in terms of environmental quality" (Dietz & Stern, 2008, p. ES1). Still, in many cases, barriers prevent the meaningful involvement of citizens in decisions affecting their communities or the natural environment. As a result, a number of studies have scrutinized government agencies in the United States and other nations to identify the opportunities for—and barriers to—the participation of ordinary citizens, as well as environmentalists and scientists, in an agency's decision making.

Studies of public participation have covered topics such as citizens' comments about forest management plans, public access to information about pollution in local communities (Beierle & Cayford, 2002), and the barriers that citizens in India faced in gaining information and the privileging of technical discourse about a proposed hydropower (dam) project in their region (Martin, 2007). (We take up the study of public participation in Chapter 12.)

3. *Environmental collaboration and conflict resolution.* Dissatisfaction with some of the adversarial forms of public participation has led practitioners and scholars to explore other models of resolving environmental conflicts. They draw inspiration from communities that have discovered ways to bring quarreling parties together. For instance, groups that had been in conflict for years over logging in Canada's coastal Great Bear Rainforest reached agreement recently to protect 5 million acres of temperate rainforest.

At the center of these modes of conflict resolution is the ideal of collaboration, a mode of communication that invites stakeholders to engage in problem solving rather than advocacy. Collaboration has been defined as "constructive, open, civil communication, generally as dialogue; a focus on the future; an emphasis on learning; and some degree of power sharing and leveling of the playing field" (Walker, 2004, p. 123). (We describe collaboration further in Chapter 13.)

4. *Media and environmental journalism.* In many ways, the study of environmental media is its own subfield. Research in this area focuses on ways in which news, advertising, and commercial programming portray nature and environmental problems as well as the effects of media on public attitudes. Subjects include the **agenda-setting** role of news media (media's ability to affect the public's perception of the importance of an issue); journalist values of objectivity and balance; and media frames—the central organizing themes that connect the different elements of a news story (headlines, quotes, etc.) into a coherent whole.

Studies in environmental media also explore online news and social media in engaging environmental concerns. These studies range widely, from interviews with the editors of *Inside Climate News* (insideclimatenews.org), a small website that won a Pulitzer Prize in 2013 for its reporting on the dangers posed by poorly regulated oil

©iStockphoto.com/urbancow

Photo 1.1	How do news, advertising, and other media affect our perceptions and attitudes toward the natural world or our understanding of environmental issues?

pipelines in the United States, to study of activists' uses of Twitter profile feeds, hyperlinks, and community-generated hashtags at a recent United Nations climate summit (Segerberg & Bennett, 2011). (We describe environmental journalism and uses of new/social media in Chapters 5 and 9.)

5. *Representations of nature in advertising and popular culture.* The use of nature images in film, television, photography, music, and commercial advertising is hardly new or surprising. What is new is the growing number of studies of how such popular culture images influence our attitudes about the environment. Scholars explore a range of cultural products—film, green advertising, SUV ads, wildlife films, supermarket tabloids, and more.

Scholars also are mapping some of the ways in which popular media sustain attitudes of cultural dominance or exploitation of the natural world. For example, Todd (2010) examines how the photographic images and travel narratives in *National Geographic* magazines depict Africa's landscapes through "anthropocentric distance," as "a wilderness theme park, and a part of the global scenery" (p. 206). (We'll look at the role of images in Chapter 4.)

6. *Advocacy campaigns and message construction.* A growing area has been the study of public information and advocacy campaigns by environmental groups, corporations, and scientists informing the public about climate change. These campaigns attempt to educate, change attitudes, and mobilize public audiences—for example, campaigns to

educate homeowners about energy savings from "smart" thermostats, campaigns to mobilize support for a wilderness area, or a corporate accountability campaign to persuade building supply stores to buy lumber only from sustainable forests.

In this area, media and communication scholars have documented the challenge of communicating the dangers from climate change as well as barriers to the public's sense of urgency (Moser, 2010). And others have found that, as environmental groups turn to new/social media, their visibility and effectiveness have changed (Lester & Hutchins, 2012). (We'll look at a range of campaigns, including uses of digital media, in Chapters 8, 9, 10, and 11.)

7. *Science and risk communication.* How effective are signs warning that fish caught from a lake may be contaminated with toxic chemicals? Did regulators ignore warnings about the risks from deep water oil drilling in the Gulf of Mexico? How are consumers to judge the risks of the chemical BPH in plastic water bottles? Such questions illustrate the study and practice of environmental science and risk communication—the ways environmental risks can be communicated to affected publics.

Science and risk communication includes a range of practices—from news media reports of the risk of pollution from hydraulic fracturing ("fracking") to government reports of the likelihood of developing cancer from exposure to the spraying of pesticides on agricultural fields. (We'll describe case studies of science and risk communication in Chapters 6 and 7.)

Defining Environmental Communication

With such a diverse range of topics, the field can appear at first glance to be confusing. If we define environmental communication as simply *talk* or the transmission of information about environmental topics—water pollution or grizzly bear habitat—our definitions will be as varied as the many topics.

A clearer definition takes into account the distinctive roles of language, visual images, protests, music, or even scientific reports as different forms of **symbolic action**. This term comes from Kenneth Burke (1966), a rhetorical theorist. In his book *Language as Symbolic Action,* Burke stated that even the most unemotional language is necessarily persuasive. This is because our language and other symbolic acts *do* something, as well as *say* something. Language actively shapes our understanding, creates meaning, and orients us to a wider world.

The view of communication as a form of symbolic action might be clearer if we contrast it with an earlier view, the **Shannon–Weaver model of communication.** Shortly after World War II, Claude Shannon and Warren Weaver (1949) proposed a model that defined human communication as simply the transmission of information from a source to a receiver. There was little effort in this model to account for meaning or for the ways in which communication acts on or shapes our awareness. Unlike the Shannon–Weaver model, symbolic action assumes that language and symbols do more than transmit information. Burke (1966) went so far as to claim that "much that we take as observations about 'reality' may be but the spinning out of possibilities implicit in our particular choice of terms" (p. 46).

If we focus on symbolic action, then we can offer a richer definition. In this book, we use the phrase **environmental communication** to mean *the pragmatic and constitutive vehicle for our understanding of the environment as well as our relationships to the natural world*; it is the symbolic medium that we use in constructing environmental problems and in negotiating society's different responses to them. Defined this way, environmental communication serves two different functions:

1. *Environmental communication is **pragmatic.*** It educates, alerts, persuades, and helps us solve environmental problems. It is this instrumental sense of communication that probably occurs to us initially. It is the vehicle or means that we use in problem solving and is often part of public education campaigns. For example, a pragmatic function of communication occurs when an environmental group educates its supporters and rallies support for protecting a wilderness area or when the electric utility industry attempts to change public perceptions of coal by buying TV ads promoting "clean coal" as an energy source.

2. *Environmental communication is **constitutive.*** Embedded within the pragmatic role of language and other forms of symbolic action is a subtler level. By constitutive, we mean that our communication also helps us construct or compose representations of nature and environmental problems as subjects for our understanding. Such communication invites a particular perspective, evokes certain values (and not others), and thus creates conscious referents for our attention. For example, different images of nature may invite us to perceive forests and rivers as natural resources for use or exploitation, or as vital life support systems (something to protect). A campaign to protect a wilderness area may educate and rally supporters (pragmatic), but at the same time, its advocates may also tap into cultural resonances that invite us to perceive "wilderness" as a pristine or unspoiled nature, thus constructing or composing nature in new ways for our understanding.

Communication as constitutive also assists us in defining certain subjects as "problems." For example, when climate scientists call our attention to tipping points, they are naming thresholds beyond which warming "could trigger a runaway thaw of Greenland's ice sheet and other abrupt shifts such as a dieback of the Amazon rainforest" (Doyle, 2008). Such communication orients our consciousness of the possibility of an abrupt shift in climate and its effects; it therefore constitutes, or raises, this possibility as a subject for our understanding.

Act Locally!

Pragmatic and Constitutive Communication in Messages About Climate Change

Examples of communication about climate change occur daily in news media, websites, blogs, TV ads, and other sources. Select one of these that interests you. It might be a news report about rising sea levels, food scarcity, or acidification of oceans; a YouTube video about the impacts of climate change on the Arctic; or a TV ad about coal or natural gas as a form of "clean energy."

The message or image you've chosen undoubtedly uses both pragmatic and constitutive functions of communication; that is, it may educate, alert, or persuade while also subtly creating meaning and orienting your consciousness to a wider world. After reflecting on this message, answer these questions:

1. What pragmatic function does this communication serve? Who is its intended audience? What is it trying to persuade this audience to think or do? How?

2. Does this message draw on constitutive functions, as well, in its use of certain words or visual images? How do these words or images create referents for your attention and understanding, invite a particular perspective, or orient you to a set of concerns?

Environmental communication as a pragmatic and constitutive vehicle serves as the framework for the chapters in this book and builds on the three core principles:

1. Human communication is a form of symbolic action.

2. Our beliefs, attitudes, and behaviors relating to nature and environmental problems are mediated or influenced by communication.

3. The public sphere emerges as a discursive space in which diverse voices engage the attention of others about environmental concerns.

These principles obviously overlap (see Figure 1.1). As we've noted, our communication (as symbolic action) actively shapes our perceptions when we see the natural world through myriad words, images, or narratives. And when we communicate publicly with others, we share these understandings and invite reactions to our views.

Nature, Communication, and the Public Sphere

Let's explore the three principles that organize the chapters in this book. We'll introduce and illustrate these briefly here and then draw on them in each of the remaining chapters.

Human Communication as Symbolic Action

Earlier, we defined environmental communication as a form of symbolic action. Our language and other symbolic acts *do* something. Films, online sites and social media, photographs, popular magazines, and other forms of human symbolic behavior act upon us. They invite us to view the world this way rather than that way to affirm these values and not those. Our stories and words warn us, but they also invite us to celebrate.

Language that invites us to celebrate also leads to real-world outcomes. Consider the American gray wolf. In 2010, a federal judge restored protection to gray wolves in the Northern Rocky Mountains under the nation's Endangered Species Act. But it was not always this way. Wolves had been extirpated from the region by the mid-20th

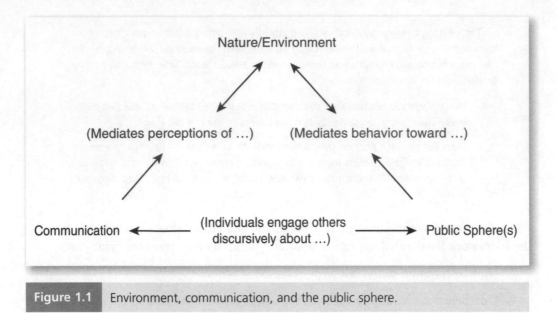

Nature/Environment

(Mediates perceptions of …)   (Mediates behavior toward …)

Communication ← (Individuals engage others discursively about …) → Public Sphere(s)

| **Figure 1.1** | Environment, communication, and the public sphere. |

century through intensive "predator control" (trapping, poisoning, or shooting). It was not until the mid-1990s that the U.S. Fish and Wildlife Service initiated a restoration plan for wolves.

In 1995, Secretary of Interior Bruce Babbitt celebrated the return of the first American gray wolf to Yellowstone National Park in a speech marking the event. Earlier that year, he had helped carry and release the wolf into the transition area in the park where she would mate with other wolves also being returned. After setting her down, Babbitt recalled, "I looked . . . into the green eyes of this magnificent creature, within this spectacular landscape, and was profoundly moved by the elevating nature of America's conservation laws: laws with the power to make creation whole" (para. 3).

Babbitt's purpose in speaking that day was to support the beleaguered Endangered Species Act, under attack in the Congress at the time. In recalling the biblical story of the flood, Babbitt evoked a powerful narrative for revaluing wolves and other endangered species. In retelling this ancient story, he invited the public to embrace a similar ethic in the present day:

> And when the waters receded, and the dove flew off to dry land, God set all the creatures free, commanding them to multiply upon the earth.
>
> Then, in the words of the covenant with Noah, "when the rainbow appears in the clouds, I will see it and remember the everlasting covenant between me and all living things on earth."
>
> Thus we are instructed that this everlasting covenant was made to protect the whole of creation. . . . We are living between the flood and the rainbow: between the threats to creation on the one side and God's covenant to protect life on the other. (Babbitt, 1995, paras. 34–36, 56)

U.S. National Park Service

Photo 1.2	Secretary of the Interior Bruce Babbitt, releasing the first American gray wolf into Yellowstone National Park in 1995.

Communication enables us to make sense about our world; it orients us toward events, people, and yes, wildlife. And, because different individuals (and generations) may value nature in diverse ways, we find our voices to be a part of a conversation with others about this world. Secretary Babbitt invoked an ancient story of survival to invite the American public to appreciate anew the Endangered Species Act. So, too, our communication mediates or helps us make sense of our own relationships with nature, what we value, and how we shall act.[1]

Human communication, therefore, is symbolic action because we draw upon language and other symbols to construct a framework for understanding and valuing and to bring the wider world to others' attention. (We explore this aspect of communication more closely in Chapters 2 and 3.)

Mediating "Nature"

It may seem odd to place "nature" in quotation marks. The natural world definitely exists: Forests are logged or remain standing; streams may be polluted or clean; and large glaciers in Antarctica are calving into the Southern Ocean. So what's going on? As one of our students asked, "What does communication have to do with nature or the study of environmental problems?" Our answer to her question takes us into the heart of this book.

Simply put, whatever else the environment may be, it is deeply entangled with our very human ways of interacting with, and knowing, the wider world. As Norwegian environmentalist Arne Naess (2000) once exclaimed, "Having been taken at least twice by avalanches, I have never felt them to be social constructions. But every word I utter about them may have social origins and the same applies to the meanings of these words" (p. 335). At a basic level, our beliefs, attitudes, and behaviors toward nature are mediated by human ways of representing the world—through our language, television, photos, art, and contemplation. *Mediating* is another way of saying that our acts of pointing to and naming something are our means for recognizing and understanding it. "Pointing and naming generate certain kinds of ecocultural knowledge that constitute aspects of nature as considered, unique, sorted, or marked" (Milstein, 2011, p. 4).

When we name the natural world, we also orient ourselves in this world. We become located or interested in it; we have a view onto this world. As Christine Oravec (2004) observed in her essay on Utah's Cedar Breaks National Monument, this act of naming is not only a mode by which we socially construct and know the natural world, but it orients us and thus "influences our interaction with it" (p. 3). For instance, is *wilderness* a place of primeval beauty, or is it a territory that is dark, dangerous, and alien to humans? Or is it something else? Early settlers in New England viewed North American forests as forbidding and dangerous. Puritan writer Michael Wigglesworth named or described the region as

A waste and howling wilderness,

Where none inhabited

But hellish fiends, and brutish men

That Devils worshiped. (quoted in Nash, 2001, p. 36)

As a result of these different orientations to the natural world, writers, citizens, conservationists, poets, scientists, and business lobbyists have fought for centuries over whether forests should be logged, rivers dammed, air quality regulated, and endangered species protected.

Consider the weather (and climate):

Periodically, winters in the United States and other parts of the world are bitingly cold, with record low temperatures and blizzards. For some, cold, snowy weather invites sarcastic remarks: "Where's that global warming?" In fact, after a recent cold winter, the percentage of Americans who believed global warming was occurring "dropped 7 points" from the previous fall ("After Cold Winter," 2013, para. 1). During cold winters, you're very likely to hear competing claims in the media or online about the reality of global warming. One skeptic, for example, quipped, "Um . . . if the globe is warming why is my car buried under all this snow?" (Beck, 2011, para. 1).

Climate scientists, on the other hand, distinguish *weather*—changing every few days—and *climate*, measured in longer, 20-year periods. For example, during the winter of 2013–2014, the Intergovernmental Panel on Climate Change (IPCC)—a body of over 2,000 scientists from 154 countries—issued a summary of recent research. It concluded: "Warming of the climate system is unequivocal, and since the 1950s, many of the observed changes are unprecedented over decades to millennia" (IPCC, 2013, p. 3). In their own ways, climate scientists, news reports, and climate change deniers offer differing views or constructions of weather/climate and what they mean. This is what we meant earlier in saying our beliefs, attitudes, and behaviors relating are mediated by communication.

Our point is that, although nature invites different responses from us, it is, in itself, politically silent. Ultimately, it is we—through our symbolic actions—who invest its seasons and species with meaning and value.

Public Sphere as Discursive Space

A third theme central to this book is the idea of the public sphere or, more accurately, public spheres. Earlier, we defined a public sphere as a realm of influence that is created when individuals engage others in communication—through conversation, argument, debate, or questioning—about subjects of shared concern or topics that affect a wider community. The public comes into being in our everyday conversations as well as in more formal interactions when we talk about the environment. And the public sphere is not just words: Visual and nonverbal symbolic actions, such as marches, banners, and photographs, also have prompted debate and questioning of environmental policy as readily as editorials, speeches, and TV newscasts.

The German social theorist Jürgen Habermas (1974) offered a similar definition when he observed that "a portion of the public sphere comes into being in every conversation in which private individuals assemble to form a public body" (p. 49). As we engage with others, we translate our private concerns into public matters and thus create circles of influence that affect how we and others view the environment and our relation to it. Such translations of private concerns into public matters occur in a range of forums and practices that give rise to something akin to an environmental public sphere—from a talk at a campus environmental forum to a scientist's testimony before a congressional committee. In public hearings, newspaper editorials, blog posts, speeches at rallies, street festivals, and on countless other occasions in which we engage others in conversation or debate, the public sphere emerges as a potential sphere of influence.

But private concerns are not always translated into public action, and technical information about the environment may remain in scientific journals, proprietary files of corporations, or other private sources. Therefore, it is important to note that two other spheres of influence exist parallel to the public sphere. Communication scholar Thomas Goodnight (1982) named these areas of influence the *personal* and *technical* spheres. For example, two strangers arguing at an airport bar is a relatively private affair, whereas the technical findings of biology that influenced Rachel Carson's (1962) discussion of the insecticide dichlorodiphenyltrichloroethane (DDT) in *Silent Spring* were originally limited to technical journals. Yet Carson's book presented this information in a way that engaged the attention—and debate—of millions of readers and scores of public officials. In doing this, *Silent Spring* gave rise to a sphere of influence as she translated technical matters into subjects of public interest.

The idea of the public sphere itself, however, can be misunderstood. Three misconceptions occur—the beliefs that the public sphere is (a) only an official site or forum for government decision making, (b) a monolithic or ideal collection of all citizens, and (c) a form of "rational" or technical communication. Each of these ideas is a misunderstanding of the public sphere.

First, the public sphere is not only, or even primarily, an official space. Although there are officially sponsored spaces such as public hearings that invite citizens to communicate about the environment, these official sites do not exhaust the public sphere. In fact, discussion and debate about environmental concerns more often occur outside of government meeting rooms and courts. The early fifth-century (BCE) Greeks called these meeting spaces of everyday life *agoras,* the public squares or marketplaces where citizens gathered to exchange ideas about the life of their community. Similarly, we find everyday spaces and opportunities today, publicly, to voice our concerns and influence the judgment of others about environmental concerns.

Second, the public sphere is neither monolithic nor a uniform assemblage of all citizens in the abstract. As the realm of influence that is created when individuals engage others discursively, the public sphere assumes concrete and local forms: They include calls to talk radio programs, blogs, letters to the editors of newspapers, or local meetings where citizens question public officials, for example, about risks to their health from contaminated well water. As Habermas (1974) reminds us, the public sphere comes into existence whenever individuals share, question, argue, mourn, or celebrate with others about their shared concerns.

Third, far from elite conversation or "rational" forms of communication, the public sphere is most often the arena in which popular, passionate, and democratic communication occurs, as well as reasoned or technical discourse. Such a view of the public sphere acknowledges the diverse voices and styles that characterize a robust, participatory democracy. In fact, in this book, we introduce the voices of ordinary citizens and the special challenges they face in gaining a hearing about matters of environmental and personal survival in their communities.

Diverse Voices in a "Green" Public Sphere

The landscape of communication about environmental concerns is diverse, complex, and often colorful, like an Amazonian rainforest or the Galapagos Islands' ecology. Whether in local community centers, on blogs, at rallies, or in corporate-sponsored TV ads, individuals and groups speaking about the environment appear in diverse sites and public spaces.

In this final section, we describe some of the voices you may hear in the public sphere communicating about environmental issues. These include the voices of

1. Citizens and community groups

2. Environmental groups

3. Scientists and scientific discourse

4. Corporations and lobbyists

5. News media and environmental journalists

6. Student and campus groups

7. Anti-environmentalist and climate change critics

Individuals in these seven groups take on multiple communication roles—writers, press officers, group spokespersons, community or campus organizers, information technology specialists, communication directors, marketing and campaign consultants, and more.

Citizens and Community Groups

Residents who complain to public officials about an environmental problem in their community—such as air pollution, asbestos in their children's school, or contaminated well water—and who organize their neighbors to take action are the most common sources of environmental change. Some are motivated by urban sprawl or development projects that destroy their homes as well as green spaces in their cities. Others, who may live near an oil refinery or chemical plant, may be motivated by noxious fumes to organize resistance to the industry's lax air quality permit.

In 1978, Lois Gibbs and her neighbors in the working-class community of Love Canal in upstate New York became concerned when, after they noticed odors and oily substances surfacing in the local school's playground, their children developed headaches and became sick. Gibbs also had read a newspaper report that Hooker Chemical Company, a subsidiary of Occidental Petroleum, had buried dangerous chemicals on land it later sold to the school board (Center for Health, Environment, and Justice, 2003).

Despite an initial denial of the problem by state officials, Gibbs and her neighbors sought media coverage, carried symbolic coffins to the state capital, marched on Mother's Day, and pressed health officials to take their concerns seriously. Finally, in 1982, the residents succeeded in persuading the federal government to relocate those who wanted to leave Love Canal. The U.S. Justice Department also prosecuted Hooker Chemical Company, imposing large fines (Shabecoff, 2003, pp. 227–229). As a result, Love Canal became a symbol of toxic waste sites and fueled a citizens' anti-toxics movement in the United States.

Lois Gibbs's story is not unique. In rural parishes in Louisiana, in inner-city neighborhoods in Detroit and Los Angeles, on Native American reservations, and in communities in India, China, Europe, the Philippines, Latin America, Africa, and throughout the world, community groups have launched campaigns to protest smog and pollution, halt toxic runoff from mining operations, and stop illegal logging of community forests. As they do, activists and residents face the challenge of finding their voice and overcoming barriers to express their concerns and persuade others to join them in demanding accountability of public officials.

Environmental Groups

Environmental organizations are among the most visible sources of communication about the environment. These groups come in a wide array of organizational types and networks, online and on the ground. They range, in the United States, from grassroots groups in local communities to nationwide organizations like the Nature Conservancy (nature.org), Sierra Club (sierraclub.org), Environmental Defense Fund (edf.org), and National Wildlife Federation (nwf.org). And there are similar groups in almost every country working for environmental protection, biological diversity, and sustainability— groups like *Navdanya,* meaning "nine seeds" (navdanya.org) in India, a women-centered movement for protecting native seeds and biological diversity, and the African Conservation Foundation (africanconservation.org), a continent-wide effort to protect Africa's endangered wildlife and their habitats. Other groups, such as Conservation International (conservation.org) and Greenpeace (greenpeace.org), are organized on an international scale, while global networks like 350.org and Avaaz.org link groups world-wide in the fight against climate change and other concerns.

These groups address a diversity of issues and often differ in their modes of advocacy. For example, the Sierra Club and Natural Resources Defense Council (nrdc.org) focus on climate change through their advocacy campaigns and lobbying the U.S. Congress on energy policy, while the Nature Conservancy and local conservancy groups protect endangered habitat on private lands by purchasing the properties themselves. Other groups such as Greenpeace and Rainforest Action Network (ran.org) use "image events" (DeLuca, 1999) to shine the spotlight of media attention on concerns as diverse as climate change, illegal whaling, and destruction of tropical rainforests.

Scientists and Scientific Discourse

The warming of the Earth's atmosphere first came to the U.S. public's attention when climate scientists testified before Congress in 1988. Since then, scientific reports, such as the periodic assessments of the Intergovernmental Panel on Climate Change (IPCC), have prompted spirited public debate over appropriate steps that governments should take to prevent what the United Nations Framework Convention on Climate Change (1994) called a "dangerous anthropogenic interference" with the global climate (para. 37). As we shall see in succeeding chapters, the work of climate scientists has become a fiercely contested site in today's public sphere, as environmentalists, public health officials, ideological skeptics, political adversaries, and others question, dispute, or urge action by Congress to adopt clean energy policies.

As with climate change, scientific reports have led to other important investigations of, and debate in the public sphere about, environmental problems affecting human health and Earth's biodiversity. From asthma in children caused by air pollution and neurological illnesses from mercury poisoning by eating contaminated fish to an accelerating loss of species of plants and animals, the early warnings of scientists have contributed substantially to public awareness, debate, and corrective actions. In Chapter 6, we'll describe the importance of science communication as well as the ways in which climate and other environmental sciences themselves have become a site of controversy in some quarters.

Corporations and Lobbyists

Environmental historian Samuel Hays (2000) reported that, as new environmental sciences in the 1960s began to document the environmental and health risks from industrial products, the affected businesses challenged the science "at every step, questioning both the methods and research designs that were used and the conclusions that were drawn" (p. 222).

Corporate opposition to environmental standards has developed for two reasons: (1) restrictions on the traditional uses of land (for example, mining, logging, or oil and gas drilling) and (2) perceived threats to the economic interests of industries such as petrochemicals, energy production, computers, and transportation. Worried

The Buffalo News

| Photo 1.3 | Local residents like Lois Gibbs of Love Canal, New York, who complain to public officials about pollution or other environmental problems and who organize their neighbors to take action are the most common and effective sources of environmental change. |

by the threat of tighter limits on air and water discharges from factories and refineries, affected corporations formed trade associations such as the Business Round Table and the Chemical Manufacturers Association to conduct public relations campaigns and lobby Congress on behalf of their industries.

On the other hand, some corporations recently have begun to go "green"— improving their operations and committing to standards for sustainability (lower energy use and lower impact on natural resources) in their operations. (We explore some of these efforts in Chapter 11.) Others, however, have skillfully adopted practices of "greenwashing," misleading advertising that claims a product promotes environmental values.

News Media and Environmental Journalists

It would be difficult to overstate the impact of news media—both "old" and new— on our understanding of environmental concerns. News media not only report events but act as conduits for other voices—scientists, public officials, corporate spokespersons, environmentalists—seeking to influence public attitudes. Media also exert influence through their agenda-setting role—that is, the ability to influence the public's perception of the salience or importance of an issue. As journalism scholar Bernard Cohen (1963) first explained, the news media filter or select issues for attention and therefore set the public's agenda, telling people not *what* to think but *what to think about*. For example, the public's concern about tropical contagious diseases soared after extensive news coverage of the Ebola outbreak in West Africa in 2014.

The Ebola news stories focused on a single, dramatic event that was newsworthy, but many environmental topics, such as toxic poisoning or species loss, may be less visible. As a result, the environment is often underreported by traditional news media. In Chapter 5, we'll look at the ways that news media shape our awareness of environmental problems.

Student and Campus Groups

Since Earth Day 1970, when interest in the environment first exploded on U.S. campuses, students and campus groups have been at the forefront of environmental reform. Today, students and campus groups are starting sustainability programs, opposing coal-burning power plants, and organizing forums on global environmental justice—urging responsibility to those who will bear the worst impacts from climate change. And in late 2014, student groups were in the forefront of the estimated 400,000 individuals attending the People's Climate March in New York City; others marched in cities globally.

On many campuses in the United States, student environmental activists are coordinating with wider networks and environmental organizations like the Sierra Student Coalition's "Beyond Coal" campaign (ssc.org) and 350.org's push for divestment from fossil fuel companies. (In Chapter 11, we'll look at recent student-led initiatives for sustainability on college campuses.)

Anti-Environmentalists and Climate Change Critics

Although it may be difficult to conceive of groups that are opposed to protection of the environment (clean air, healthy forests, safe drinking water, and so on), a backlash against government regulations and even environmental science has arisen, particularly, in the United States This is often fueled by the perception that environmental regulations harm economic growth and jobs.

One expression of this opposition, beginning in the 1990s was **Wise Use groups** or *property rights* groups. These groups objected to restrictions on the use of their property for such purposes as protection of wetlands or habitat for endangered species. They include groups like Ron Arnold's Center for the Defense of Free Enterprise (which is opposed to environmental regulations generally). Arnold, a controversial figure in the anti-environmentalist movement, once told a reporter, "Our goal is to destroy environmentalism once and for all" (Rawe & Field, 1992, quoted in Helvarg, 2004, p. 7).

More recently, climate change deniers have questioned whether global warming is occurring at all, or whether human activities (such as burning of fossil fuels) are a contributing cause of warming. Using online sites, talk radio, conservative think tanks, and films like *The Great Global Warming Swindle,* such skeptics have fueled debate and sometimes stalled government action on climate change in the United States.

Global Study of Environmental Communication

The many diverse sources we've identified (above) are speaking about environmental concerns in nearly every country in the world. In 2011, scholars and practitioners established the International Environmental Communication Association (theieca. org) to coordinate research and other activities worldwide. Interest has grown not only in North America, the United Kingdom, and Europe, where "environmental communication has grown substantially as a field" (Carvalho, 2009, para. 1), but also in China, Southeast Asia, India, the Middle East, Australia, Russia, Korea, Africa, and Latin America. We will return to some of these voices and the practices of environmental communication by diverse interests in the following chapters.

SUMMARY

This chapter defined environmental communication, its major areas of study, and the principal concepts around which the chapters of this book will be organized:

- The field of environmental communication includes several major areas: environmental rhetoric and the social–symbolic "construction" of nature, public participation in environmental decision making, environmental collaboration and conflict resolution, media and environmental journalism, representations of nature in corporate advertising and popular culture, advocacy campaigns and message construction, and science and risk communication.

- The term *environmental communication* itself was defined as the pragmatic and constitutive vehicle for our understanding of the environment as well as our relationships to the natural world; it is the symbolic medium that we use in constructing environmental problems and in negotiating society's different responses to them.
- Using this definition, the framework for the chapters in this book builds on three core principles:
  1. Human communication is a form of symbolic action.
  2. Our beliefs, attitudes, and behaviors relating to nature and environmental problems are mediated or influenced by communication.
  3. The public sphere emerges as a discursive space for communication about the environment.

Now that you've learned something about the field of environmental communication, we hope you're ready to engage the range of topics—from the challenge of communicating about climate change to your right to know about pollution in your community—that make up the practice of speaking for and about the environment. And along the way, we hope you'll feel inspired to join the public conversations about environmental concerns happening today.

SUGGESTED RESOURCES

- Depoe, S., & Peeples, J. (2014). *Voice and Environmental Communication.* New York, NY: Palgrave. Explores how people give voice to, and listen to the voices of, the natural environment.
- Hansen, A. (2010). *Environment, Media and Communication.* London, England, and New York, NY: Routledge.
- Henry, J. (2010). *Communication and the Natural World.* State College, PA: Starta.
- Follow or subscribe to an environmental daily news site like Environmental News Network (enn.com), the Grist (grist.org), and huffingtonpost.com/green.

KEY TERMS

Agenda setting   13

Constitutive   16

Environmental
    communication   16

Pragmatic   16

Shannon–Weaver model of
    communication   15

Symbolic action   15

Wise Use groups   27

DISCUSSION QUESTIONS

1. Is nature ethically and politically silent? What does this mean? If nature is politically silent, does this mean it has no value apart from human meaning?

2. The rhetorical theorist Kenneth Burke (1966) claims that "much that we take as observations about 'reality' may be but the spinning out of possibilities implicit in our particular choice of terms" (p. 46). Does this mean we cannot know "reality" outside of the words we use to describe it? What did Burke mean by this?

3. In our society, whose voices are heard most often about environmental issues? What influence do corporations, TV personalities, and partisan blogs have in the political process? Are there still openings for ordinary citizens to be heard?

NOTE

1. There is an update to the story of the gray wolf. Since 2011, U.S. Fish and Wildlife Service has begun removing the American gray wolf from the endangered species list and its protections and turning over management to the Western states in the wolves' territories. Since 2011, these states have permitted hunting of the wolves, resulting in a decline in wolf populations (UPI, 2013) and fueling renewed debate over viable populations of the wolf.

"Many people assume that I must have been inordinately brave to face down the thugs and police during the campaign for Karura Forest. The truth is that I simply did not understand why anyone would want to violate the rights of others or ruin the environment. Why would someone destroy the only forest left in the city and give it to friends and political supporters to build expensive houses and golf courses?" (Maathai, 2008, p. 272).

—Wangari Maathai (1940–2011),
founder of the Green Belt Movement in
Kenya and recipient of the 2004 Nobel Peace Prize

CHAPTER 2

Contested Meanings
of Environment

Nature might well be thought of as the original Rorschach test.

—Jan E. Dizard, *Going Wild: Hunting,
Animal Rights, and the Contested
Meaning of Nature* (1994, p. 160)

We need different ideas because we need different relationships.

—Raymond Williams, *Problems in
Materialism and Culture* (1980, p. 85)

Since the latter half of the 20th century, few words have acquired the symbolic currency of *environment*. No matter which culture or time period one studies, it is important to realize that our beliefs about the environment and how we communicate about them are highly contingent. Like Rorschach inkblot tests used by psychologists to determine one's state of mind, the hopes and fears we feel in relation to the environment reflect a good deal about ourselves in a specific place and moment of time, in addition to the environment we are describing. To illustrate this dynamic relationship and significant legacies that continue to shape perceptions today, this chapter traces some of the more notably contested meanings of environment in the United States and, to a lesser extent, globally.

Chapter Preview

In this chapter, we'll describe five pivotal historical periods through which individuals and new movements contested the dominant attitudes about the environment and what society accepted as an environmental problem or solution.

- The first section describes early American beliefs that fostered a more favorable relationship with what they understood as nature: romanticism, nationalism, and transcendentalism.
- The second section defines two early movements in the United States—a 19th-century preservationist movement that challenged dominant views about the exploitation of wilderness and an early 20th-century ethic of conservation of natural resources.
- The third section describes the rise in the 20th century of an ecological sensibility and a challenge to urban pollution for the sake of human health, as well as the environmental commons.
- The fourth section describes the discourse of environmental justice, which contests a view of nature as a place void of humans' everyday lives.
- Finally, the fifth section engages contemporary discourses of sustainability, which challenges business as usual, and climate justice, which invites us to imagine environmental connections from our backyards to our atmosphere.

Throughout this chapter, we imagine each of these moments through notable changes in **discourse**, or a pattern of knowledge and power communicated through linguistic and nonlinguistic human expression (Foucault, 1970). One way to analyze discourses is to identify their conditions of possibility or how they both reflect previous attitudes and emerging **antagonisms** of a culture in a particular period of history. In everyday language, the term *antagonism* means a *conflict* or *disagreement*. Here, we are using the term more specifically to signal the cultural recognition of the *limit* of an idea, a widely shared viewpoint, or ideology (Laclau & Mouffe, 2001). A conceptual limit is recognized when questioning or criticism reveals a prevailing view to be inadequate or unresponsive to new demands. Recognizing this inadequacy creates an opening for alternative voices—and ideas to redefine what is appropriate, wise, or ethical within a specific context—in this case, the changing relationship between people and the environment.

In this chapter, we underscore major antagonisms that have defined environmental discourse when new voices challenged prevailing cultural values. This is not an exhaustive list from the United States or globally, but highlighting these antagonisms is meant to illustrate how contested meanings of environmental discourses are shaped by culture and shape culture in return.

Overall, this chapter contextualizes and defines a cluster of words that often serve as synonyms for the environment, but symbolize distinct meanings and power relations: *nature*, *wilderness*, *natural resources*, *ecology*, *public health*, *the commons*, *environmental justice*, and *climate justice*. Following the definition of environmental communication in Chapter 1, each of the following discourses about these terms is born of *pragmatic* exigencies and *constitutes* different ways of relating with the environment.

Learning to Love Nature

Many early European colonists did not immediately value nature in North America. Colonist Michael Wigglesworth (1662), for example, described the dark forests of his era as "a waste and howling wilderness" (Nash, 2001, p. 36). "Progress" often was defined by dominating nature and indigenous peoples to make way for colonial farms and cities. Writing from a European perspective of "the New World" at Plymouth in 1620, William Bradford incredulously asked, "What could they see but a hideous and desolate wilderness, full of wild beasts & wild men?" With that phrase, he began what environmental historian Roderick Nash (2001) called an American "tradition of repugnance" for nature and people associated with it (p. 24).

Eventually, voices in art, literature, and on the lecture circuits began to challenge the colonial view of nature solely as alien and exploitable through the championing of wilderness. In his classic study, *Wilderness and the American Mind*, Roderick Nash (2001) identifies multiple sources of this cultural shift away from early colonial attitudes including:

1. *Romantic aesthetics:* "Appreciation of wilderness," Nash argues, "began in cities" (p. 44). In the 18th and early 19th centuries, English nature poets and aestheticians, such as William Gilpin, "inspired a rhetorical style for articulating [an] appreciation of uncivilized nature" (p. 46). These urban dwellers were removed from the day-to-day hardships of living in rural areas and fostered, in American art and literature, an ideal of sublimity in wild nature. The **sublime** was an aesthetic category that associated God's influence with the feelings of awe and exultation that some experienced in the presence of wilderness. "Combined with the primitivistic idealization of a life closer to nature, these ideas fed the Romantic movement which had far-reaching implications for wilderness" (p. 44). Carleton Watkins's 1861 photographs of Yosemite were pivotal to establishing the area as the nation's first protected land and in fostering admiration for the environment (DeLuca & Demo, 2000).

2. *American national identity:* Believing that the new nation could not match the reverence many felt for Europe's illustrious monuments and cathedrals, advocates of a uniquely American identity championed the distinctive characteristics of its natural landscape. "Nationalists argued that far from being a liability, wilderness was actually an American asset" (Nash, p. 67). Writers and artists of the Hudson River school, such as Thomas Cole, celebrated the wonders of the American wilderness by defining a nationalistic style in fiction, poetry, painting, and eventually photography. In his 1835 "Essay on American Scenery," for example, Cole argued, "American scenery . . . has features, and glorious ones, unknown to Europe. The most distinctive, and perhaps the most impressive, characteristic of American scenery is its wildness" (quoted in Nash, 2001, pp. 80–81).

3. *Transcendentalist ideals:* The 19th-century philosophy of **transcendentalism** also proved an important impetus for revaluing wild nature. Transcendentalists held that "natural objects assumed importance because, if rightly seen, they reflected

universal spiritual truth" (Nash, 2001, p. 85). Among those who drew upon such beliefs to challenge older discourses about wilderness was the writer and philosopher Henry David Thoreau. Thoreau (1893/1932) argued that "in Wildness is the preservation of the World"; and that there exists "a subtle magnetism in Nature, which, if we unconsciously yield to it, will direct us aright" (pp. 251, 265). The Union of Concerned Scientists and Penguin Classic Books (2009) have launched an interactive digital book titled *Thoreau's Legacies: American Stories About Global Warming* in tribute to Thoreau's keen observations of the world around him and ability to inspire environmental activists, including John Muir and Rachel Carson, both of whom we discuss further in this chapter.

With the articulation of each of these discourses, though they vary in many ways, the focus primarily is on constituting the environment as **nature** or the physical world that generally exceeds human creation (trees, birds, bears, clouds, rainbows, oceans, seashells, and so forth). Of course, today, from practices in landscape architecture that radically transform the Earth (such as New York City's Central Park) to the capability of genetically cloning animals, this distinction between what humans can create and what we cannot is more complicated, which perhaps is why *environment* has become a more prominent term than *nature* today.

Nevertheless, how humans relate with this definition of the environment remains an ongoing cultural anxiety. Consider, for example, Richard Louv's (2008) best-selling book, *Last Child in the Woods: Saving Our Children From Nature Deficit Disorder*, in which he argues in an age of increased technology that we must remember how direct exposure to nature is essential for emotionally and physically healthy human development and our ability to respond to current environmental crises. Those concerns seem to resonate with early beliefs that the salvation of urban dwellers would be found in nature.

Wilderness Preservation Versus Natural Resource Conservation

As more people began to imagine the value of the environment, diverging viewpoints came about as to its use. Should we set spaces aside where humans tread lightly in order to enable nature to thrive? Or should we find ways to cultivate nature efficiently for increasing human demands for wood, paper, drinking water, and more?

John Muir and the Wilderness Preservation Movement

By the 1880s, key figures had begun to argue explicitly for the **preservation** of wilderness areas, that is, to protect from harm and to maintain certain places.[1] Arising out of these efforts were campaigns to designate spectacular regions of natural scenery as preservation areas, such as Yosemite Valley and Mariposa Giant Tree Grove in California. The discourse of preservation was invoked to ban commercial use of these areas, instead keeping them for appreciation, study, and low-impact outdoor recreation.

One of the leaders of the U.S. preservation movement was Scottish immigrant John Muir, who was influenced by Thoreau and whose own literary essays in the 1870s and 1880s did much to arouse national sentiment for preserving Yosemite Valley. Communication scholar Christine Oravec (1981) has observed that Muir's essays evoked a Romantic **sublime response** from his readers through his description of the rugged mountains and valleys of the Sierra Nevada. This response on the part of readers was characterized by (a) an immediate awareness of a sublime object (such as Yosemite Valley), (b) a sense of overwhelming personal insignificance and awe in the object's presence, and (c) ultimately a feeling of spiritual exaltation (p. 248).

Muir's influence and the support of others led to a long-term national campaign to preserve Yosemite Valley, including the art of George Catlin (Mackintosh, 1999) and the landscape photographs of Carleton Watkins (DeLuca & Demo, 2000). By 1890, these combined efforts resulted in the U.S. Congress's creation of Yosemite National Park, "the first successful proposal for preservation of natural scenery to gain widespread national attention and support" from the public (Oravec, 1981, p. 256).

Logging of giant redwood trees along California's coast in the 1880s also fueled interest in the preservation movement. Laura White and the California Federation of Women's Clubs were among those who led successful campaigns to protect redwood groves in the late 19th century (Merchant, 2005). As a result of these early campaigns, groups dedicated to wilderness and wildlife preservation began to appear: John Muir's Sierra Club (1892), Audubon Society (1905), Save the Redwoods League (1918), National Parks and Conservation Association (1919), Wilderness Society (1935), and National Wildlife Federation (1936). In the 20th century, these groups launched other preservation campaigns that challenged exploitation of these wild lands. The National Parks Act of 1916 established a national system of parks that continues to expand today. Other designations of

Library of Congress

Photo 2.1

In 1903, John Muir led U.S. president Theodore Roosevelt into Yosemite Valley (posing on Overhanging Rock at the top of Glacier Point) as part of his continuing efforts to preserve wilderness areas.

parks, wildlife refuges, and wild and scenic rivers would follow into the 21st century. Preservationists' most significant victory was the 1964 Wilderness Act, which authorized Congress to designate **wilderness** areas using the following definition:

> A wilderness, in contrast with those areas where man [sic] and his own works domi-
> nate the landscape, is hereby recognized as an area where the earth and its community
> of life are untrammeled by man, where man himself is a visitor who does not remain.

Another Viewpoint: Western Conquest and the Construction of "Wilderness"

The idea of wilderness as a concept created by lone European male heroes is a convenient way to share this history, but it is contested (see, for example, more on the role of women in Merchant, 1995).

Communication studies scholar and wilderness advocate Kevin DeLuca (2001) persua-sively has argued that contemporary concerns over the role of the private sector in environ-mental organizations and matters has an ahistorical and mythic perspective of history: "The concern is predicated on imagining environmentalism to be as pristine as the wilderness it valorizes" (p. 634). In contrast, DeLuca points out:

> In 1851, Captain James Savage and the Mariposa Battalion stumbled upon Yosemite Valley in pursuit of their genocidal goal of cleansing the region of Native Americans. For the Ahwahneechee, Yosemite was not wilderness but home. The campaign of the Mariposa Battalion in the 1850s literally and figuratively cleared the ground for the construction of Yosemite as pristine wilderness. One of the soldiers, Lafayette Bunnell, admired the scenery and, recognizing the tourist potential, established a toll road in 1856. In less than a decade, Yosemite Valley passed from Ahwahneechee home to tourist attraction and wilderness icon. (p. 638)

> DeLuca argues that the fact that wilderness is a social construction with a bloody history should not deter us from defending its preservation, but "preservation must rest on the rec-ognition that wilderness is not a divine text but a significant social achievement. The preser-vation and expansion of that achievement depends on making arguments about the worth of wilderness" (p. 649).

Gifford Pinchot and the Conservation of Natural Resources

Muir's ethic of wilderness preservation clashed with a competing vision that sought to manage America's forests more like a "natural resource" that needed to be culti-vated and harvested. Influenced by the philosophy of **utilitarianism,** the idea of the greatest good for the greatest number, some in the early 20th century began to pro-mote a new conservation discourse, which promoted economic gain as the primary value to arbitrate contested environmental decisions. Associated principally with Gifford Pinchot, President Theodore Roosevelt's chief of the Division of Forestry (now the U.S. Forest Service), the term **conservation** interpreted the most valuable relationship with the environment to be "the wise and efficient use of natural resources" (Merchant, 2005, p. 128). That is, while conservationists tended to enjoy the outdoors for hunting, fishing, hiking, and more, they believed human relationships with the environment ultimately should be determined by economic demands. For example, in managing public forest lands as a source of timber, Pinchot instituted a sustained yield policy, according to which logged timberlands were to be reforested after cutting, to ensure future timber supplies (Hays, 1989; Merchant, 2005.) In the following decades, Pinchot's conservation approach strongly influenced the management of natural resources by U.S. government agencies such as the Forest Service and the Bureau of Land Management (BLM).

The debate between Muir's ethic of preservation and Pinchot's conservation approach came to a head in the fierce controversy over the building of a dam in Hetch Hetchy Valley in Yosemite National Park. In 1901, the City of San Francisco's proposal to dam the river running through this valley as a source for its residents' water supply sparked a multiyear dispute over the purpose of the new park with the Sierra Club launching a grassroots campaign to stop the dam. This conflict over the value of national parks and of the environment more broadly defined would continue long after conservationists won and the Hetch Hetchy dam was approved in 1913. (For more information, see www.sierraclub.org/ca/hetchhetchy/history.asp.)

Cultivating an Ecological Consciousness

The tension between the discourses of wilderness preservation and conservation of natural resources continues to be a feature in some current debates. As early as the 1980s, a split had developed in the U.S. movement as a result of the perceived failure of mainstream environmental groups to preserve more wild lands. Disillusioned wilderness activists formed the radical group Earth First! to engage in **direct action,** or physical acts of protests, such as road blockades, sit-ins, and **tree spiking**[2] to prevent logging in old growth forests. Their first action was to dramatize another dam debate by draping cloth down the side of the Glen Canyon dam in Arizona to make it appear cracked. Other groups, such as the Earth Liberation Front, turned for a while to arson and property damage in a controversial effort to protect the shrinking habitat of endangered species.

Internationally, environmental, indigenous, and climate justice groups have also campaigned for the protection of remaining large areas of forests, both for their importance for habitat and the homes of indigenous peoples and also for their importance in mitigating or slowing climate change. For example, scientists, environmental activists, and advocates for indigenous peoples like the Coalition of Rainforest Nations have succeeded in gaining some measure of protection through the United Nation's program, Reducing Emissions from Deforestation and Forest Degradation (REDD). Using financial incentives, REDD encourages developing nations not to cut their forests. Although a major goal is to reduce carbon dioxide (CO_2) emissions from deforestation, the protection of these areas also preserves biodiversity, natural resources, and communal areas for forest communities. In addition, Wangari Maathai (pictured at the beginning of this chapter) founded the Greenbelt Movement in

Kenya in the 1970s with the belief that planting trees, peace, and women's rights could be intertwined. In recognition of the significance of her transformative efforts, in 2004, she became the first African woman to receive the Nobel Peace Prize. Overall, the contemporary movement to preserve and conserve nature as wilderness or as a natural resource consists of a broad and diverse range of both voices and strategies that set the stage for a new ecological sensibility.

The prefix *eco* in *ecology* (and *economics*) has roots in ancient Greece with the term *oikos*, meaning a house or dwelling. Yet it wasn't until the turn of the 20th century that a German scientist and artist, Ernst Haeckel (1904), coined the modern term **ecology** as the study of how an organism relates with its exterior world. The modern environmental movement remains heavily influenced by early 20th-century ecologists and core terms they have identified, such as **resilience**, an organism's ability to adapt and to persist at the same time. This perspective assumes the environment always is dynamic or changing but also recognizes limits to a species' ability to adapt before failing to thrive.

In the mid-20th century, an ecological consciousness began to be articulated, including Aldo Leopold's 1949 classic *A Sand County Almanac*, where he defined a land ethic as:

> (1) The land is not merely soil. (2) That native plants and animals kept the energy circuit open; others may or may not. [and] (3) That man-made [sic] changes are of a different order than evolutionary changes, and have effects more comprehensive than is intended or foreseen. (p. 255)

This growing sense of nature as dynamic carried into urban spheres and human communities as well, and it is to this third antagonism that we now turn.

Public Health and the Ecology Movement

By the 1960s in the United States much was changing. Among the many social justice movements mobilizing during this time, urban activists fostered another pivotal antagonism based on an ecological perspective. At a time when environmental protections for the **commons** (resources accessible to all people and not privately owned, such as air, water, and the Earth) were weak or nonexistent, everyday people in the United States joined growing concerns among ecological scientists and began to question the effects of urban pollution, nuclear fallout, and pesticides on human health. Their concerns included the air and water emissions from factories and refineries, abandoned toxic waste sites, exposure to chemicals used to control agricultural pests, and the radioactive fallout from aboveground nuclear testing.

Rachel Carson and the Public Health Movement

Often, biologist and writer Rachel Carson is credited for voicing the first nationally recognized public challenge to business practices that affect the environment, including public health or the prevention of disease and prolonging human life. In her book *Silent Spring*, Carson (1962) wrote, "We are adding a . . . new kind of havoc—the direct killing of birds, mammals, fishes, and indeed, practically every form of wildlife by chemical insecticides indiscriminately sprayed on the land" (p. 83). Fearful of the ecological consequences of insecticides like DDT (dichlorodiphenyltrichloroethane), she warned that modern agribusiness had "armed itself with the most modern and terrible weapons, and that in turning them against the insects it has also turned them against the earth" (p. 262). With her prescient writings, public advocacy on behalf of national toxic legislation, and untimely death from cancer, Rachel Carson is widely considered the founder of the modern environmental movement.

Although *Silent Spring* did prefigure a popular movement, earlier voices from the 1880s through the 1920s had warned of dangers to human health from poor sanitation and occupational exposures to lead and other chemicals. Trade unions, *sanitarians* (reformers from Jane Addams's Hull House in Chicago), and public health advocates had warned of hazards to both workplace and urban life: "contaminated water supplies, inadequate waste and sewage collection disposal, poor ventilation and polluted and smoke-filled air, [and] overcrowded neighborhoods and tenements" (Gottlieb, 1993, p. 55).

Ellen Swallow Richards was the first woman admitted to MIT, earning her degree in 1873. Richards was trained as a chemist; although she is considered the founder of home economics, she is an underrecognized early thinker on what has become known as "ecology" through her work on water and pollution, as well as other areas of research (Pezzullo, 2014). Dr. Alice Hamilton also was "a powerful environmental advocate in an era when the term had yet to be invented" (Gottlieb, 1993, p. 51) who worked in the 1920s to reform the "dangerous trades" of urban workplaces. With the publication of *Industrial Poisons in the United States* and her work with the Women's Health Bureau, Hamilton (1925) became "the country's most powerful and effective voice for exploring the environmental consequences of industrial activity" (p. 51) including the impacts of occupational hazards on women and minorities in the workplace.

Still, until *Silent Spring* in 1962, there was no such thing as an "environmental movement" in the United States in the sense of a "concerted, populous, vocal, influential, active" force (Sale, 1993, p. 6). However, by the late 1960s, news coverage of air pollution, nuclear fallout, fires on the Cuyahoga River near Cleveland, Ohio, when its polluted surface ignited, and oil spills off the coast of Santa Barbara, California, fueled a public outcry for greater protection of the environment.

Earth Day and Legislative Landmarks

By the first **Earth Day** on April 22, 1970, students, public health workers, activist groups, and urban workers had coalesced into a movement to champion environmental controls on industrial pollution. Some 20 million people took part in protests, teach-ins, and festivals throughout the country in one of the largest demonstrations in American history.

At the same time, new nongovernmental organizations arose to address the relationship between human health and the environment. Among the earliest were the Environmental Defense Fund (1967), Environmental Action (1970), and the Natural Resources Defense Council (1970). The popularity of the paradigm shift of ecology, expressed through the discourse of the modern environmental movement, led lawmakers to enact bold, new federal legislation in 1970, including the National Environmental Policy Act (NEPA), the Clean Air Act, the Occupational Safety and Health Act (OSHA), and the creation of the U.S. Environmental Protection Agency.

Despite these landmarks, struggles to hold polluting entities accountable for their waste and impact on everyday people continue in the United States and globally. (See photo 2.2.) Communities became increasingly worried by the chemical contamination of their air, drinking water, soil, and school grounds. For example, the small New York community of Love Canal became a symbol of the nation's widening consciousness of the hazards of chemicals. (See the brief description of the Love Canal case in Chapter 1.) Ordinary citizens felt themselves surrounded by what Hays (1989) termed the "toxic 'sea around us'" (p. 171) and began to organize in community-based groups to demand cleanup of their neighborhoods and stricter accountability of corporate polluters.

Prompted by the toxic waste scandals at Love Canal and other places, such as Times Beach, Missouri, the U.S. Congress passed the Superfund law of 1980, which authorized

©iStockphoto.com/Hung_Chung_Chih

| Photo 2.2 | Smog emergency shuts down Beijing, China. According to the World Health Organization, the air pollution in other cities may be worse. How does wearing a face mask to breathe every day change a child's or an adult's perception of the meaning of *environment*? |

the Environmental Protection Agency (EPA) to clean up toxic sites and take action against the polluting parties. Local citizens also took advantage of new federal laws such as the Clean Water Act to participate in agencies' issuance of air and water permits for businesses. Both structural change of public participation in public venues and individual consumer habits still continue to warrant attention. (We discuss legal spheres further in Chapters 12 and 14 and individual negotiations of risk more in Chapter 7.)

Environmental Justice: Linking Social Justice and Environmental Quality

Even as the U.S. environmental movement widened its concerns in the 1960s to include public health along with wilderness preservation, there remained a definition of the environment that provided contradictory accounts of humans' place in nature and assumed a "long-standing separation of the social from the ecological" (Gottlieb, 2002, p. 5). By the 1980s, however, new activists from low-income groups and communities of color within and outside the United States had begun to gain ground in challenging the dominant perceptions of nature as a wild place apart from people's everyday lives, disclosing another notable antagonism in prevailing cultural views of the environment: environmental justice.

Redefining the Meaning of "Environment"

Despite earlier efforts to bring environmentalist, labor, civil rights, and religious leaders together in the 1960s and 1970s,[3] U.S. environmental organizations largely failed to recognize the problems of urban residents and people of color communities. Interdisciplinary scholar Giovanna Di Chiro (1996) reports, for example, that in the mid-1980s, residents in south central Los Angeles who were trying to stop a solid waste incinerator from being located in their neighborhood discovered that "these issues were not deemed adequately 'environmental' by local environmental groups" (p. 299). Identifying a salient antagonistic limit in communication, activists in communities of color

AP Photo/LM Otero

| Photo 2.3 | The environmental justice movement reminds us that humans are part of the environment and that the environment matters to who we are, everything we do, and everywhere we go. |

were particularly vocal in criticizing mainstream environmental organizations for being "reluctant to address issues of equity and social justice, within the context of the environment" (Alston, 1990, p. 23).

In a historically significant move, the movement proposed to redefine the word environment to mean "where we live, where we work, where we play, and where we learn" (Lee, 1996, p. 6). The discourse articulated a new perspective for environmental matters in the United States and beyond in which U.S. civil rights and environmental advocacy became intertwined in their values, tactics, and solutions.

For example, a key moment in the launching of this new movement occurred in 1982 with the protests by residents of the largely African American community of Warren County, North Carolina. Local residents and leaders of national civil rights groups tried to halt the state's plans to locate a toxic waste landfill in this rural community by sitting in roads to block 6,000 trucks carrying soil contaminated by toxins known as PCBs (polychlorinated biphenyls).[4] More than 500 arrests occurred from this act of **civil disobedience**, a peaceful form of protest that violates laws and accepts legal consequences (such as arrest) in order to point out ongoing injustices. Sociologists Robert Bullard and Beverly Hendrix Wright (1987) called this moment in Warren County "the first national attempt by blacks to link environmental issues (hazardous waste and pollution) to the mainstream civil rights agenda" (p. 32). (For more on the significance of this event as a *story of origin* in the movement, see Pezzullo, 2001.)

With similar struggles in other parts of the nation and reports of the heavy concentration of hazardous facilities in minority neighborhoods, some charged that these communities suffered from a form of **environmental racism** (Sandler & Pezzullo, 2007, p. 4) or, more broadly, environmental injustice (Roberts, 2007, p. 289). Residents and critics alike began to speak of being poisoned and "dumped upon" and of certain communities targeted as "sacrifice zones" (Schueler, 1992, p. 45). Importantly for these critics, the term *environmental racism* meant not only threats to their health from hazardous waste landfills, incinerators, agricultural pesticides, sweatshops, and polluting factories but also the disproportionate burden that these practices placed on people of color and the workers and residents of low-income communities. The discourse of environmental injustice, therefore, both described what people were experiencing and helped constitute new possibilities.

Defining Environmental Justice

Emerging from these grassroots struggles, for example, was a robust discourse of **environmental justice.** For most activists, this term connected the safety and quality of the environments where people lived and worked with concerns for social and economic justice. Residents and movement activists insisted that environmental justice referred to the basic right of all people to be free of poisons and other hazards. At its core, environmental justice also was a vision of the democratic inclusion of people and communities in the decisions that affect their health and well-being.

Many people criticized decision-making processes that failed to provide meaningful participation "for those most burdened by environmental decisions" and called for at-risk communities to share more fully in those decisions that adversely impact their communities (Cole & Foster, 2001, p. 16; also see Gerrard & Foster, 2008).

The demand for environmental justice received significant publicity in 1991, when delegates from local communities, along with national leaders of civil rights, religious, and environmental groups, convened in Washington, DC, for the first People of Color Environmental Leadership Summit. For the first time, different strands of the emerging movement for environmental justice came together to challenge mainstream environmental discourse. The delegates also adopted a powerful set of **Principles of Environmental Justice** that enumerated a series of rights, including "the fundamental right to political, economic, cultural, and environmental self-determination of all peoples" (*Proceedings,* 1991, p. viii).

☞ FYI   Principles of Environmental Justice (1991)

The following are excerpts from the 1991 text adopted by delegates to the First People of Color Environmental Leadership Summit, Washington, DC.

WE, THE PEOPLE OF COLOR, gathered together at this multinational People of Color Environmental Leadership Summit, to begin to build a national and international movement of all peoples of color to fight the destruction and taking of our lands and communities . . . do affirm and adopt these Principles of Environmental Justice:

1. Environmental Justice affirms the sacredness of Mother Earth, ecological unity and the interdependence of all species, and the right to be free from ecological destruction.

2. Environmental Justice demands that public policy be based on mutual respect and justice for all peoples, free from any form of discrimination or bias. . . .

3. Environmental Justice demands the right to participate as equal partners at every level of decision making, including needs assessment, planning, implementation, enforcement and evaluation.

4. Environmental Justice affirms the right of all workers to a safe and healthy work environment without being forced to choose between an unsafe livelihood and unemployment. It also affirms the right of those who work at home to be free from environmental hazards.

5. Environmental Justice protects the right of victims of environmental injustice to receive full compensation and reparations for damages as well as quality health care. . . .

6. Environmental Justice affirms the need for urban and rural ecological policies to clean up and rebuild our cities and rural areas in balance with nature, honoring the cultural integrity of all our communities, and provides fair access for all to the full range of resources. . . .

SOURCE: "Principles of Environmental Justice" at: www.ejnet.org/ej/principles.html.

In 1994, the environmental justice movement achieved an important goal when President Clinton issued an executive order directing each federal agency to "make achieving environmental justice part of its mission by identifying and addressing . . . disproportionately high and adverse human health or environmental effects of its programs, policies, and activities on minority populations and low-income populations in the United States" (Executive Order No. 12898, 1994). Nevertheless, the movement continues to face real-world, on-the-ground challenges to building more just and ecologically sound communities within the United States and globally. (We describe the contemporary movement for environmental justice in more detail in Chapter 10.)

Movements for Sustainability and Climate Justice

Since at least the late 1980s, a diverse movement has grown in many countries. Throughout Europe, Asia, North and South America, Africa, Australia, and the Pacific Islands, countless numbers of local and regional groups have challenged environmentally dangerous and inequitable practices in their societies. These challenges often are similar to the antagonisms described earlier—efforts to protect natural systems, safeguard human health, and secure social justice. In *Blessed Unrest,* Paul Hawken (2007) describes a remarkable sample of these diverse efforts:

> It is composed of families in India, students in Australia, farmers in France, the landless in Brazil, the Bananeras of Honduras, the "poors" of Durban, villagers in Irían Jaya, indigenous tribes of Bolivia, and housewives in Japan. . . . These groups defend against corrupt politics and climate change, corporate predation and the death of oceans, governmental indifference and pandemic poverty, industrial forestry and farming, and depletion of soil and water. (pp. 11, 164–165)

These and thousands of other efforts reflect diverse agendas and methods of communication. Yet running through all is the growing recognition of a "limit" or the unsustainability of "the currently disruptive relationship between earth's two most complex systems—human culture and the living world" (Hawken, 2007, p. 172).

Introducing Sustainability

In response to these diverse signs of crisis, a new antagonism has begun to emerge—the challenge of building a more *sustainable* world in the face of disruptive or unsustainable social and economic systems. *Sustainability* is admittedly an elusive term. For some, the sustainability movement "is nothing less than a rethinking and remaking of our role in the natural world. It is a recalibration of human intentions to coincide with the way the biophysical world works" (Edwards, 2005, p. xiv). For many, the .

movement encompasses three goals or aspirations—Environmental protection, Economic health, and Equity or social justice—often called the Three Es (Edwards, 2005, p. 17). The rhetorical appeal of sustainability, as Tarla Rai Peterson (1997) argues, "lies in its philosophical ambiguity and range" (p. 36). For our purposes, a working definition of **sustainability** will be the capacity to negotiate environmental, social, and economic needs and desires for current and future generations. We will discuss this movement more in Chapter 11.

Moving Toward Climate Justice

Sustainability efforts implicate the local and the global. The **climate** refers to average atmosphere changes over a long period of time. **Climate justice** is a global movement that recognizes the intertwined relationship between global warming and social injustice. That is, climate justice activists point out that (1) the root causes of our current climate crisis are primarily due to First World fossil fuel consumption and (2) low-income communities and communities of color around the globe are the people who are most significantly impacted first. Thus, solutions to the climate crisis are to be found in movements for increased democratic public participation and accountability across economic and national boundaries.

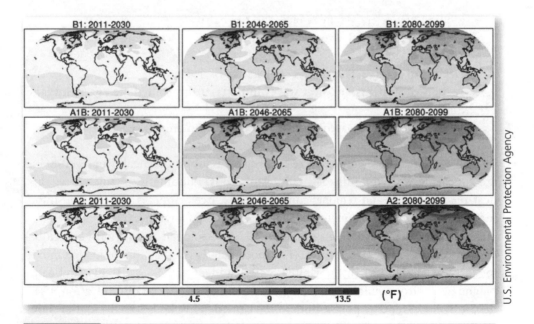

U.S. Environmental Protection Agency

| Photo 2.4 | The climate is a long-term assessment of Earth's changing atmosphere. Weather is much more short-term and place specific; for this reason, some of us may feel confused about global "warming," when we can see snow out the back window! Above is the Intergovernmental Panel on Climate Change's (IPPC) projected average global temperatures, for three scenarios. |

More specifically, both sustainability movements and climate justice movements are critical of the unrestrained growth of carbon-based economies, often called a **business as usual (BAU)** scenario of the future. Carbon is the energy source—primarily fossil fuels such as oil, coal, and natural gas—used by human societies to produce electricity, fuel transportation, and sustain other dimensions of life. Questioning these energy sources has been important because climate scientists believe there is an increased warming of the Earth due principally to the emissions (CO_2) caused by combustion of fossil fuels as well as other so-called greenhouse gases.

That the Earth is warming is no longer in doubt. The most recent IPCC (2013) determined that warming of the Earth's climate is now "unequivocal" (p. 1). Furthermore, the evidence for anthropogenic or human influence in this warming has grown. The latest IPCC report concluded, "It is *extremely likely* that human influence has been the dominant cause of the observed warming since the mid-20th century" (p. 15).

As a result, growing numbers of scientists, health officials, students, and many grassroots groups are questioning the BAU or unsustainable approach to economic growth. This is because the impacts of a rapidly warming climate are already occurring in many regions. These include the following:

- Melting of glaciers, Arctic sea ice, and permafrost
- Decline in snow-fed rivers and water sources in North and South America, Asia, and Africa
- Decline of coral reefs and increased acidity of oceans
- Sustained droughts and heat waves in parts of the United States, Africa, and other areas
- Migration (and increased extinction) of plant and animal species

And with greater warming, severe consequences for human health and well-being are appearing: spreading disease, failures of crops, flooding of populated regions, heat-related deaths, and regional conflicts over scarce resources. (For further information, see the IPCC's *Climate Change 2013* reports at www.ipcc.ch/.)

In response, a movement for climate justice has emerged, urging local and global leaders to take immediate action. (For more details on the climate justice movement, see Chapter 10.) In the United States, student activism, TV ads, and campaigns by environmental organizations have raised the alarm and set off a public debate over U.S. energy policy. Both in the United States and globally, tens of thousands of activists and local groups are connecting through social networking sites such as 350.org, itsgettinghotinhere.org, and the Climate Action Network (climate-network.org). And scientists, concerned about the slowness of governmental response, have spoken publicly. In a series of reports, the U.S. National Research Council advised the U.S. Congress that, "climate change . . . poses significant risks for a broad range of human and natural systems" ("Top," 2010, para. 2).

Act Locally!

Local Resources and Antagonisms

There are undoubtedly many resources on your campus or in your community—faculty, community leaders, activists, and other professionals—who are knowledgeable about one or more of the antagonisms described in this chapter: nature as a romantic, national, or transcendental resource; wilderness preservation; natural resource conservation; an ecological perspective that includes public health and the environmental commons; environmental justice; sustainability; and climate justice.

Invite someone with knowledge about or personal experience in one of these areas to visit your class or campus organization. Ask him or her to speak about the specific antagonism you've chosen—that is, speak about the efforts by individuals, groups, or movements to contest prevailing views and raise awareness among wider publics about the protection of natural areas, the effects of air or water pollution on public health, the persistence of environmental racism in society, and so forth. Among the questions you might ask these individuals, consider these:

- How did you come to study (or be engaged with) this issue? What do you find most interesting?
- What progress has been made in addressing this problem, and what are the most difficult challenges remaining?
- What role has communication played in either perpetuating the problem or in finding solutions?
- What resources exist locally? How can we become more involved with this issue?

SUMMARY

John Muir famously said: "When we try to pick out anything by itself, we find it hitched to everything else in the Universe."[5] Indeed, throughout the history of environmental values and actions, antagonisms are debated over how we can grapple with the interconnectedness of human and nonhuman life, as well as humans with other humans. As we have seen, "the environment" is subject to redefinition as new voices and interests contest prevailing understandings. The core of these challenges is a distinctly rhetorical process, through which humans influence, question, and persuade.

One way to analyze discourses of individuals, activists, movements, and community members is to identify emerging antagonisms of a culture in a particular period of history. We hope this chapter has helped illustrate the ways in which we can identify these pivotal cultural moments through discourse and how a culture's prevailing values may be transformed.

In this chapter, we've described five historical periods in which individuals, organizations, and movements rhetorically contested dominant cultural ideas about the environment and constituted new ones through discourse. Each moment is defined by antagonisms, the recognition of the limit of this idea, viewpoint, or ideology. A limit is recognized when questioning or contesting a prevailing view reveals it to be inadequate or unresponsive to new demand and, therefore, creates an opening for alternative voices and ideas. In U.S. history, we have identified five key moments of historical transformation in our perceptions of the environment:

1. American values from the 1600s to 1900s that began to foster a more favorable relationship with nature, as Americans understood it, included romanticism, nationalism, and transcendentalism.

2. Two movements in the United States challenged dominant colonial views about nature solely as something to be conquered and abhorred: a 19th-century preservationist movement that valued nature as wilderness and an early 20th-century ethic of conservation of natural resources.

3. The pollution of the 20th century gave birth to an ecological sensibility and a challenge to urban pollution that made possible a concept of the environmental commons and a renewed commitment to public health.

4. A grassroots movement for environmental justice arose in the late 1970s that challenged mainstream views of nature as a place apart from where people work, live, learn, and play.

5. In the late 1980s, global movements for sustainability and climate justice began to challenge business-as-usual carbon economy models and make connections between social justice and environmental quality from our backyards to our atmosphere.

These antagonisms reveal the highly contingent nature of our understanding and viewpoints about the environment. Pivotal to these cultural shifts and our ability to contest them is a powerful process of social–symbolic construction, which we describe in more detail in the next chapter.

SUGGESTED RESOURCES

- Bullard, R. (2013). The father of environmental justice. Retrieved from http://drrobertbullard.com/
- Goodwin, N. (Writer & Director). (2007). *Rachel Carson's* Silent Spring [Television series single episode]. In J. Crichton (Executive producer), *American experience.* Arlington, VA: PBS.
- Intergovernmental Panel on Climate Change. (2013). *Climate Change 2013* reports. Retrieved from http://www.ipcc.ch
- Mackintosh, B. (1999). The National Park Service: A brief history. Retrieved from http://www.cr.nps.gov/history/hisnps/npshistory/npshisto.htm

- Philippine movement for climate justice. (2013). Retrieved from http://climate justice.ph/
- The Green Belt Movement. (2014). Wangari Maathai: Key speeches and articles. Retrieved from http://www.greenbeltmovement.org/wangari-maathai/key-speeches-and-articles
- *Wrenched*. (2014). [Motion Picture]. Retrieved from http://wrenched-the-movie .com/

KEY TERMS

Antagonism   32

Business as
usual (BAU)   46

Civil disobedience   42

Climate   45

Climate justice   45

Commons   38

Conservation   36

Direct action   37

Discourse   32

Earth Day   39

Ecology   38

Environmental justice   42

Environmental racism   42

Nature   34

Preservation   34

Principles of Environmental
Justice   43

Resilience   38

Sublime   33

Sublime response   35

Sustainability   45

Transcendentalism   33

Tree spiking   37

Utilitarianism   36

Wilderness   36

DISCUSSION QUESTIONS

1. European colonialists initially were scared of and threatened by nature, but a love of nature arose once European settlers felt a distance from nature in their everyday lives. Do you feel more comfortable in the woods or downtown in a city? Why do you think that is so? Does your comfortability shape what you value?

2. Is wilderness merely a symbolic construction? Does this matter to whether or not you want to protect it?

3. Many people have become motivated to care about ecology (particularly air quality and water quality) because they have or know someone with asthma, cancer, or another illness that is environmentally triggered. Do you or someone you know have health concerns that shape your relationship with the environment?

4. Do you believe the environmental justice movement invites a distinct understanding of the environment or an overlapping one? Re-read the Principles of Environmental Justice. What similarities and differences do you perceive between the attitudes we describe of early European colonizers and contemporary environmental justice advocates?

5. When you hear the word *climate*, what other words come to mind? Many find it overwhelming to try to imagine something on the scale of the climate, let alone a climate crisis. Which label do you prefer to describe the atmospheric challenges we face: *climate change*, *climate crisis*, *global warming*, *climate disruption*, or something else? Explain your choice.

NOTES

1. In 1872, President Ulysses S. Grant signed a law designating 2 million acres for Yellowstone National Park, the nation's first national park. And 13 years later, the state of New York set aside 715,000 acres in its Adirondack Mountains. Still, these first acts were less motivated by an aesthetic or spiritual appreciation of wilderness than by a desire to protect against land speculation and a need to protect forest watersheds for New York City's drinking water (Nash, 2001, p. 108).

2. Tree spiking is the practice of driving spikes or long nails into some trees in an area that is scheduled to be logged. The metal—or sometimes plastic—spikes do not actually hurt the trees. Tree spiking can discourage loggers from cutting the trees because of possible damage to their chain saws or, later, to the machinery in timber mills when saw blades strike the hidden spikes.

3. The 1971 Urban Environment Conference (UEC) was one of the first successful efforts to link environmental and social justice concerns. The UEC sought "to help broaden the way the public defined environmental issues and to focus on the particular environmental problems of urban minorities" (Gottlieb, 1993, pp. 262–263; Kazis & Grossman, 1991, p. 247).

4. Although the State of North Carolina completed the landfill in Warren County in 1982, local activists persisted in calling for its detoxification. Two decades later, in 2004, their efforts finally paid off when the state cleaned up the landfill (Pezzullo, 2001).

5. This statement by Muir is often misquoted. Given the popularity of online research today, we recommend you read this website about popular Muir misquotations and how the Sierra Club judges the most accurate language used: www.sierraclub.org/john_muir_exhibit/writings/misquotes.aspx.

PART II

Constructions of the Environment

Our language (terministic screens) powerfully shapes or mediates our experiences—what is selected for notice, what is deflected from notice, and therefore how we perceive the environment. "Organic" apples generally are those grown and minimally processed without synthetic pesticides, fertilizers, artificial ingredients, or preservatives. "Conventional" apples are those sprayed, as in this image, with more pesticides than any other fruit or vegetable. How do these labels shape our perceptions and choices?

CHAPTER 3

Symbolic Constructions of Environment

Symbolic and natural systems are mutually constituted.

—Depoe (2006, p. vii)

I think people have this insatiable desire to point at and name things. "Oh, that's a . . ." or "What is that?"

—Whale tour boat captain (quoted in Milstein, 2011, p. 4)

As we saw in the in the last chapter, cultural perceptions of the environment may change as new voices and interests arise to contest or challenge prevailing discourses. The core of these challenges is a human process of communication, including symbolic construction and public negotiations. In this chapter, we want to build on our definition of environmental communication in Chapter 1 as both a pragmatic and constitutive vehicle for our understanding of and with the environment to establish a more in-depth perspective of how symbolic constructions and the environment are intertwined.

Chapter Preview

In this chapter, we introduce a symbolic approach to environmental studies in order to better understand the different ways in which "environment" is constructed and negotiated in the public sphere. To do so, we define several key terms:

1. A rhetorical perspective, particularly through the use of terministic screens and naming, helps remind us of the importance of language in shaping our world.

(Continued)

(Continued)

2. The rhetorical situation is an effective way to analyze the ways we construct environmental exigencies, audiences, and constraints.

3. Tropes and genres are useful to interpret human dramas about the environment.

4. Communication frames direct our attention to certain meanings or stories about the environment and offer a guide for our understanding.

5. Dominant and critical discourses provide a means to analyze prevalent beliefs, values, and actions, as well as the voices that challenge these.

By the late 20th century, scholars such as Donna Haraway (1991), Andrew Ross (1994), Klaus Eder (1996b), Bruno Latour (2004), and Neil Evernden (1992) had begun to describe the discursive constructions that shape our views of the environment. Herndl and Brown (1996), in fact, argue that "environment" is

> a concept and an associated set of cultural values that we have constructed through the way we use language. In a very real sense, there is no objective environment in the phenomenal world, no environment separate from the words we use to represent it. (p. 3)

This **symbolic perspective** focuses on the cultural sources that construct our perceptions of the world.

This is not to suggest that there is no material world. Of course there is. Try holding your breath while reading this chapter, and you soon will be reminded how important clean air is to your ability to function. But it is through different symbolic modes that we understand and engage this world, infuse it with significance, and act toward it. Holding our breath to swim underwater for fun is quite different from holding our breath because the pollution in an area might cause an asthma attack. Deciding whether or not we should actively protect clean water and air is a decision we make daily that has material impacts well beyond our own bodies. As rhetorical scholar Stephen P. Depoe (2006), founding editor of the journal *Environmental Communication: A Journal of Nature and Culture,* bluntly put it, "Symbolic and natural systems are mutually constituted" (p. vii). The environment affects us, but our language and other symbolic action also have the capacity to affect or construct our perceptions of the environment itself.

The environment is something that we know, at least partly, through symbols. In Chapter 1, we discussed the pragmatic and constitutive dimensions of environmental communication. The history we reviewed in Chapter 2 also made clear that different choices are possible and that the discourses surrounding these choices construct diverse meanings for the worlds we know. As a result, some scholars adopt a more specifically rhetorical perspective to study the different language choices by which journalists, scientists, corporations, environmentalists, and citizens attempt to influence our perceptions and behaviors toward the environment. The remaining sections of this chapter aim to provide an even wider range of concepts from rhetorical studies.

A Rhetorical Perspective

The study of rhetoric traces its origins to classical Greek philosopher teachers such as Isocrates (436–338 BCE) and Aristotle (384–322 BCE), who taught the arts of citizenship to political leaders in democratic city–states such as Athens. The practice in these city–states was for citizens to speak publicly in law courts and the political assembly, where each citizen represented his own interests. (In Athens and other cities, civic speech was limited principally to native-born, property-owning, male citizens.) As a result, competency in public speaking, debate, and persuasion was vital for conducting civic business—war and peace, taxes, construction of public monuments, property claims, and so forth.

It was during this period that Aristotle instructed his students in the art of civic speaking when he defined **rhetoric** as "the faculty of discovering in any given case the available means of persuasion" (Herrick, 2009, p. 77). This art of rhetoric rested not simply on skillful delivery but on the ability to discover the resources for persuasion that were available in a specific situation. This draws our attention to rhetoric as a purposeful choice among the available means useful in accomplishing some effect or outcome. As a result, we can say that a **rhetorical perspective** focuses on purposeful and consequential efforts to influence society's attitudes and behavior through communication, including public debate, protests, advertising, and other modes of symbolic action (Campbell & Huxman, 2008). This definition also underscores rhetoric as a perspective that is social or publically oriented and contextual. By *contextual* we mean that, in addition to being publicly oriented, rhetoric also addresses *particular* situations, audiences, and concerns about the environment that are not universal. What might be relevant or compelling in one place with one audience regarding a particular topic might not be in another. (Consider, for example, how you might make an appeal to your parent or guardian to go on a trip abroad compared to persuading a peer.)

Although rhetoric traditionally has been viewed as an instrumental or pragmatic activity—persuading others—its use clearly has a second function as noted in Chapter 1: The purposeful use of language also helps to constitute our perception of the world itself. A rhetorical perspective, then, invites us to be sensitive to this deeper meaning, even as we identify the more pragmatic, available means used in communication about the environment. Let us now consider some of these rhetorical resources.

Terministic Screens and Naming

A recognition of this constitutive function appears within the field of rhetorical studies. For example, Kenneth Burke (1966) used the metaphor of screens to describe the way our language orients us to see certain things, some aspects of the world, and not others: "If any given terminology is a reflection of reality, by its very nature as a terminology it must be a selection of reality; and to this extent it must function also as a deflection of reality" (p. 45). That is, our **terministic screens** powerfully shape or mediate our experiences—what is selected for notice, what is deflected from notice, and therefore how we perceive our world. As a result, whenever we communicate, we actively participate in constituting our world.

One important dimension of terministic screens is **naming**—the means by which we socially represent objects or people and therefore know the world, including the natural world. The act of pointing and naming something out there in the world, as Milstein (2011) reminds us, is "a foundational act"; pointing and naming are "the basic entry to socially discerning and categorizing parts of nature" (p. 4). And in doing so, naming also indicates "an orientation" to the world and thus "influences our interaction with it" (Oravec, 2004, p. 3).

Consider, as an example, the massive loss of biodiversity we are experiencing globally. A few persist at calling this trend "natural" (assuming species always have gone extinct due to evolutionary trends), others an "act of God" (believing humans have little agency because God must have a plan that includes species going extinct), others believe it is due to a "climate crisis" (noting that mass extinctions have happened before but rarely turn out well for species wanting to survive them), and the list could go on. While some are aware of these labels as reflections of their value systems, others may not realize how choosing one terministic screen might deflect attention away from another.

Another striking example of the use of naming in a deliberate way was the successful campaign by a trade association that works with sewage treatment plants, the Water Environment Federation (WEF). WEF decided to rename sewage sludge as *biosolids*. In the process of treating sewage, a sludge-like material is typically left over. This substance (which sometimes contains life-threatening, toxic chemicals, such as dioxin) is often used as fertilizer on agricultural fields. As a result, many environmental and health groups have raised concerns about the risk from the chemicals found in sewage sludge.

Industry critic Sheldon Rampton described the efforts of WEF in the 1990s to coin another term for this toxic sludge "in hope of escaping its negative connation" and, in the words of a WEF spokesperson, "to win public acceptance for the beneficial use of biosolids" (quoted in Rampton, 2002, p. 348). The campaign also succeeded in having biosolids placed in the *Merriam-Webster Dictionary*. A WEF official reported at the time that he was "pleased that the term 'sludge' will not appear in the definition of biosolids" (quoted in Rampton, 2002, p. 349). Today, WEF's website (www.wef.org) describes biosolids as "the nutrient-rich organic materials resulting from the treatment of domestic waste at a wastewater treatment facility."

☞ **FYI**   ## Naming Water Cultures Versus Water Commodities

Below is an excerpt from a 2008 interview between media studies scholar Andy Opel and Vandana Shiva, a physicist, ecologist, activist, editor, and author of many books. In India, she has established Navdanya, a movement for biodiversity conservation and farmers' rights. The excerpt below is an answer Shiva provides to illustrate the stakes of naming water an economic commodity versus a culture.

Shiva:

When the awareness and consciousness of our living in the water cycle dies, that is when water culture dies. To me, water culture is the consciousness of water, the consciousness of being immersed in a water cycle, the consciousness of knowing that we are 70% water, and that the planet is 70% water, and to tread extremely lightly to ensure that water balance is not destroyed. Heightened water awareness creates water culture and water cultures build into them cultures like the sacredness of India's rivers. If Indians could have such a long-term evolution of civilization in the Ganges basin, it is because the Ganges was related to as a sacred mother nourishing the entire basin. The culture that creates is extremely different from the culture which sees water running into the sea as wasted and sees rivers as wild women to be tamed and creates the most violent technologies for rerouting rivers, imprisoning rivers and drying out rivers. That idea of control that develops technologies that disrupt the water cycle and impair the water culture goes hand in hand and are leading to the current thinking that water is just another commodity on the planet, you don't have to give it any special respect. And every right wing think tank that is promoting and supporting water privatization repeatedly states that water is just another commodity. (Shiva, 2008, pp. 498–499)

Constructing an Environmental Problem: The "Rhetorical Situation"

As we've just seen, we may be alarmed (toxic sludge) or reassured (biosolids) as a result of the selective naming that constructs the issue or problem at hand. We also might worry about global access to clean water or take for granted the idea that some can buy it bottled from a store. German sociologist Klaus Eder (1996a) explains that often it is "the methods of communicating environmental conditions and ideas, and not the state of deterioration itself, which explain . . . the emergence of a public discourse on the environment" (p. 209). This is particularly important when communication names an environmental problem. Political scientist Deborah Stone (2002) observed that problems "are not given, out there in the world waiting for smart analysts to come along and define them correctly. They are created in the minds of citizens by other citizens, leaders, organizations, and government agencies" (p. 156).

The symbolic construction of the environment arises from this ability to characterize certain facts or conditions one way rather than another and, therefore, to name it as a problem or not a problem. Such an ability also recognizes the *constitutive* function of communication we described in Chapter 1. It is for precisely this reason that questions of "how and why certain environmental issues become identified as 'problems'" are such an important part of environmental communication (Tindall, 1995, p. 49). For example, climate skeptics who don't believe humans have any influence on global warming deny it is a problem; instead, they attribute any warming to natural causes: "Climate is always changing. We have had ice ages and warmer periods when alligators were found in Spitzbergen" (Lindzen, 2009, para. 1).

An environmental problem, as discussed in Chapter 2, might be helpfully analyzed through three terms that define a **rhetorical situation**: (1) **exigency,** a set of conditions that have been constituted as a "problem," grievance, or crisis that becomes marked by a sense of urgency; (2) **audience**, the people being addressed, their beliefs, actions, and larger cultural understandings; and (3) **constraints**, the cultural limitations and possibilities of the context.[1] For people making decisions about the environment, considering how an exigency is constructed and then addressed within a particular context is vital to how we make judgments and invent responses. We will revisit these terms in more depth when discussing advocacy campaigns in Chapter 8. For now, let us elaborate on rhetoric.

Tropes and Genres

The efforts of citizens, environmental groups, and others to educate, persuade, and mobilize draw on other resources of language as well. Rhetorical scholars explore a range of such resources—argumentation, narrative accounts, emotional appeals, tropes, and rhetorical genres—in websites, films, campaign materials,

Courtesy of Surfrider Foundation

WHAT GOES IN THE OCEAN GOES IN YOU.

Photo 3.1   Environmental advocates long have tried to create a sense of exigency about plastic pollution in the oceans. This ad by the Surfrider Foundation attempts to create a sense of exigency about the issue by appealing to people who eat sushi. Which rhetorical constraints does this ad address, or fail to address?

ads, and other communication in environmental controversies. Let's look briefly at two of these resources: tropes and genres.

Among the most ubiquitous resources of language are **tropes.** Tropes refer to the use of words that turn a meaning from its original sense in a new direction for a persuasive purpose. Tropes draw on a basic function of language itself. There is "no un-rhetorical 'naturalness' of language," 19th century philosopher Friedrich Nietzsche observed. "All words in themselves and from the start are, in terms of their meaning, tropes" (Mayer, 2009, p. 37).

Over time, specific tropes have acquired familiar names, such as **synecdoche** or the part standing for the whole, as in references simply to melting glaciers to signal the wider impacts caused by global warming. Another common trope is **irony** or the use of language that is the opposite of one's belief, often for humor. The humorous online environmental news source *Grist* especially is adept at reporting on ironic events. For example, in a 2011 story by Christopher Mims on a solar-powered oil field, he notes that the project appears to be run on both "sunshine" and "irony":

> Basically, Glasspoint Energy is using the solar panels to make steam, which can help extract "heavy" oil from old fields. It's like using a knife to get the last bits of ketchup out of the bottle . . . if the knife were made out of anti-ketchup. (para. 2)

Metaphor, one of the major tropes, abounds in the landscape of environmental communication: Mother Nature, population bomb, Spaceship Earth, and carbon footprint are just a few examples. Metaphor's role is to invite a comparison by "talking about one thing in terms of another" (Jasinski, 2001, p. 550).

Newer metaphors have arisen as scientists, environmentalists, and others bring attention to new problems. Chris Russill (2008), for example, identified climate scientists' use of "tipping point" as a metaphor to forewarn the public about irreversible and catastrophic occurrences if global warming continues.

In their efforts to shape public attitudes or agendas about environmental issues, advocates may turn to metaphors as a way of communicating their concerns or goals. For example, many environmentalists use the carbon "footprint" metaphor to help people grasp the impact of our choices (regarding transportation, home size, family size, diet, and more) on the amount of greenhouse gases emitted into our atmosphere. Calculators are readily found online for each one of us to consider these otherwise less tangible environmental impacts. Interestingly, oil companies also use the metaphor of a footprint to influence news stories. Consider, for example, the opening of the Arctic National Wildlife Refuge in Alaska to oil drilling. In 2001, as a vote neared in the U.S. Congress, oil company officials evoked this image to suggest that the drilling would have little impact on the environment. Touting advances in technology, industry spokespeople insisted, "With sideways drilling and other advances, the oil beneath the 1.5 million-acre coastal plain can be tapped with a 'footprint' on the surface no larger than 2,000 acres" (Spiess & Ruskin, 2001, para. 1). The *Anchorage Daily News* reported that the oil industry's footprint metaphor "proved to be a potent piece of rhetoric," implying that drilling would affect less than 1 percent of the coastal plain" (para. 18).

NASA

Photo 3.2	"Blue Marble" is the official name of the first photo of Earth taken from outer space (in 1972 by the *Apollo 8* crew on their flight to the moon). What does the metaphor of "marble" signify compared to other choices? What name would you give this image?

☞ FYI   Metaphor of Spaceship Earth

The practice of speaking of the Earth as a "spaceship" became widespread after astronauts took the first photos of it from space. The photos of the blue-green Earth against a dark universe invited a concern for the precarious existence of this small planet. U.S. Ambassador Adlai Stevenson famously evoked the metaphor of Spaceship Earth when he addressed the United Nations on July 9, 1965. He spoke about UN delegates traveling as passengers on "a little space ship, dependent on its vulnerable reserves of air and soil" (Park, 2001, p. 99). The metaphor was further popularized in the late 1960s by architect Buckminster Fuller (1968) in his *Operating Manual for Spaceship Earth.*

Economist Kenneth Boulding (1965) invoked the most prescient use of the Spaceship Earth metaphor in a 1965 address: "Once we begin to look at earth as a space ship, the appalling extent of our ignorance about it is almost frightening. This is true of the level of every science. We know practically nothing, for instance, about the long-run dynamics even of the physical system of the earth. . . . We do not even know whether the activities of man *[sic]* are going to make the earth warm up or cool off" (para. 7).

Second, environmental actors often rely on different genres to influence perceptions of an issue or problem. Though studied throughout the humanities, **rhetorical genres** are generally defined as distinct forms or types of speech that "share characteristics distinguishing them" from other types of speech (Jamieson & Stromer-Galley, 2001, p. 361). In the last chapter, for example, we observed John Muir's use of the genre of the sublime in his nature writing in the 19th century to evoke feelings of spiritual exaltation. Current examples of genres include *apocalyptic* rhetoric, the *jeremiad,* and what Schwarze (2006) has termed *environmental melodrama.*

For example, Paul Ehrlich (1968) and Rachel Carson (1962), in their classic books *The Population Bomb* and *Silent Spring,* appropriated **apocalyptic narrative** literary styles to warn of impending and severe ecological crises. Literary critics Jimmie Killingsworth and Jacqueline Palmer (1996) explained that

> in depicting the end of the world as a result of the overweening desire to control nature, [these authors] have discovered a rhetorical means of contesting their opponents' claims for the idea of progress with its ascendant narratives of human victory over nature. (p. 21)

More recently, scientists such as James Lovelock have evoked apocalyptic images to warn of the potentially catastrophic effects of global warming on civilization. For example, Lovelock (2006) cautioned that "before this century is over billions of us will die and the few breeding pairs of people that survive will be in the Arctic where the climate remains tolerable" (para. 7). Yet reliance on apocalyptic rhetoric to address climate crises may generate feelings of hopelessness or erode one's credibility since many environmental catastrophes cannot be proven until it is too late to change course. Scientists therefore face a dilemma: How do you raise awareness of future, serious effects from climate changes—rising sea levels, regional conflicts, and so on—without causing depression or paralysis?

A similar genre to apocalypse is jeremiad. Communication scholar Dylan Wolfe (2008) draws on the genre of the jeremiad to sound an environmental alarm in Dr. Seuss's *The Lorax.* The genre, originally named for the lamentations of the Hebrew prophet Jeremiah, has been a recurring genre of American public address (Bercovitch, 1978). **Jeremiad** refers to speech or writing that laments or denounces the behavior of a people or society and warns of future consequences if society does not change its ways. *The Lorax,* of course, is a fable but one that is also a jeremiad in which the Lorax speaks for the trees, whose fate is imperiled by the Once-ler, a symbol of industrial capitalist society.

Other rhetorical critics such as Steven Schwarze (2006) and William Kinsella (2008) have analyzed the genre of an **environmental melodrama** to clarify issues of power and the ways advocates moralize an environmental conflict. As a genre, melodrama "generates stark, polarizing distinctions between social actors and infuses those distinctions with moral gravity and pathos" and is therefore "a powerful resource for rhetorical invention" (Schwarze, 2006, p. 239). Schwarze says that melodrama, by identifying key social actors and where the "public interest" lies, can "remoralize situations" that have been obscured by inaccuracies and "the reassuring rhetoric of technical reason" (p. 250). He offers the example of Bill Moyers's PBS

Courtesy of Utah Moms for Clean Air

| Photo 3.3 | Utah Moms for Clean Air draws on environmental melodrama by using the trope of motherhood and the symbolism of children in their campaigns. In December 2013, they launched a 12 Days of Christmas campaign to further moralize the horrible air pollution inversions residents have been facing annually. For more information see http://blog.utahmomsforcleanair.org/ |

documentary *Trade Secrets* (Moyers & Jones, 2001) about the health dangers of the vinyl chloride chemical industry:

> *Trade Secrets* shuttles between images of confidential company memos describing toxic workplace exposure in scientific language, and episodes of workers on hospital beds or widows tearily recalling their spouse's suffering. These melodramatic juxtapositions offer a clear moral framework for interpreting the actions of company decision makers. They characterize officials as being knowledgeable about toxic hazards in scientific terms, but utterly indifferent to the human suffering that resulted from those hazards.... Melodrama puts the inaccuracy of scientific language on display and highlights its potential blindspots. (Schwarze, 2006, p. 251)

As this example makes clear, melodrama can serve pragmatic purposes: public education and criticism of the chemical industry, for example. But it also illustrates rhetoric's constitutive function, a reordering of public consciousness and specifically the restoration of a moral frame in judging the actions of this industry. This reference to a moral frame brings up another rhetorical resource—the use of communication frames that mediate or affect our understanding of environmental concerns.

Communication Frames

The term *frame* was first popularized by sociologist Erving Goffman (1974) in his book *Frame Analysis*. He defined **frames** as the cognitive maps or patterns of interpretation that people use to organize their understanding of reality. A frame, then, is *constitutive*; it helps to construct a particular view or orientation to some aspect of reality. It also may be *persuasive*, because one frame might make a person's worldview more compelling over another's.

For example, the U.S. Food and Drug Administration (FDA) recently put into place a new policy restricting the use of antibiotics in industrial farms to spur growth and weight gain in cattle, pigs, and chickens. A *New York Times* story reported that experts have warned the indiscriminate use of these antibiotics "has endangered human health by fueling the growing epidemic of antibiotic resistance" (Tavernise, 2013, p. A1). The story also quoted former FDA commissioner David Kessler: "This is the first significant step in dealing with this important public health concern in 20 years. No one should underestimate how big a lift this has been in changing widespread and long entrenched industry practices" (p. A1). By invoking a public health frame, the *Times* highlighted the possible danger to the public while downplaying the economic benefits of growth antibiotics to the agricultural industry. Framing the issue in this way enabled the newspaper to add its support for the FDA's proposed ban on these antibiotics.

Framing can shape or construct how we perceive both problems and solutions and attempt to persuade others of a particular perspective. As media and international affairs scholar Robert Entman (1993) explained, "To frame is to select some aspects of a perceived reality and make them more salient . . . in such a way as to promote a particular problem definition, causal interpretation, moral evaluation, and/or treatment recommendation for the item described" (p. 56).

For example, President Obama has argued that clean energy initiatives in the United States will help to grow the economy and create jobs. The president used this job creation frame in his weekly address to the nation, speaking from a hybrid bus plant in Indianapolis:

> The clean energy jobs at this plant are the jobs of the future—jobs that pay well right here in America. And, in the years ahead, it's clean energy companies like this one that will keep our economy growing, create new jobs, and make sure America remains the most prosperous nation in the world. (Headapohl, 2011, para. 3)

On the other hand, the Institute for Energy Research (2010), a free market research group, invoked a jobs-versus-environment frame in the debate over U.S. energy policy. In a report on its website, the institute claimed a cap on greenhouse gas emissions from oil or coal-burning plants "would reduce U.S. employment by roughly 522,000 jobs in 2015, rising to over 5.1 million jobs by 2050" (para. 4). In each case, the competing uses of a jobs frame by President Obama and the institute construct different meanings for the public in the debate over energy policy and the economy, as well as attempt to persuade audiences of particular beliefs and actions as a result of a particular frame. We address frames more in Chapters 5 and 8.

Dominant and Critical Discourses

The rhetorical claims about a loss of jobs draw on a broader discourse in U.S. politics that asserts there is a trade-off of jobs versus the environment. This concept of *discourse* reminds us that rhetorical resources are broader than any single metaphor, frame, or utterance. As we noted in the last chapter, a discourse is a pattern of knowledge and power communicated through linguistic and nonlinguistic human expression; as a result, it functions to "circulate a coherent set of meanings about an important topic" (Fiske, 1987, p. 14). Such meanings often influence our understanding of how the world works or should work.

When a discourse gains a broad or taken-for-granted status in a culture (for example, growth is good for the economy) or when its meanings help to legitimize certain practices, it can be said to be a **dominant discourse**. Often, these discourses are invisible in the sense that they express naturalized or taken-for-granted assumptions and values about how the world is or should be organized.

Perhaps the best example of a dominant environmental discourse is what biologists Dennis Pirages and Paul Ehrlich (1974) called the **Dominant Social Paradigm (DSP)**. Communication scholars would point out that a dominant social paradigm is a specific discursive tradition. As expressed in political speeches, advertising, movies, and so forth, today's DSP affirms society's "belief in abundance and progress, our devotion to growth and prosperity, our faith in science and technology, and our commitment to a laissez-faire economy, limited government planning, and private property rights" (Dunlap & Van Liere, 1978, p. 10). In everyday terms, this dominant discourse is recognized in references to free markets as the source of prosperity and the wise use of natural resources to build a strong economy, and so forth.

Other discourses may question society's dominant discourses. These alternative ways of speaking, writing, or portraying the environment in art, music, and photographs illustrate **critical discourses**. These are recurring ways of speaking that challenge society's taken-for-granted assumptions and offer alternatives to prevailing discourses. In some ages, critical discourses are muted or absent, whereas in other periods, they may be boisterous and widespread. In our own time, critical discourses have proliferated in mainstream media and online, questioning dominant assumptions about growth and the environment. For example, we saw (earlier) the emergence of a new antagonism, opening space for a discourse of sustainability by scientists, environmentalists, and students on many campuses who are questioning the use of fossil fuels and calling for "climate justice" (see Photo 3.4).

Today, food also continues to be a contested topic, from dominant social paradigms of how to grow, produce, and consume which foods when, as well as critical discourses that attempt to challenge our taken-for-granted assumptions. In Italy, Carlos Petrini helped mobilize an international food movement in the late 1980s. Two events helped solidify the exigency that Petrini felt about Italian culture, food, and economy transforming in unsustainable ways. First, the opening of the first McDonald's fast-food store in Piazza di Spagna, a well-known piazza or public square in Rome, he argued, was a sign of increased multinational influence on eating habits and local economies. Second, the poisoning of hundreds of Italians in northern Italy, including the death of 19, as a result of cheap wine cut with methanol, underscored the high risk of becoming alienated from one's food sources and the laws that govern them.

James Ennis/Flickr. Used under Creative Commons license
https://creativecommons.org/licenses/by/2.0/

Photo 3.4	The growing movement on college campuses for divestment from the fossil fuel industry, and the call for "climate justice," is a direct challenge to the dominant (taken-for-granted) discourse undergirding U.S. and global energy policy.

Petrini has since helped foster an international, critical discourse that resists the dominant discourse of fast, cheap food. Petrini and his allies launched the **Slow Food Movement**, which promotes environmental sustainability through local food cultures by promoting "good, clean, and fair food." This movement has built an international coalition with La Via Campesina, a self-declared international peasant movement. La Via Campesina challenges dominant corporate-driven transnational agriculture in defense of small farmers and **food sovereignty**, the right of everyday people, farmworker unions, and sovereign countries to public participation in agricultural and food policy.

SUMMARY

In this chapter, we've described a perspective on communication that emphasizes the symbolic construction of our views about the environment. We introduced several rhetorical concepts in this chapter that bear relevance to environmental communication:

- A rhetorical perspective, particularly the use of terministic screens and naming, helps remind us of the important of socio-symbolic language in shaping our world.

- The rhetorical situation is a helpful way to analyze the ways we construct environmental problems, including exigencies, audiences, and constraints.
- Rhetorical tropes and genres are useful to interpret human dramas about the environment in ways that both persuade and constitute environmental attitudes and actions.
- Framing helps us to appreciate the ways in which language choices direct our attention and shape our views of certain topics or meanings.
- Dominant and critical discourses provide a means to analyze both prevalent and counterintuitive beliefs, values, and actions that sustain and challenge a culture's view of the environment.

While the examples in this chapter broadly illustrate socio-symbolic choices related to the environment, discourses are rehearsed, performed, and negotiated in concrete contexts in the public sphere. In the next chapter, we expand on this view of environmental communication by looking at visual and popular culture contexts.

SUGGESTED RESOURCES

- Milstein, T. (2011). Nature identification: The power of pointing and naming. *Environmental Communication: A Journal of Nature and Culture, 5,* 3–24.
- Shiva, V. (2002). *Water wars: Privatization, pollution, and profit.* Cambridge, MA: South End Press.
- Irena Salina (Director). (2008). *FLOW: For love of water* [Film]. USA: Group Entertainment.
- Moyers, B., & Jones, B. (2001). *Trade secrets.* Retrieved from http://www.pbs.org/tradesecrets
- Stauber, J., & Rampton, S. (2002). *Toxic sludge is good for you: Lies, damn lies and the public relations industry.* Monroe, ME: Common Courage Press.
- Slow Food International. http://www.slowfood.com/

KEY TERMS

Apocalyptic narrative 61

Audience 58

Constraints 58

Critical discourse 64

Dominant discourse 64

Dominant Social Paradigm
   (DSP) 64

Environmental melodrama 61

Exigency 57

Food sovereignty 65

Frames 62

Irony 59

Jeremiad 61

Metaphor 59

Naming 56

Rhetoric 55

Rhetorical genres 61

DISCUSSION QUESTIONS

1. What do Herndl and Brown (1996) mean when they claim that "in a very real sense, there is no objective environment in the phenomenal world, no environment separate from the words we use to represent it" (p. 3)? Do you agree with this assertion?

2. Are apocalyptic warnings about global warming effective, or do such warnings create problems of credibility or paralyze action? How can scientists raise awareness of future, serious effects from climate changes—rising sea levels, deaths from prolonged droughts, and so on—without relying on some vision of catastrophic events?

3. There are many debates over food today, from whether or not we should reduce pesticide consumption to what we should serve kids at lunch to the rights of food producers to participate in shaping food cultures. We have noted terms such as "organic," "conventional," and "real food." What other words have you noticed in food debates? Which ones have moved you to take action or change your daily practices? Which do you find unpersuasive?

4. Since our understanding of the environment is *symbolically constructed* (by words, images, etc.), does that mean there is no inherently "right" or "wrong" view or attitude toward the environment? Where do our ethics of environmental behavior and attitudes come from?

NOTE

1. For more on the rhetorical situation, see Bitzer (1968) and Vatz (1973).

Avatar (2009) commonly is estimated to be the top-grossing film in the world. Writer and director James Cameron's self-declared goal was to create a film with a powerful environmental message inviting audiences to become warriors for the Earth. The unsustainability of the film industry from the massive carbon footprint of new technologies to the waste of disposable 3-D glasses remains a challenge that Cameron will attempt to address in his sequels, including using solar power (Braude, 2012).

CHAPTER 4

The Environment in/of Visual and Popular Culture

Climate change really is a made-for-TV story. It has all the drama of Hollywood, with real-life villains and heroes thrown in. We scientists struggle everyday to communicate the importance of climate change to the world. It is great to see communication experts come in and accomplish what scientists alone cannot.

—John Abraham (2014)

While images are often said to embody complexity (being worth the prover-bial thousand words), media theory tells us that they also reduce complexity by providing interpretive frames or narratives that selectively blend fact and emotion.

—Darryn Anne DiFrancesco and Nathan Young (2010, p. 518)

Communication about the environment, of course, is not limited to verbal communication. For most of us, it is challenging to separate our understanding of the environment from the images we associate with it. Since ancient times, across the globe from China to Greece to Egypt, initial art forms (such as pottery vases, stone etchings, and hieroglyphics) included depictions of animals, food, trees, and the places where people lived. Today, we continue to communicate about the environment explicitly and implicitly across a range of media, including but not limited to phones, computers, television, and films. How we interact with our environments and what we do as part of the environment, in other words, long have been and continue to be intertwined with sight. Like words, images can both enhance our imaginations and stifle them.[1]

Chapter Preview

The first section of this chapter on the environment and popular culture describes the ways in which dominant mass-consumed media about the environment might be analyzed. We highlight three approaches:

1. How audiences respond to environmental media through *dominant*, *oppositional*, and *negotiated* positions

2. How the environmental impact of communication technologies may include an assessment of how they are made, used, and disposed

3. How a *circuit of culture* approach helps provide a more nuanced environmental analysis of media than just what we see on a screen or page

The second part of this chapter moves into specific examples of communication technologies and what they do or do not "afford" or enable us to "see" about the environment. In particular, we look at two notable examples:

1. The ways in which the *visual rhetoric* of media such as paintings and still photographs can affect attitudes and behavior toward nature

2. How a range of media that show moving images teach us something about environmental embodiment and the witnessing of ecological disasters, in particular

The third section of this chapter moves into the fields of green art, marketing, and graphic design, respectively. We discuss the following:

1. The ways environmental artists use nature as a medium of expression and are inspired to respond poetically to environmental problems

2. Two competing metaphors to consider the ways digital technologies are transforming how we construct, send, and receive green messages

3. Finally, how graphic design (such as infographics and interactive maps) provides opportunities to communicate technical environmental information and data

Given the significance of sight and media technologies to how the environment is imagined and expressed in culture, in this chapter we introduce several key communication concepts of visual and popular cultures. Although we will address a range of technologies, media specificity is important to appreciating the communicative dimensions of these objects. That is, we do not believe a television show enables the same communicative experiences as a photograph, and so forth. Political economist Harold Innis (2008) argued that this is the **bias of communication**, the ways any given medium creates and limits conditions of possibility across space and time in a particular culture. Throughout this chapter, we'll attempt to underscore various constraints (limitations and possibilities) that various media enable without guaranteeing results. That is, while a television show is different than a photograph, watching a television show does not determine the environmental consequences of one's experience.

Further, we do not believe that focusing on visual and popular media means that our relationships with them solely are limited to our eyes. Borrowing from psychologist James Jerome Gibson's (1986) classic *The Ecological Approach to Visual Perception,* therefore, this chapter will describe which **affordances** are enabled by various media and which are not. "The *affordances* of the environment," Gibson explains, "are what it *offers* the animal, what it provides or *furnishes,* either for good or ill" (p. 127). Approaching visual and popular media communication through affordances entails considering both physical characteristics (a horizontal versus a slanted surface, stationary versus portable, et cetera) and how those characteristics interact with the person using it (a knee-high seat for an adult, Gibson points out, is not knee-high for a child). We explore a range of visual and popular culture in this chapter (although we address advertising in Chapter 11 and images are significant throughout the book).

Many media scholars study **aesthetics,** or the role of art or taste. Applied aesthetics of popular culture include but are not limited to the position of the camera, the use of light, the role of sound, design of costume and sets, the pace of a moving picture, and the application of special effects. We'll elaborate on the affordances various media provide aesthetically in the second part of this chapter, but first we want to provide context for thinking about how environmental communication is interpreted and circulates.

The Environment and Popular Culture

Popular or mass media abound in the First World across more technologies than we almost can count and the environment is a frequent theme. Environmental communication does not just occur when two or more people converse on a hike in the woods with each other, it also happens through the radio, graffiti, posters, bumper stickers, phones, Skype calls, televisions, movie theatres, computers, and more. Further, any given medium may enable multiple messages, commercial film screenings open with advertisements, phones include texts, tweets, e-mails, and calls, et cetera. But, how do we navigate these mediums? And what can we assume is their communicative significance?

Encoding/Decoding Environmental Media

In a seminal essay, "Encoding and Decoding in the Television Discourse," cultural studies scholar Stuart Hall (1973) provided a useful way to describe the way audiences (and users) interact with media. To begin, he helps remind us that any medium is both **encoding,** or created with a message, and **decoding,** or interpreted by audiences or receivers. Consider, for example, a popular show like the longest-running prime time show on television, *The Simpsons.* For any given episode, the writers of that show will create a storyline, *encoding* it with a plot, side jokes, and more. Yet audiences may interpret or *decode* that particular episode in divergent ways. While some may laugh with the mischievous and self-centered Bart, others may identify with his vegetarian and environmentally conscious sister, Lisa. The intent of the authors, therefore, does not determine the reactions of audiences, though they can make certain responses more or less imaginable.

For example, a *Simpsons* episode like "Treehouse of Horror XXIV" (from Season 25, originally aired on October 6, 2013), introduces common environmental elements of the show, such as the nuclear plant where Homer Simpson works, toxic pollution, and kids playing outdoors. Afterward, one might expect that at least some viewers could become more skeptical about nuclear power as a safe energy option, worried about sources of toxic pollution in their lives, and curious about building a treehouse; yet if you finished watching it and launched into a debate about national parks, those watching with you might be surprised since the writers did not include anything about national parks in the episode. (For a more extended engagement with the series, see Todd, 2002.)

Any given popular cultural artifact might enable a range of responses. Hall (1984) identifies three ways a media consumer might decode a cultural text:

1. *Dominant position:* when the consumer agrees with the text's cultural biases

2. *Oppositional position:* when the consumer rejects the text's cultural biases

3. *Negotiated position:* when the consumer accepts some of the text's cultural biases, but rejects others

How do these different positions influence the communicative possibilities of popular culture? These positions are a useful way to consider a range of popular culture practices that exceed the visual. For example, eco-celebrity initiatives are announced regularly. In 2011, international musical artist and producer will.i.am and The Coca-Cola Company launched a new brand, *EKOCYCLE*™, that encourages zero waste through recycled fashion. A dominant position in relation to this news could be to believe that both partners genuinely care about environmental sustainability and, perhaps, to be excited that one can purchase clothes that might not only be hip, but also reduce waste. An oppositional position could be to act exasperated and cynical about this news, believing anything a celebrity does in the name of the environment is just another form of self-promotion and that a multinational corporation like Coca-Cola, known for water abuses and sugar promotion, could not possibly be initiating anything worthwhile. A negotiated position might believe that will.i.am is interested in self-promotion and that Coca-Cola is not the healthiest corporation to ever shape our planet, but also acknowledge that it is possible that their collaboration could reduce clothing waste.

As we provide popular culture examples throughout this chapter and you consider ones that you think have been most significant, try to identify not only the aesthetics and content of the cultural text, but also the values that you believe might enable someone to decode it from a particular position.

Media's Lifecycle and the Circuit of Culture

In addition to studying what media technologies afford us to communicate or not and the ways that we respond to those communications, media have environmental impacts through the contexts in which they are made and disposed, as well as circulated. Let's now turn to two ways to think about that approach to the environmental impact of media.

Most of us were taught the life cycle of a species in our first biology course. The life cycle is the changes a species goes through in its life from birth to death. Usually, teachers choose frogs to illustrate this biological process, because their life cycles are pretty easy to see as distinct stages; frogs transform from an egg to a tadpole to a tadpole with legs to a froglet to an adult frog. Environmentalists sometimes refer to a **life-cycle-assessment (LCA)** as a way of studying how a product's "life" develops from-cradle-to-grave. An LCA might include raw material extraction, production, distribution, use, and disposal. Something like an iPad, for example, is made from metals and chemicals, developed as components that then are combined and shipped, used through electric power for a few years, and then usually disposed of or recycled, as best as possible. Nevertheless, digital media often are promoted as a more environmentally sustainable way to communicate than printing on paper.

Another framework to study the various environmental impacts of media beyond aesthetics is called the *circuit of culture*, which initially mapped five elements and the relations between them: regulation, production, consumption, representation, and identity. The **circuit of culture** is a flexible framework that reminds us of three characteristics of media: (1) culture always is changing and moving as part of broader networks or contexts; (2) to study it, we must choose which elements will or will not be the focus of our analysis; and (3) people involved may or may not be involved in more than one element (Pezzullo, 2011).

This framework helps remind us that representation is only one way to assess visual media. The production, consumption, and disposal of communication technologies also have environmental impacts. Apple computers, for example, provide environmental reports of all of their products, none of which focus on green or antigreen representations shared on their devices but focus instead on four categories: climate change (focusing on the carbon footprint), restricted substances (use of toxic chemicals), energy efficiency (of use), and material efficiency (products used to make components and how they break down). (See apple.com/environment/reports/.)

Once again, while an applied aesthetic perspective might focus on what one sees on a screen or a representation, a circuit of culture approach reminds us that environmental messages and values also are enabled and constrained throughout a medium's life cycle. Now that we have considered the environmental impact of visual media technologies, let us turn to some of the more popular uses of visual communication technologies to engage environmental topics.

Looking at the Environment

The circuit of culture framework reminds us that visual images do not exist by themselves. They are not simply isolated images of something, such as an eagle, mountain, or polluting factory. Rather, a specific image may evoke other images and texts and, therefore, a multitude of associations and meanings. It is important then to understand "how that image fits into the larger ecosystem of images and texts" (Dobrin & Morey, 2009, p. 10). Studying how various media interact with each other and people is called **intermediation**.

Now that we have established an aesthetic and cultural studies approach to media contexts, let us return to specific affordances of communication technologies and what they do or do not enable. How do the visual media both persuade and constitute our perceptions of environmental problems and possibilities?

Visual Rhetoric and Nature

Given the prevalence of visual media in most of our everyday lives, rhetorical scholars have begun to look more closely at the significance of visual images in the public sphere. As communication scholars Lester Olson, Cara Finnegan, and Diane S. Hope (2008) point out in their study *Visual Rhetoric,* "public images often work in ways that are rhetorical; that is, *they function to persuade* [emphasis added]" (p. 1). Rhetorical scholars Sidney I. Dobrin and Sean Morey (2009) also have called for the study of *Ecosee*, the "study and the production of the visual (re)presentation of space, environment, ecology, and nature in photographs, paintings, television, film, video games, computer media, and other forms of image-based media" (p. 2). As such, we can say that **visual rhetoric** functions both pragmatically—to persuade—and constitutively, to construct or challenge a particular "seeing" of nature or what constitutes an *environmental problem*.

While what we see firsthand in our environments surely affects us, our symbolic actions—including images—also affect our perceptions of the environment. All images, of course, are artifacts; they are human made. They select certain aspects of the world (and not others), certain angles of vision, frames, and ways of composing this larger reality. As a result, visual culture influences meaning; it suggests an orientation to the world.

Let's pursue this insight by looking at two ways visual rhetoric have had a significant influence on environmental communication in the United States: the 18th- and 19th-century paintings of the American West and 2003 photographs of the Arctic National Wildlife Refuge. What is seen in these paintings and photos, and what meaning or orientation do they suggest?

Seeing the American West

In Chapter 2, we saw that 18th- and 19th-century artists such as Thomas Cole, Albert Bierstadt, and the Hudson River School painters were significant sources of the public's awareness of the American West. Equally important were the photographers who followed military expeditions and surveyors into western territories and who were among the first to portray the West to many people who lived in eastern cities and towns. Photographs of Yosemite Valley, Yellowstone, the Rocky Mountains, and the Grand Canyon not only popularized these sites but, as they became broadly available in the media, "were factors in building public support for preserving the areas" (DeLuca & Demo, 2000, p. 245).

With such popularization, however, came an orientation and also ideological disposition toward nature and human relationships with the land. On the one hand, the paintings of the Hudson River School and, in the 20th century, photographs of environmental advocates like Ansel Adams aided in constituting natural areas as pristine and as objects of the sublime. Yet rhetorical scholars Gregory Clark, Michael Halloran, and Allison Woodford (1996) have argued that such portrayals of wilderness depicted nature as separate from human culture; the viewpoint of paintings distanced the human observer by viewing the landscape from above or in control of nature. They concluded that, although expressing a reverence for the land, such depictions functioned "rhetorically to fuel a process of conquest" (p. 274).

More recently, rhetorical critics Kevin DeLuca and Anne Demo (2000) have argued that what was left out of landscape photographs of the West may be as important as what was included. They gave the example of early photos of Yosemite Valley taken in the 1860s by the photographer Carleton Watkins. DeLuca and Demo (2000) wrote that, when Watkins portrayed Yosemite Valley as wilderness, devoid of humans, he also helped to construct a national myth of pristine nature that was harmful. In a critique of the implicit rhetoric of such scenes, they argued that the "ability of whites to rhapsodize about Yosemite as paradise, the original Garden of Eden, depended on the forced removal and forgetting of the indigenous inhabitants of the area for the past 3,500 years" (p. 254). Writer Rebecca Solnit (1992) has pointed out,

> The West wasn't empty, it was emptied—literally by expeditions like the Mariposa Battalion [which killed and/or relocated the native inhabitants of Yosemite Valley in the 1850s], and figuratively by the sublime images of a virgin paradise created by so many painters, poets, and photographers. (p. 56, quoted in DeLuca & Demo, 2000, p. 256)

Whether or not one agrees with DeLuca and Demo's claim about the impact of Watkins's photos, it is important to note that visual images often play pivotal roles in shaping perceptions of natural areas as well as our awareness of the impacts of pollution and toxic waste on human communities. As DeLuca and Demo (2000) have argued, visual portrayals often are "enmeshed in a turbulent stream of multiple and conflictual discourses that shape what these images mean in particular contexts"; indeed, in many ways, such pictures constitute "the context in which a politics takes place—they are creating a reality" (p. 242).

☞ **FYI**   **Documenting the Environment**

The U.S. Environmental Protection Agency (EPA) hired freelance photographers to document images of "environmental problems, EPA activities, and everyday life in the 1970s." This effort became known as the *Documerica Project* (1971–1977). Photographers chose to document a range of sites, including pictures of African Americans in Chicago enjoying parks, swimming outdoors, and music; European Americans working in a Tennessee coal plant; Navajo and Hopi reservations; power lines; birds; and much more. The collection includes over 15,000 images, which can be found today in its entirety at the National Archives (http://arcweb.archives.gov/arc/action/ExternalIdSearch?id=542493&jScript=true) and partially on flickr (www.flickr.com/photos/usnationalarchives/collections/72157620729903309/). Many of these images might challenge assumptions about how narrowly some believe the word *environment* was imagined in the 1970s.

Today, Google Earth makes the Documerica Project archive appear incredibly small. Originally built by the U.S. Central Intelligence Agency (CIA) and bought by Google in 2004, it uses satellite imagery, aerial photography, and computer graphics to show cities, skies, oceans, the moon, and Mars. You can visually move laterally or zoom in or zoom out on more places than ever before. Such unprecedented visual access has ambiguous environmental impacts, from discoveries in biodiversity to military surveillance (www.google.com/earth/).

Picturing the Arctic National Wildlife Refuge

In October 2000, a 33-year-old physicist named Subhankar Banerjee began a two-year project to photograph the seasons and the biodiversity of Alaska's Arctic National Wildlife Refuge. His project, which took him on a 4,000-mile journey by foot, kayak, and snowmobile through the wildlife refuge, culminated in stunning photographs, published in his book *Arctic National Wildlife Refuge: Seasons of Life and Land* (Banerjee, 2003). (For a sample of the photographs, see www.subhankarbanerjee.org/.)

Banerjee hoped that his photographs would educate the public about threats to Alaska's remote refuge. The Smithsonian Museum in Washington, DC, had scheduled a major exhibition of Banerjee's photos for 2003. However, the young scientist–photographer suddenly found his photos and the exhibit caught in the midst of a political controversy. During a March 18, 2003, debate in the U.S. Senate about oil drilling in the Arctic National Wildlife Refuge, Senator Barbara Boxer of California urged every Senator to visit Banerjee's exhibit at the Smithsonian "before calling the refuge a frozen wasteland" (Egan, 2003, p. A20). The vote to open the refuge to oil drilling later failed by four votes.

Although Banerjee's photos were not the only influence on the Senate's vote, the controversy over his photos caused a political firestorm and helped to create a context for debate over the refuge itself. Journalist Timothy Egan (2003) reported that the Smithsonian told Banerjee "the museum had been pressured to cancel or sharply revise the exhibit" (p. A20). Documents from the museum give an idea of the revised exhibit. For example, Egan reported that the original caption for a photo of the Romanzof Mountains quoted Banerjee as saying, "The refuge has the most beautiful landscape I have ever seen and is so remote and untamed that many peaks, valleys and lakes are still without names." However, the new caption simply reads, "Unnamed Peak, Romanzof Mountains" (p. A20). After the failed vote, the Smithsonian "sent a letter to the publisher of Banerjee's book, saying that the Smithsonian no longer had any connection to Mr. Banerjee's work" (p. A20).

The Smithsonian's criticism of Banerjee's photographs was revealing. Photographs can be powerful rhetorical statements. As DeLuca and Demo (2000) argued, they can constitute a context for understanding and judgment. Especially when accompanied by captions that encourage a particular meaning, photos can embody a range of symbolic resources that sustain or challenge prevailing viewpoints. Some observers felt that Banerjee's photos of Alaska's wilderness had this potential. A reviewer for the *Planet* in Jackson Hole, Wyoming, observed, "Sometimes pictures have a chance to change history by creating a larger understanding of a subject, thus enlightening the public and bringing greater awareness to an issue" (Review, 2003).

Photo 4.1 Former U.S. vice president Al Gore's and the Intergovernmental Panel on Climate Change's joint film and multi-media campaign *An Inconvenient Truth* remains an exemplar of an eco-cinema documentary and contemporary environmental advocacy. What fewer know is that Gore traces his own career-long belief in the importance of environmental communication as far back as his senior undergraduate thesis on visual rhetoric's impact on political debates (Gore, 2007).

Moving Images of Disasters

Visual media do not always stand still. How does an advocacy organization or educational group decide to use photographs or video? What does looking at images that move tell us about the affordances provided by a medium that moves beyond a single moment?

Witnessing Ecological Crises

Through various modes of media, more people appear more connected through communication technologies than ever before. Of course, these connections are not equal across the globe or even within the United States. Nevertheless, environmental advocates have used these interconnections to afford opportunities to witness ecological crises. In general, **witnessing** traditionally has been defined as an act of hearing and seeing oral and written evidence through firsthand experience. "Bearing witness," as visual communication and drama scholar Jan Cohen-Cruz (1998) points out, "uses heightened means to direct attention onto actions of social magnitude, often at sites where they actually occur, and from a perspective that would otherwise be missing" (p. 65). As such, social movements have found witnessing to be a persuasive mode of communication, either in the form of people "on the ground" bearing witness to their own situation and experiences or by inviting those who live farther away from particular sites and conditions to witness "a perspective that would otherwise be missing." At minimum, therefore, the potential rhetorical efficacy of providing opportunities to witness is to suggest alternative modes of viewing and acting in the world.

The goal of witnessing is to build communities and connections between people. As media scholars Bhaskar Sarkar and Janet Walker (2010) write in their collection, *Documentary Testimonies: Global Archives of Suffering,*

> In our latter twentieth and twenty-first century "era of witness," media testimonial initiatives—be they official, grassroots, guerilla, transitory, insistent, or any combination thereof—participate in the creation of ethical communities by bringing testifiers and testimonial witnesses together at the audiovisual interface. (p. 1)

The communicative goal usually is to hope that these "ethical communities" will be transformed through the experience and, after the viewing is over, take additional actions.

In the environmental movement, Greenpeace perhaps is best known for its use of witnessing as a tactic of resistance. Started in 1971, the organization articulates or links "green" concerns to a pacifist ethic of "peace." The landmark event that catapulted them into the international spotlight drew on the Quaker tradition of "bearing witness" by chartering a boat to protest a nuclear test site on the island Amchitka in the North West Pacific. This approach to social change involved "expressing opposition simply by turning up and being seen at the site of the activity to which they object."[2] There are risks to these acts of witnessing that are not risks taken by those who look at the images from the comfort of their homes. Greenpeace activists have been arrested in Russia, Japan, and elsewhere for these acts of witnessing. We discuss this communication tactic more in the next chapter.

Visual communication scholar Diane S. Hope (2009) reminds us that environmental perspectives sometimes are easily communicated through images, while others are not:

> Visual reportage and photographic documentation of environmental degradation have been notoriously difficult. While individual instances of ecological crises such as large oil spills, heavy smog, and mountaintop removal are "photogenic," persistent deterioration of air, water, and land is largely invisible to the eye and to the camera lens. Although measures of climate change and levels of polluting toxins may be visualized through a variety of imaging technologies, such images need significant expertise to interpret. (p. 33)

Let's turn now to climate change and oil spills to consider how various crises pose rhetorical constraints—limitations and possibilities—for environmental advocates to visually portray exigency.

Polar Bears as Condensation Symbols

As media scholar Anders Hansen (2010) explains, the significance we associate with a public issue such as climate change is the result of a great deal of communicative work. For example,

> images of melting ice-floes, Arctic/Antarctic landscapes, glaciers, etc. become synonymous with—come to mean or signify—"threatened environments" and ultimately "global warming" or climate change, where in the past they would have signified something quite different such as "challenge" or a test of human endeavour . . . or simply "pristine" and aesthetically pleasing environments, as yet untouched and unspoilt. (p. 3)

"If anything symbolizes the Arctic," writes Tim Flannery (2005), author of *The Weather Makers,* "it is surely *nanuk,* the great white bear" (p. 100). Seen on greeting cards, in environmental groups' appeals, and featured as Lorek Byrnison, the great armored ice bear who aided Lyra in the 2007 film *The Golden Compass,* the images of the polar bear abound in popular culture. Recently, compelling images of polar bears struggling for survival have emerged as a powerful symbol of global warming. As early as 2005, scientists were finding evidence that polar bears have been drowning in the Arctic Sea due to the melting of ice floes from climate change. Polar bears feed from these ice floes, and as they drift farther apart, the bears are being forced to swim longer distances.

In the summer of 2008, observers flying for a whale survey over the Chukchi Sea spotted polar bears swimming in open water. The bears were 15 to 65 miles off the Alaskan shore, "some swimming north, apparently trying to reach the polar ice edge, which on that day was 400 miles away" ("As Arctic Sea Ice Melts", 2008, p. A16). "Although polar bears are strong swimmers, they are adapted for swimming close to the shore. Their sea journeys leave them vulnerable to exhaustion, hypothermia or being swamped by waves" (Iredale, 2005, para. 3).

News reports about global warming, therefore, constitute a very different context in which images of polar bears appear than do the contexts of greeting cards or a fantasy–adventure film. Within the context of images of melting ice and news of climate change, images of polar bears function as a visual **condensation symbol**. A condensation symbol is a word or phrase (or, in this case, an image) that "stirs vivid impressions involving the listeners' most basic values" (Graber, 1976, p. 289). Political scientist Murray Edelman (1964) stressed that such symbols are able to "condense into one symbolic event or sign" powerful emotions, memories, or anxieties about some event or situation (p. 6). Images of vulnerable polar bears may be one condensation symbol for our anxieties about the planet's warming. As such, they also help to construct what we understand to be an environmental problem itself.

©iStockphoto.com/EdStock

Photo 4.2	Within the context of images of melting ice floes and news of climate change, polar bears function as a visual *condensation symbol* for climate crisis. Yet what difference would it make if an image of a climate change refugee became the new condensation symbol? Do you think polar bears on broken ice are more visually resonant than people walking through floods? As more humans are impacted by climate disruptions, do you think the condensation symbol will change?

Pollution in Real Time

Visual media of disasters also move. Increasingly, international journalists are using drones to record large-scale ruins or hazardous places that are unsafe for humans to visit. After the 2013 Fukushima disaster in Japan, the Namie city mayor invited Google Earth to take images of their community to which residents believe they may never be able to return.[3] People also are watching more real-time video of environmental disasters as they occur, whether provided through private citizens posting video online or corporations sharing videos of cleanup efforts.

The multinational corporation, BP, provides one notable example of moving images worth considering for its environmental implications. On April 20, 2010, a fiery explosion erupted on the *Deepwater Horizon* oil platform in the Gulf of Mexico, killing 11 workers. The collapse of the rig and failure of BP's "blowout preventer" allowed crude oil a mile beneath the surface to spew into the Gulf waters. The oil gushed for 87 days until BP temporarily blocked the wellhead. Although BP initially claimed the rate of oil flow was low, U.S. government scientists determined as much as 1.5 to 2.5 million gallons of oil a day were spewing into the Gulf (CNN Wire Staff, June 16, 2010). By September 18, 2010, when BP permanently sealed the well, nearly 5 million barrels or 206 million gallons of oil had spilled into Gulf waters. Oil sheen and tar balls ultimately reached into wetlands, marshy bays, and onto beaches in Louisiana, Mississippi, Alabama, and Florida. The U.S. government declared a fishery disaster off the coasts of Louisiana, Mississippi, and Alabama, and hotels and restaurants reported that business had fallen off sharply for the summer season. The U.S. official in charge of overseeing the cleanup called it "the worst oil spill in U.S. history" (Hayes, 2010, para. 1).

From the day of the oil rig explosion, visual images framed the public's sense of the event as an economic and ecological problem: dramatic film showing the collapse of the Deep Horizon oil rig; satellite images of oil spreading over 40,000 square miles on the surface; interviews with unemployed Louisiana shrimpers; and emotionally gripping photos of oil-soaked seabirds, turtles, and dolphins. TV cameras showed oil-tainted wetlands and workers in white hazmat suits scooping up tar balls that had washed onto beaches.

Despite these compelling images, some worried they were not seeing the real extent of the problem. "I think there's an enormous amount of oil below the surface that unfortunately we can't see," one scientist told an oversight committee (quoted in Froomkin, 2010, para. 7). A partial answer to "seeing" the problem would come from the live video from BP's Remotely Operated Vehicles (ROVs) of the oil gushing from the underwater wellhead. Initially, BP had wanted to block this live feed to news outlets. Under intense pressure from federal officials, however, the company relented and agreed to "make a live video feed from the source of the leak continuously available to the public" (Froomkin, 2010, p. 1).

The real-time images of the oil gushing from the ruptured well were compelling: "The sight of an ecological catastrophe unfolding in real time has gripped the public more than the series finale of 'Lost'" (Fermino, 2010, p.1). Within days, millions of people had viewed the video, with at least 3,000 websites using the live feed (Jonsson, 2010). One reporter observed, "The glimpse into the deep has proved mesmerizing" (Jonsson, 2010, para. 2). The images had become so riveting that "when President Obama spoke about the historic spill, TV news channels showed a split screen of Obama and the gushing oil" (Fermino, 2010, p. 1).

Green Art, Marketing, and Graphic Design

One of the relevant affordances of environmental media is how they enable or constrain humans to enable or constrain our agency or ability to act, respond, or make a difference. In this last section of the chapter, we want to highlight various metaphors used to think about the ways digital media enable agency, when they can fail, and how sometimes more abstract images make more powerful statements.

Environmental Art

As we said previously, the environment has provided inspiration for humans' greatest fears and fantasy as long as humans have walked the Earth. Totem poles, tapestries, masks, baskets, sculptures, graffiti, and more have been made from and symbolized human interpretation of our environments. Some of the best-known art in the world focuses on nature, including Utagawa Hirshige's *Mountains and River on the Kiso Road*, Claude Monet's *Water Lilies*, and Georgia O'Keefe's *Summer Days*.

Today, **environmental art** usually references two types of artists: (1) artists who foreground particular materials in their work—such as sticks, stones, flowers, and mud—and whose work is displayed in the outdoors and (2) artists that aspire to communicate about environment problems through their work.

An exemplar of the former type of environmental artist is Bavarian Nils-Udo. For decades, he has been using locally found materials wherever he plans to create what he imagines as "potential utopias," such as giant nests. As such, his work resonates with the Romanticism we discussed in Chapter 2; yet Nils-Udo believes most of us have become disconnected from nature. Key to this type of art is inviting an embodied, sensuous site-specific experience to heighten our awareness of nature's poetics, as well as the work's ephemerality. As he (2002) notes in an artist's statement,

> Natural space experienced through hearing, seeing, smelling, tasting and touching. By means of the smallest possible interventions, living, three-dimensional natural space is re-organized, unlocked and put under tension. Reorganisation, of course for a finite period of time. One day, the intervention is wiped away, undone by nature without leaving a trace. (greenmuseum.org)

Nils-Udo not only has made environmental art throughout Europe, but also in India, Japan, Israel, and Mexico.

While some environmental artists focus on nonhuman nature, others find inspiration in the often-fraught relationship between humans and the environment. Perhaps one of the best-known contemporary environmental artists in the United States who communicates on a number of environmental problems is Seattle-based Chris Jordan. He uses photographs and objects of mass consumption to visualize consumption. For example, on his webpage *Running the Numbers II: Portraits of Global Mass Culture* (2009–present), Jordan includes a 2009 image titled "Gyre" that at first appears to be three vertical panels of Japanese art of the ocean, but once one zooms in closer, one can see that it "[d]epicts 2.4 million pieces of plastic, equal to the estimated number of pounds of plastic pollution that enter the world's oceans every hour. All of the plastic in this image was collected from the Pacific Ocean" (Jordan, 2009–present).

Art, of course, need not be in a museum or a large-scale production. Contemporary environmental protests often include creative signs and puppets. Another relatively common environmental advocacy campaign is to paint trashcans. Seventy-five-year-old Lebanese Elie Saba launched the "Ana Ma Bkeb" (I Don't Litter) campaign, which included the painting of metal barrels with trees, suns, and other natural imagery to raise the profile of trash in the streets. Saba said, "People forget the street is a public place and belongs to everyone and must be kept clean so we can have a clean environment" (Topalian, 2013).

Act Locally!

Another exemplar of environmental art is the Maine-based Beehive Design Collective that draws complex environmental relations based on collaborative work with environmental advocates and artists. In 2010, for example, they launched the True Cost of Coal graphics campaign with Appalachian grassroots organizers.

Poet and author Susanne Antonetta (2012) wrote of them: "The visual power of the banner offers a clear and intricate story that draws the eye everywhere at once, fascinating in its detail . . . and overwhelming in its breadth. With its symbolism and visual density. . . . I find my skepticism gone by the end of the presentation, replaced by the excitement of viewing artwork that feels more like an experience than an image, communicating time, change, story, and possibility" (orionmagazine.org).

You can visit the Beehive Design Collective's website to invite them to your campus and to download the poster (and this image) to use in your own campaigns.

Viral Marketing

There are many metaphors people use to describe how digital media work. One that often is used for green advertising and campaigns is *viral*. To do so suggests the nonlinear and accumulative way that digital media (and viruses) work. That is, the **viral media** metaphor suggests that communication spreads digitally in random patterns (will you share MisterEpicMan's *How Animals Eat Their Food* YouTube video from an acquaintance with one friend or ten or your cousin?) and exponentially (you send the Taylor Swift *Goat Duet* YouTube video to your friend who sends it to two friends who tweet it to one hundred, and so forth).

In advertising, **viral marketing** is increasingly popular. This approach to selling products and brands adopts a word-of-mouth campaign often through digital media, involves little or unpaid services, and often is targeted at an audience in a way to appear organic and grassroots rather than top-down. A common college town example is students who are paid to go to a bar to order a drink in the hopes on the part of a liquor company that they can create a buzz around a particular drink using their product and increase their sales. We revisit this concept in Chapter 9.

Failed Persuasion

Digital media, obviously, sometimes fail. In 2010, the British climate campaign 10:10 (http://www.1010global.org/) commissioned the well-known comic screenwriter Richard Curtis (*Love Actually, Bridget Jones' Diary, Four Weddings and a Funeral, Notting Hill,* etc.) to write and produce a mini-movie. Shot on 35 millimeter by a professional crew, the partnership showed the promise of adding a digital media component to the group's broader efforts. Yet when the mini-movie *No Pressure* came out, audiences felt they had gone too far.

In an attempt to be comedic, the film involved scenes congratulating those who were joining efforts to address climate change and literally blowing up those who are not with gory blood explosions. The controversy cause the organization to take the video down from its site (though it is available online if you search for it) and to issue an apology in which they wrote: "At 10:10 we're all about trying new and creative ways of getting people to take action on climate change. Unfortunately in this instance we missed the mark. Oh well, we live and learn" (*Sorry,* n.d.).

This example shows how the humor often championed by digital media as a contemporary affordance of media-savvy consumers can backfire when one is not in touch with one's intended audiences.

Green Graphic Design

Most of the examples in this chapter show how visual media provide more detailed documentation of the world we live in or imagine. Yet one of the ways digital media is inventive is that it also has led to a proliferation of **infographics**, or visual interpretations of information or data such as a table or chart that explain something (like a report, concept, or set of connections). These images are not valued for their detail, but their abstraction. The assumption is that more people will be

Jacobs School of Engineering

| Photo 4.3 | Given the increase of air pollution in certain cities and (the not unrelated) popularity of mobile communication technologies, more companies and governments are designing air pollution apps for people to monitor up-to-date and localized air quality information. |

moved by information or data if it can be summarized in a single frame, with compelling connections and/or quantification illustrated quickly. They often use arrows to show how points are linked, clip art to simplify a point (of a dollar bill or a light bulb, for example), and bars or other objects to represent scale (like how many disposable diapers can be replaced with how many cloth ones—hint: the answer is in the thousands).

Data visualizations illustrate numerical data through an image. Examples include bar charts or three-dimensional representations. For example, Visual.ly regularly updates an infographic about endangered rhinos that uses multiple data visualizations, such as graphs of rhinos killed and the overall reduction in population, numbers linked to particular species, a map of world demand, and various sizes of circles showing populations by country (Visual.ly, 2013).

Given how technical a lot of environmental information can be, it is perhaps not a surprise that many of the most successful infographics today are focused on environmental communication. The technology magazine Wired.com lists the following topics among the top infographics of 2013: a U.S. population map, how 85 common dog breeds can be traced genetically to four kinds of dogs, where tornados touch down in the United States, the most common travel paths in New York City based on geotagged Tweets, and the microbiome that exists in human bodies (Vanhemert, 2013).

Increasingly popular also are **interactive maps**, two-way electronic communications that synthesize complex data through a cartographic representation and respond to user activity. One exemplar is the World Resources Institute's Aqueduct global water risk mapping tool. The self-declared goal of this interactive infographic is to help "companies, investors, governments, and other users understand where and how water risks and opportunities are emerging worldwide" (World Resources Institute, n.d., para. 1). The risks they map include physical quantitative risks, physical qualitative risks, and regulatory and reputational risks. By going to their website, one can click on any location in the world and find out more information about the water present, threatened, or lacking in an area (http://www.wri.org/our-work/project/aqueduct/aqueduct-atlas). Interactive maps such as this afford the producer to summarize a great deal of data, while allowing the user to interact with the information in a way that is more specific to her or his interests. (We elaborate further on how environmental activists use new media in Chapter 9.)

SUMMARY

- Environmental communication increasingly involves images. This chapter offers an overview of the complicated and changing visual landscape we live in.
- We considered various approaches to assessing and studying the relationship between the environment and popular culture. Although some focus on aesthetics or representations encoded or decoded in media, we introduced three additional approaches: (1) studying how audiences respond to environmental media texts through "dominant," "oppositional," and "negotiated" positions; (2) analyzing the life cycle of communication technologies, including production, use, and disposal/reuse; and (3) mapping a "circuit of culture" approach that can account for environmental communication constituted through representation, as well as how that relates to broader contexts.
- Having established different approaches to popular culture, we then turned to how to analyze visual texts. First, we considered the ways visual rhetoric in paintings and still photographs can shape attitudes and behavior toward nature. Second, we explored how moving images afford particular relationships with environmental embodiment and the witnessing of ecological disaster, including constituting condensation symbols and real-time images.
- Finally, we turned to the fields of green art, marketing, and graphic design to reflect on what they do or do not afford. We described how environmental artists sometimes use nature as a medium of expression and, at other times, use art to respond poetically to environmental crises. For marketing, we focused on two competing metaphors to consider the ways digital technologies are transforming how we construct, send, and receive green messages. We also noted how graphic

design, particularly infographics and interactive maps, increasingly afford new possibilities for communicating technical environmental information and data.

- In the next chapter, we turn to how news media use images and words to communicate about the environment in the public sphere.

SUGGESTED RESOURCES

- To see quintessential photographs of Yosemite National Park and other stunning places on Earth, you can visit the park or see the online Ansel Adams Gallery at http://www.anseladams.com/
- Krista Bryson's (2014) West Virginia Water Crisis blog shows how one student can develop a transmedia website in response to an environmental crisis: http://west virginiawatercrisis.wordpress.com/
- Google Earth provides interactive digital maps of Fukushima, Japan, at http://www.nationsonline.org/oneworld/map/google_map_Fukushima.htm
- Louie Schwartzberg, founder of Blacklight films' TEDx talk (and 10-munute, time-lapse film of the natural world) is at http://www.youtube.com/embed/gXD MoiEkyuQ
- To further explore the creative choices of visual metaphors, art, and sustainability, see Brian Cozen (2013) "Mobilizing Artists: Green Patriot Posters, Visual Metaphors, and Climate Change Activism."
- For stunning art that looks like photographs, but is mostly pastels, see Zaria Forman's work. Her series on the Maldives, which are particularly vulnerable to sea levels rising, is notable: http://www.zariaforman.com
- On the 350.org "Do the Math" tour, environmental advocates take a "family photo" at every location to show how the movement to address climate change is growing. See, for example, http://www.rollingstone.com/politics/pictures/bill-mckibben-and-350-org-on-the-do-the-math-tour-20121120/we-have-a-little-tradition-for-every-show-0347561

KEY TERMS

Aesthetics 71

Affordances 71

Bias of communication 70

Circuit of culture 73

Condensation symbol 79

Data visualization 85

Encoding and decoding 71

Environmental art 82

Infographics 84

Interactive maps 86

Intermediation 74

Life-cycle-assessment (LCA) 73

Viral marketing 84

Viral media 84

Visual rhetoric 74

Witnessing 78

DISCUSSION QUESTIONS

1. Some climate scientists and journalists have complained that the public cannot "see" global warming. How would you solve this problem? Which medium do you think can most compellingly express the impacts of prolonged drought, rising sea levels, disease, and so forth?

2. How many phones have you owned in your life? What do you do with the old one when you buy a new one: trash, donate, or recycle it? Where do recycled phones go?

3. Find three environmentally focused photographs about animals online: one you share the dominant message with, one you resist completely, and one you think shows both persuasive and unpersuasive points. How do the images alter your opinion of the topic they are portraying, if at all? Are you more influenced by aesthetic choices of the photographer or something else?

4. What is your favorite movie with an environmental message? Why? Do you think films that show what you value and do not want harmed (such as beautiful sunsets at the beach or healthy children playing at a park) or document a problem (such as people walking through apocalyptic floods or dirty water coming out of someone's faucet) or portray a fictional time and place motivate people more? Why?

5. Draw an infographic about climate change. How are numbers used, if at all? What choices did you make to have complicated data appear quicker and easier to understand? Why do you think it is a compelling interpretation?

NOTES

1. Both authors of this textbook have been influenced by what we have seen or not seen in the world, though (and sometimes perhaps because) we long have worn glasses and contacts. Since we live in an ocularcentric or seeing-centered culture, we think this chapter is important to include in this edition. Nevertheless, people who are severely seeing impaired or blind would not have the heavy reliance on sight that we underscore in this chapter.

2. Mark Warford (Ed.). *Greenpeace Witness: Twenty-Five Years on the Environmental Front Line* (London: Andre Deutsch, 1997). We are grateful to Kevin DeLuca for pointing out this connection. For an elaboration on the argument in this paragraph about witnessing as an environmental advocacy tactic of the past and present, see Pezzullo (2007).

3. We are grateful to Joshua Trey Barnett for first telling us of this invitation.

Environmental activists, left, hold a press conference on beach during the United Nations Climate Change Conference in Cancun, Mexico, 2010.

CHAPTER 5

News Media and Environmental Journalism (Old and New)

News about the environment, environmental disasters, and environmental issues or problems does not happen by itself, but is . . . "produced," "manufactured," or "constructed."

—Hansen (2010, p. 72)

Digital and social media sources have had a tsunami-like impact in providing environmental information that cannot be discounted, or more importantly, should not be overlooked, . . . in light of major declines in traditional media coverage of environmental issues.

—Friedman (2015)

By now, we've seen that our perceptions of nature and environmental issues are mediated by many sources—political debate, art, documentaries, and so forth. Among the important sources of information about the environment have been journalists and the news media. In this chapter, we explore the emergence and changing nature of environmental journalism, the forces that influence the "production" of environmental news, and some of the effects of media on readers and viewers.

In referring to *news media,* it is important to distinguish between *traditional news media* and news publicized through *digital technologies.* By **traditional news media**, we mean the researching, writing, and broadcasting of reports for newspapers or magazines in print, television, or radio, usually by professionals trained as journalists. In this chapter, we describe the significant role of these traditional sources in covering environmental news stories. Then, we explore the explosive growth of **digital technologies**, communication media that are based on computational devices (blogs, social media, mobile devices, etc.), to address how they have served as supplements to, and sometimes substitutes for, traditional news media's role in producing information about the environment.

Apart from technological advances for audiences and professional changes for journalists, there remains another reason to distinguish traditional news media from digital technologies: content. At the heart of traditional environmental journalism is a dilemma. Journalism professor Sharon Friedman (2004) has observed that environmental journalists working in traditional media still must deal with a "shrinking news hole while facing a growing need to tell longer, complicated and more in-depth stories" (p. 176). In journalistic parlance, a **news hole** is the amount of space that is available in a newspaper or TV news story relative to other demands for the same space. Friedman explains that competition for shrinking news space has increased the pressure on journalists to simplify or dramatize issues to ensure that a story gets printed or aired. As a result, in addition to pressures to use popular digital technologies (such as Twitter), many journalists also are using blogs associated with their newspapers, which often offer greater freedom and, notably, space to tell more in-depth stories.

When you've finished this chapter, you should be aware of some of the factors influencing the production of news and information about the environment in the public sphere. You should also be able to raise questions about the possible influence of news media in shaping our perceptions and behavior about important environmental concerns.

Growth and Changes in Environmental News

By the 1960s, news stories and visual images of environmental concerns began to appear prominently in U.S. news media—from the photo of the Earth taken by astronauts on *Apollo 8* in 1968 to TV film of an oil spill off the coast of Santa Barbara to *Time* magazine's story of Ohio's Cuyahoga River bursting into flames from pollution in 1969. During the next decades, the media's interest would periodically expand and wane, portraying environmental problems in multidimensional and often dramatic ways.

In this section, we describe the origins of environmental reporting in the United States and some of the characteristics of news coverage of the environment. We also describe the "perfect storm" that's led to the closing or downsizing of many newspapers and its impact on environmental journalism in recent years.

Emergence and Cycles in Environmental News

As environmentalism became a formidable force after Rachel Carson's *Silent Spring* (1962), "environmental journalism grew with it." Some newspapers began an environmental "beat," but beat or no beat, reporters found themselves covering issues like dioxin, smog, and endangered species, as well as oil spills, air pollution, and nuclear fallout (Palen, 1998, para. 1). In 1990, the field of environmental journalism was given a boost by the creation of the Society of Environmental Journalists (SEJ), whose mission "is to strengthen the quality, reach and viability of journalism across all media to advance public understanding of environmental issues" (www.sej.org). By the first decade of the 21st century, more than 1,400 journalists identified as environmental reporters in the United States, with more than 7,500 journalists in other countries covering the environment (Wyss, 2008, p. ix).

Even as news coverage of environmental issues grew during this period, the characteristics of this coverage—and the demands on journalists—are not unlike other areas of journalism. That is, news stories about the environment often are driven by specific events—oil spills or nuclear plant accidents—and, like other stories, they must compete with news of war, unemployment, terrorism, and other breaking news.

Event-Driven Coverage

Inevitably, contemporary news "is largely event focused and event driven," and it is this norm that is important in determining "which environmental issues get news coverage and which don't" (Hansen, 2010, p. 95). Indeed, environmental news thrives on dramatic events such as oil spills, forest fires, hurricanes, and accidents at nuclear power plants. Yet, as veteran environmental journalist Bob Wyss (2008) points out, environmental issues rarely are as dramatic as a hurricane. "Rather than striking with a fury, some [stories] ooze, seep, or bubble silently . . . or the story might change so imperceptibly as to be nearly invisible, such as the disappearance of another animal or plant

©iStockphoto.com/USO

Photo 5.1	Environmental journalists travel the world to provide eyewitness evidence of ecological atrocities and solutions.

species" (p. 8). The long-term effects of global climate change are not "immediately observable"; nor is the link between pollution and respiratory diseases obvious to observers (Hansen, 2010, pp. 95–96).

This invisibility of many environmental phenomena presents a challenge for journalists. As media scholar Anders Hansen (2010) reminds us,

> The timescale of most environmental problems is ill-suited to the 24-hour cycle of news production: Many environmental problems take a long time to develop; there is often uncertainty for years about the causes and wider effects of environmental problems . . . and even where a scientific and political consensus may emerge, the "visualization" for a wider public audience of what is happening requires a great deal of communicative "work." (p. 96)

As a result, many environmental news stories—like other news stories—present a *snapshot,* a specific moment, event, or action from a larger phenomenon. When such stories appear, they are usually driven by dramatic visuals, such as the illegal slaughter of elephants for their ivory tusks.

Added to this challenge is another difficulty. Many environmental reporters, themselves, lack training in the issues they're covering. These issues are often complex, ranging from depletion of the ozone layer around the Earth to the health effects of genetically modified organisms. The Knight Center for Environmental Journalism, for example, reported "only 12 percent of environmental journalists had degrees in scientific or environmental fields" (Wyss, 2008, p. 18). Still, some journalists, such as former *New York Times* journalist Andrew Revkin and the *Chicago Tribune's* staff, including Michael Hawthorne, Patricia Callahan, and Sam Roe, have excelled in reporting complex issues like climate change and flame retardants in a manner that has informed readers about the causes, long-term impacts, and even the science behind these important subjects.

The event-driven nature of news and the invisibility of many environmental issues, nevertheless, have meant that an environmental story must compete with other breaking news. Even an in-depth, but complicated story about the global shortage of water may be shoved aside in favor of a more dramatic news event. Indeed, over the years, the frequency of environmental news has risen and fallen as wars, economic recession, terrorism, and other concerns have seized TV and newspaper headlines.

Cycles in Environmental News Coverage

With the rise of an ecology movement in the late 1960s, environmental news grew in coverage and reached an early peak after Earth Day during the early 1970s. Such news coverage, however, began to disappear in the United States in the 1980s, "just as environmental issues were becoming more complex involving health effects, scientific uncertainty, economic and regulatory issues" (Friedman, 2015). As often happens, a dramatic event in 1989, the *Exxon Valdez* oil spill in Alaska, spurred new interest in the environment. Film of oil-soaked birds and otters and oil-blackened coastlines of Alaska's Prince William Sound filled TV screens nightly. The Tyndall Report, which tracks network news minutes, reported that environmental stories that year saw an unprecedented 774 minutes, combined, on the *CBS Evening News, NBC Nightly News,* and *ABC World News Tonight* (Hall, 2001).

The years following the *Exxon Valdez* oil spill, however, would see a marked difference. Without environmental disasters, or large Earth Day celebrations, traditional news media's interest in the environment waned. Shabecoff (2000) reports that, not only did environmental stories not grow in the 1990s during President Bill Clinton's administration, but the total number of news stories about the environment carried by newspapers and television networks declined substantially. The Tyndall Report tracked a low of 174 minutes in 1996 for the major TV news reports and 195 minutes in 1998 (Hall, 2001). By the end of the 1990s decade, environmental reporters were citing shrinking news holes as one of the most frequent barriers to coverage of environmental news (Sachsman, Simon, & Valenti, 2002).

A brief resurgence in environmental coverage occurred in 2001, with President George W. Bush's election, but the September 11, 2001, terrorist attack focused U.S. news on terrorism and a "war on terror" with very real wars in Iraq and Afghanistan. Friedman (2004) reported at the time that nearly all environmental journalists with whom she consulted agreed that "the events of September 11 have shrunk the [environmental] news hole even further" (p. 179).

By 2006, coverage of the environment had erupted, across all media—newspapers, TV, online news sites, film, and news magazines. *Time* magazine's cover, in April 3, 2006, illustrated the heightened interest: "Be Worried. Be Very Worried" (about global warming). Newspaper coverage of global warming, particularly, spiked during this period (Brainard, 2008b, para. 10), spurred by the riveting documentary film *An Inconvenient Truth* (2006), warning of the dangers of global climate change, and the United Nation's Intergovernmental Panel on Climate Change (IPCC) report in early 2007, concluding that increases in global temperatures since mid-20th century were "very likely" due to human-caused greenhouse gases (Sec. 2.2). By the end of this same decade, however, news coverage of the environment had begun to drop off again.

One important note about cycles of news coverage, particularly of climate change: In their comprehensive study of newspaper coverage of climate change in 27 countries between 1996–2010, Schmidt, Ivanova, and Schaefer (2013) reported that, "while coverage of climate change goes up and down in cycles, media attention to climate change has increased very significantly in an overall upward trend across all countries" (Hansen, 2015).

Overall, certain characteristics of news media coverage of the environment stand out from this early period. Hansen (2015) described two as (a) "once introduced in the 1960s, the 'environment' has consolidated itself as a news-beat and category of news coverage, and (b) . . . news coverage of the environment has gone up and down" (Hansen, 2015). As Friedman (2004) has also pointed out, "The environmental beat has never really been stable, riding a cycle of ups and downs like an elevator," often crowded-out when competing against other events—economic news, war and terrorist events, etc. (p. 177).

By the early 21st century, however, another force had also appeared, one that began what the Pew Research Center (2004) called "an epochal transition [in journalism], as momentous as the invention of the telegraph or television" (p. 4). This transition would dramatically change how news and information about the environment are communicated to the public.

A Perfect Storm: Decline of Traditional News Media

The conditions that would affect U.S., and to a similar extent, Canadian and some European news organizations, appeared early in the new century.[1] "A shrinking US economy, loss of advertising revenues, and audiences migrating to the Internet on tablets and mobile devices produced 'a perfect storm' that forced traditional mass media outlets to downsize and change the way they dealt with news." Environmental journalism was "hit hard by this downsizing" (Friedman, 2015).

As we write, the Pew Research Journalism Project is reporting that, "Print [newspaper] advertising continues its sharp decline," and while television advertising "currently remains stable, . . . the steady audience migration to the web will inevitably impact that business model, too" (Holcomb & Mitchell, 2014, para. 7). The drop in revenue has left newspapers, particularly, downsizing everything—daily circulation, the size of newspapers, space devoted to news, and the number of reporters. This has left many U.S. newsrooms with fewer experienced reporters covering the environmental beat. As Bud Ward, at the *Yale Forum on Climate Change and the Media,* said, "It's hard for reporters to focus on ambitious climate reporting . . . when their ranks are being 'carnaged'" (Ward, 2008, para. 28).

And, while online versions of newspapers are now available, these still depend on staff reporters to produce content, and therein lies a potential problem. When the Seattle *Post-Intelligencer* newspaper moved online, for example, it slashed its news staff of 165 reporters and began operating online with only 20 reporters (Yardley & Pérez-Pena, 2009). As a consequence, online daily papers increasingly depend on content aggregators for much of their national reporting.

As media cut staff reporters, there is inevitably a loss of science expertise. Some are eliminating entire beats: The San Jose *Mercury News* reported, "two decades ago nearly 150 papers had a science section. Now fewer than 20 are left, and [these]. . . usually dedicate their scarce column inches to lifestyle and health" (Daley, 2010, para. 16). Veteran TV reporter John Daley (2010) put it bluntly: "The ranks of reporters best equipped to cover . . . major environmental and climate change stories at most news outlets, particularly in local markets, are being decimated" (para. 6). In 2008, for example, CNN cut its entire science, technology, and environment news staff. A survey of news directors at radio and TV stations by the Center for Climate Change Communication at George Mason University confirmed this trend two years later:

- Television news directors are interested in running science stories, but few have staff dedicated to this beat.
- Climate change is covered relatively infrequently on local TV news.
- Most news directors are comfortable with their weathercasters reporting on climate science. (Maibach, Wilson, & Witte, 2010, pp. 4, 5)

The impacts of these changes are palpable. In 2013, the *New York Times* closed its environment desk and Green blog, claiming it was moving the environment out of its "silo" and that it was motivating every desk to cover the environment as part of what it does. Later that year, Margaret Sullivan (2013) reported that "[t]he quantity of climate change coverage decreased" and "the amount of deep, enterprising coverage of climate change in the *Times* appears to have dropped."

As traditional new media confront these challenges, some are launching their own websites and expanding links to other news sources, including online sources and news aggregators. *The Guardian*, a prominent British newspaper, for example, has increased its coverage of environmental news "many times over via the Guardian Environmental Network, a group of more than 30 content partners from across the media spectrum" (Brainard, 2015). Such trends confirm that it matters whether there are reporters paid to cover environmental news or not.

News Production and the Environment

Environmental news stories do not write themselves. As we learned in Chapter 3, an environmental issue becomes recognized as a *problem* as a result of the constitutive nature of communication itself. As Hansen (2010) reminds us, "News about the environment, environmental disasters, and environmental issues or problems does not happen by itself, but is rather 'produced,' 'manufactured,' or 'constructed'" (p. 72). In this section, we'll describe the influences that shape or constrain this "making" of news, including the demand for "newsworthiness," requirements for objectivity and "balance" in news stories, as well as other forces that may impact journalists.

Journalistic Norms and Constraints

Newsworthiness

One of the most important of the practices that affect environmental news reporting is the value or *newsworthiness* of a story. **Newsworthiness** is the ability of a news story to attract readers or viewers. In their popular guide to news reporting, *Reaching Audiences: A Guide to Media Writing,* Yopp, McAdams, and Thornburg (2009) identify criteria, found in most U.S. media guidelines, that determine the newsworthiness of a particular news story: (a) prominence, (b) timeliness, (c) proximity, (d) impact, (e) magnitude, (f) conflict, (g), oddity, and (h) emotional impact. As a result, reporters and editors feel they must strive to fit or package environmental problems according to these criteria in producing "newsworthy" stories.

Since we've seen that environmental news is event-oriented and characterized by strong visual elements (while tending to ignore environmental issues that "involve long, drawn-out processes" [Anderson, 2015]), it should not be surprising that reporters stress such criteria as impact, magnitude, and conflict in their stories. Let's look at some examples of this.

News headlines of the melting of the Antarctica ice sheets emphasized the potential impact and timeliness of the inevitable sea level rise from climate change: "Scientists Warn of Rising Oceans from Polar Melt" (Gillis & Chang, 2014), and "Ice Melt in Part of Antarctica 'Appears Unstoppable'" (Hanna, 2014). The lead paragraph in the *New York Times* elaborated on these themes:

> A large section of the mighty West Antarctica ice sheet has begun falling apart and its continued melting now appears to be unstoppable, two groups of scientists reported on Monday. If the findings hold up, they suggest that the melting could destabilize neighboring parts of the ice sheet and a rise in sea level of 10 feet or more may be unavoidable in coming centuries. (Gillis & Chang, 2014, para. 1)

While the fate of the Antarctica ice sheets captured news headlines, coverage of climate change, as we learned above, cycles up and down in U.S. media. How "newsworthy," then, are incidents like this? Check out "Act Locally!: Is Global Warming "Newsworthy"?

Act Locally!

Is Global Warming "Newsworthy"?

One measure of the "newsworthiness" is the frequency of news articles that are published in leading newspapers. How often do U.S. and other nations' leading newspapers report stories of global warming or climate change events? To answer this question, Drs. Maxwell Boykoff and Maria Mansfield track newspaper coverage of global warming in the United States and in 50 newspapers in 20 nations and on six continents (displayed in a graph that is updated monthly).

For an up-to-date look at how well the 5 largest U.S. newspapers are covering global warming, check out the latest Boykoff and Maxwell graph at http://sciencepolicy.colorado

.edu/icecaps/research/media_coverage/usa/index.html. (The five U.S. newspapers included in the tracking are *Washington Post, Wall Street Journal, New York Times, USA Today,* and the *Los Angeles Times.*)

Which newspaper reports the most stories? The least?

What factors explain the "ups and downs' in the frequency of newspaper stories about global warming in the United States?

Also, check how well the world's newspapers are covering this subject today at this same site. What country's news media report the most news about global warming?

And another example of factors that produce a "newsworthy" story is following the meltdown of several nuclear power reactors in Japan in March, 2011, this front-page headline appeared, "Japan Faces Potential Nuclear Disaster as Radiation Levels Rise." The news headline stressed the impact and magnitude of the crisis at the Fukushima Daiichi nuclear plant, which burned after an earthquake and tsunami struck Japan four days earlier. These criteria were reinforced immediately in the opening paragraph:

Japan's nuclear crisis verged toward catastrophe on Tuesday after an explosion damaged the vessel containing the nuclear core at one reactor and a fire at another spewed large amounts of radioactive material into the air, according to the statements of Japanese government and industry officials. (Tabuchi, Sanger, & Bradsher, 2011, p. A1)

Masaru Kamikura/Flickr. Used under Creative Commons license https://creativecommons.org/licenses/by/2.0/

Photo 5.2   Japanese NTV newscast during the crisis at the Fukushima Daiichi nuclear power station after the 2011 earthquake and tsunami.

Later in the Fukushima nuclear plant story, an emphasis on *proximity*, or nearness to readers, emerged to guide the story's narrative: "After an emergency cabinet meeting, the Japanese government told people living within 30 kilometers, about 18 miles, of the Daiichi plant to stay indoors, keep their windows closed and stop using air conditioning" (p. A1).

Another criterion—*conflict*—is an especially influential factor in news stories about the environment: environmentalists versus loggers, climate scientists versus global warming denialists, angry residents versus chemical companies, and so forth. And such stories of environmental conflict often are accompanied by visual elements— photos, film, etc. Indeed, some environmental groups like Greenpeace are known for their ability to generate newsworthy stories by their dramatic image events, which take advantage of news media's desire for pictures, particularly images of conflict. (See FYI, "Image Events and the Environment.")

☞ **FYI**   **Image Events and the Environment**

By offering dramatic visual events as part of a story, environmental advocates have been able to fulfill the newsworthiness criterion and have excelled in gaining photojournalism coverage for their campaigns. Greenpeace pioneered this environmental advocacy tactic in the 1970s in the Pacific Ocean with its dramatic staging of activists in small boats protesting the killing of whales next to massive ships with harpoons posed to kill them; given the enormous size of whales and their stature in our collective imagination, the comparatively small boat of resistance embodied vulnerability in the public sphere in a way that was not previously done.

Environmental communication scholar Kevin DeLuca (2005) has called these **image events,** or staged, visual events that take advantage of television's hunger for pictures, particularly images of conflict. Recognizing the shorthand images can provide, environmental advocates can organize photo opportunities accordingly. These have included film footage of large banners draped from a corporate headquarters proclaiming, "End clear-cutting of rain forests!" or the 1981 Earth First! banner draped on the side of Glen Canyon Dam, making the dam appear cracked in a photograph to announce the birth of their organization. DeLuca quotes a veteran Greenpeace campaigner who explained that such image events succeed by creating "mind bombs" and "reducing a complex set of issues to symbols that break people's comfortable equilibrium, get them asking whether there are better ways to do things" (p. 3).

Media Frames

In his classic study, *Public Opinion*, Walter Lippmann (1922) was perhaps the first to grasp a basic dilemma of news reporting when he wrote,

> The real environment is altogether too big, too complex, and too fleeting for direct acquaintance. We are not equipped to deal with so much subtlety, so much variety, so many permutations and combinations. And although we have to act in that environment, we have to reconstruct it on a simpler model before we can manage it. *To traverse the world men* [sic] *must have maps of the world* [emphasis added]. (p. 16)

As a result, journalists have sought ways to simplify, frame, or make "maps of the world" to communicate their stories.

As we noted in Chapter 3, the term frame was first popularized by Erving Goffman (1974) as the cognitive map or pattern of interpretation that people use to organize their understandings of reality. With this in mind, Entman (1993) defined the *act of framing* as the selection of "some aspects of a perceived reality and [making] them more salient in a communicating text" (, p. 56). Hansen (2010) explained, "Frames, in other words, draw attention—like a frame around a painting or photograph—to particular dimensions or perspectives, and they set the boundaries for how we should interpret or perceive what is presented" (p. 31). It is this idea that journalists rely upon as they attempt to write or "frame" a news story in a particular way.

In Chapter 1, we introduced the concept of **media frames** or the central organizing themes that connect the different elements of a news story (headlines, quotes, etc.) into a coherent whole. By providing this coherence, media frames help readers make sense of new experiences, relating them to familiar assumptions about the way the world works. Here, we want to elaborate on specific examples.

Consider the media frame that is the "central organizing theme" in a news story in the *Alaska Dispatch*, which bills itself as "news and voices from the last frontier." The story's headline, "Aerial Wolf Hunting on Kenai Peninsula Put on Hold," announced a temporary halt to the practice of shooting wolves from small planes "to help boost moose populations" ("Aerial," 2012, para. 1). A photo of a snarling wolf's face, with its fangs bared, accompanied the story. The photo, headline, quotes, and narrative structure all revealed the theme of "predator control," that is, shooting wolves, which local hunters believed would help to "improve the moose population on the Kenai" (para. 2, 3). All of the elements of the news story contributed to an "anti-wolf" frame.

Using different frames invites readers to "see" or view the world differently. And in environmental controversies, the opposing parties sometimes compete to influence the framing of a news story. Miller and Riechert (2000) explain that each side tries to gain public support for its position, often "not by offering new facts or by changing evaluations of the facts, but *by altering the frames or interpretive dimensions for evaluating the facts* [emphasis added]" (p. 45).

Now compare, for example, the *Alaska Dispatch* anti-wolf media frame with a story from the *New York Times*, "As Wolves' Numbers Rise, So Does Friction Between Guardians and Hunters." A photo of a smiling gray-haired woman, playing with a docile wolf, laying on its back, accompanied the story. The lead paragraph stated, "When people like Nancy Jo Dowler started raising wolves here decades ago, the animals were rare in Wisconsin and nearly extinct across the country" (Yaccino, 2012, para. 1). The headline, opening paragraph, and photo introduce a media frame of "conflict," with a wolf-friendly narrative that pits hunters and trappers against "local and national animal rights groups that fear the undoing of nearly four decades of work to restore a healthy number of wolves" (para. 4).

The two news stories oriented readers in very different ways. The first story organized the subject of wolf hunting around concerns for ensuring larger moose populations for hunters, depicting wolves as predators that had to be "controlled."

The second story framed wolves in light of their declining numbers—some nearing extinction—and the growing opposition to shooting them—a conflict frame, with a decidedly pro-wolf stance.

The struggle over which frames should define our understanding of the environment is a central feature in the public sphere. Indeed, the history of environmentalism in the United States and other countries can be understood to be a struggle over different but powerful frames for understanding the natural world and what constitutes an environment "problem." News media, for example, now refer more often to *rainforests* instead of *jungles* and to *wetlands* instead of *swamps*. (To test your ability to recognize different frames, select and compare competing news stories of a compelling environmental issue.)

Norms of Objectivity and Balance

The values of **objectivity and balance** have been bedrock norms of journalism for almost a century. In principle, these are the commitments by journalists to provide information that is accurate and without reporter bias and, where there is uncertainty or controversy, to balance news stories with statements from all sides of the issue. Additionally, the latter norm has been relied upon when a reporter lacks the expertise or time to determine where "truth" lies (Cunningham, 2003), or when a reporter wants to foster debate.

Objectivity

In practice, however, the norms of objectivity and balance run into difficulty. Particularly in environmental journalism, reporters struggle to maintain genuine objectivity. While a story on the shooting of wolves, for example, may be accurate on one level, a kind of bias already has occurred in the selection of this story versus others; this occurs also in its framing and the choice of sources that are interviewed (as we just saw above). In another example, environmental communication scholar Anabela Carvalho (2007) found in her study of British newspapers that the discursive constructions of climate change are "strongly entangled with ideological standpoints." That is, the ideas and values underlying different political views or orientations work as a "powerful selection device in deciding what is scientific news, i.e., what the relevant 'facts' are, and who are the authorized 'agents of definition' of science matters" (p. 223).

As a consequence, a challenge for "objective" reporting has been the need for journalists to rely on credible sources, that is, sources whose experience or insight readers and viewers will trust as the basis for "truth." Often, these individuals are those whom some have called the "authorized knowers" of society—scientists, experts in a field, government and industry leaders, and so on (Ericson, Baranek, & Chan, 1989, quoted in Hansen, 2010, p. 91). (In Chapter 7, we describe some of the difficulties that victims of environmental harms, such as residents of low-income communities, have experienced in being interviewed or recognized as "credible" sources in environmental news stories.)

Balance

Related sometimes to objectivity is the norm of balance in reporting a story. Balance usually means a responsibility to report all sides of story, particularly when there is a controversy. Yet, this also can be problematic. As Wyss (2008) points out, "Getting both sides of a story, while generally desirable in journalism, does not always work in science and environmental coverage" (p. 62). He cites the longtime science editor for the *Dallas Morning News* who quipped that, "balance in space stories would require that every story about satellites would require a comment from the Flat Earth Society" (p. 62).

Nevertheless, when environmental issues are controversial, or when reporters lack the expertise to adjudicate between conflicting claims, the tendency in journalism has been to balance stories by quoting differing viewpoints. This refers to the pairing of a disputed report or statement from a scientist or public official with an opposing viewpoint. For example, balance was a common practice in much of the early coverage of global warming. Boykoff and Boykoff (2004) cite the following example from a *Los Angeles Times* article in 1992:

> The ability to study climatic patterns has been critical to the debate over the phenomenon called "global warming." *Some scientists believe*—and some ice core studies seem to indicate—that humanity's production of carbon dioxide is leading to a potentially dangerous overheating of the planet. *But skeptics contend* there is no evidence the warning exceeds the climate's natural variations [emphasis added]. (Abramson, 1992, p. A1)

Even as late as 2010, and with mounting evidence of human-caused climate change, a survey of news directors at radio and TV stations found that "nearly all news directors (90%) believe that, like coverage of other issues, coverage of climate change must reflect a 'balance' of viewpoints" (Maibach et al., 2010, p. 4).

In recent years, the norm of balance has been sharply criticized. (See "Another Viewpoint: Objectivity and Balance in News Reporting.") Some media critics have challenged the assumption that there are always two sides of an issue, particularly when empirical data or scientific research strongly supports one side. For example, climatologist Stephen H. Schneider (2009) complained, "A mainstream, well-established consensus of hundreds of experts may be 'balanced' against the opposing views of a few special-interest PhDs. To the uninformed, each position seems equally credible" (pp. 203–204). As a result, balance will be misleading. Boykoff and Boykoff (2004) found that "balancing" in the reporting of global warming in U.S. newspapers actually led to "biased coverage" of the science of climate change and the findings of anthropogenic (human) contributions to global warming (p. 125).

The trend of balancing scientists and skeptics in reports of global warming may be changing. Brainard (2008b) reported that the news media seem to be "slowly but surely eliminating false balance when addressing human activity's role in global warming" (para. 6). Boykoff (2007) found that "balanced" coverage of global warming science has tapered off in recent years in major U.S. newspapers. Still, many newspapers continue to use "qualifying and hedging language," suggesting uncertainty, "despite ever-growing scientific agreement that human activities modify global climate" (Bailey, Giangola, & Boykoff, 2014, p. 197).

Other Influences on Environmental Journalists

Political Economy of News Media

The term **media political economy** refers to the influence of the economic interests and/or political agenda of the owners of newspapers and television networks on the news content of these media sources. In her critical study of corporate influence on the media, Australian media scholar Sharon Beder (2002) noted that most commercial media organizations are owned by multinational corporations with financial interests in other businesses (such as forestry, oil and gas, electric utilities, real estate, etc.). With increasing consolidation of media ownership, some media managers, editors, or reporters may feel pressure from owners to choose (or avoid) stories and to report news in ways that ensure a favorable political climate for these business or ideological interests. In turn, they "become the proprietor's 'voice' within the newsroom, ensuring that journalistic 'independence' conforms to the preferred editorial line" (McNair, 1994, p. 42).

Another Viewpoint: Objectivity and Balance in News Reporting

Environmental communication professional, nature writer, and Director of the Center for Environmental Communication at Loyola University, New Orleans, Robert A. Thomas describes the responsibility of journalists this way:

A reporter should use balance in researching all sides of a story, but in drawing conclusions and reporting the results of this research, a reporter must be guided by fairness and objectivity. As an example, the evidence for global climate change is solid and scientists who specialize in this arena overwhelmingly are supporters. There is a school of thought, called The Deniers, largely composed of the free market policy community and others whose best interests are served by denying the impacts of carbon on climate, who cast doubt into the equation by raising anecdotal arguments and using fragments of data sets. Good journalists will acquaint themselves with these disparate arguments, but give credence to the arguments of climate scientists who follow the scientific method in their quest for truth.

Robert A. Thomas, PhD, Director, Center for Environmental Communication, School of Mass Communication, Loyola University, New Orleans (personal communication, April 5, 2011).

Consider the example of General Electric (GE), one of the world's largest corporations and former owner of NBC television and its business channel CNBC. Beder (2002) found that GE was by no means a hands-off owner of its networks. Instead, she reported, GE officials regularly inserted the business's interests into network editorial decisions. Although GE has had environmental problems, "NBC journalists have not been particularly keen to expose GE's environmental record" (p. 224). For example, when the Environmental Protection Agency (EPA) found GE responsible for discharging more than a million pounds of toxic chemicals called polychlorinated biphenyls (PCBs) into New York's Hudson River and proposed that GE pay for a massive cleanup, "the company responded with an aggressive campaign aimed at killing the plan," spending, by its own estimates, $10 to $15 million on advertising (Mann, 2001, para. 2). (Eventually, GE agreed to fund a cleanup plan for the river.)

Finally, the sheer size of some media corporations enables their viewpoints to reach large numbers of readers or viewers. Sinclair Broadcast Group (known for its conservative political stance), for example, owns more TV stations in the United States than any other corporation. According to the *Seattle Times*, the corporation also "has a history of using its television stations to support its political views" (Heffter, 2013, para. 1, 2). For several years, newscasts on select Sinclair TV stations, for example, ended the news program with a conservative commentator who spoke of the "manmade global warming hysteria that swept the nation," and who claimed, "most people realize manmade global warming is a hoax" (Hyman, 2011, para. 1–2; see also Heffter, 2013).

Gatekeeping and Newsroom Routines

The decisions of editors and media managers to cover or not cover certain environmental stories illustrates what has been called the **gatekeeping** role of news production. Simply put, the metaphor of gatekeeping is used to suggest that certain individuals in newsrooms decide what gets through the "gate" and what stays out. White's (1950) classic study "The 'Gatekeeper': A Case Study in the Selection of News" launched the tradition in media research of tracking the routines, habits, and informal relationships among editors and reporters and among reporters' backgrounds, training, and sources.

Many editors and newsrooms find it particularly difficult to deal with the environmental beat for two reasons. First, the unobtrusive or "invisible" nature of many environmental problems makes it hard for reporters to fit these stories into conventional news formats. Second, environmental issues can be difficult to report because few reporters have training in science or knowledge of complex environmental problems such as groundwater pollution, animal waste, genetically modified crops, or cancer and disease clusters. And few news organizations have the financial resources to hire such talent. As a result, "on any given day, an environmental story may be assigned to a science specialist, a health reporter, a general assignment reporter, or even a business reporter" (Corbett, 2006, p. 217).

As a result of these constraints, reporters and editors have been turning to online news services such as ClimateWire (eenews.net/cw) or Environmental Health News (environmentalhealthnews.org). The Society of Environmental Journalists (SEJ) maintains other databases useful to reporters. (We describe these more specialized, online news sources below.)

The influences on the production of news—newsworthiness, media frames, etc.—enable, but they also limit the ability of traditional news media to report on the environment. In the next section, we go beyond this "making" of news to look at the *effects* of news media on readers and viewers.

Media Effects

In Chapter 1, we said that our understanding and behavior toward the environment depend not only on environmental sciences but also on *media representations,* as well as public debate, film, popular culture, everyday conversations with people we know, and more. But just exactly how do media shape our views of nature or our environmental behaviors? Answering these questions brings us to the ways media scholars study what are called **media effects.** By this phrase, we mean the influence of different media content, frequency, and forms of communication on audiences' attitudes, perceptions, and behaviors.

In this section, we review three major theories of the effects of news coverage on audiences' attitudes and behavior: (a) agenda setting, (b) narrative framing, and (c) cultivation theory. While these approaches provide little evidence of direct, causal effects, they do suggest media's influence is *cumulative and a part of a wider context of social influences* that help to construct our interest in, and our understanding of, the environment.

Agenda Setting

Perhaps the most influential theory of media effects that applies to environmental news is "agenda setting." Although Walter Lippmann (1922) first suggested the idea of mass media influences on the public, Bernard Cohen (1963) coined the term agenda setting to distinguish between individual opinion (what people believe) and the public's perception of the salience or importance of an issue. News reporting, he said, "may not be successful much of the time in telling people *what to think,* but it is stunningly successful in telling its readers *what to think about* [emphasis added]" (p. 13; see also McCombs & Shaw, 1972). In their study of television, Iyengar and Kinder (1987) defined agenda setting this way: "Those problems that receive prominent attention on the national news become the problems the viewing public regards as the nation's most important" (p. 16).

The agenda setting hypothesis has been influential in much environmental communication research. Such studies "have confirmed that the media can play a potentially powerful role in setting the agenda for public concern about and awareness of environmental issues" (Hansen, 2011, p. 18; Soroka, 2002). And Eyal, Winter, and DeGeorge (1981), Ader (1995), and Soroka (2002) discovered that the agenda setting effect is especially strong for *unobtrusive* issues or those to which readers or viewers have little personal access and that this effect is most apparent in the public's perceptions of risk or danger from environmental sources.

Environmental communication scholars have, nevertheless, cautioned that agenda setting studies sometimes have been conflicting. In an attempt to clarify agenda setting theory, journalism scholar Christine R. Ader (1995) investigated the influence of what she calls "real-world conditions," public opinion, and the media's agenda in a study of news reports about the environment in the *New York Times* from 1970 to 1990. Since some consider this a definitive study, let's explore it further.

Ader asked two questions: (1) Is the public's concern about environmental problems driven by real-world conditions rather than media coverage and (2) do public attitudes influence the amount of media coverage of an issue rather than the media's setting of the agenda?

Using data from Gallup polls during this period, Ader identified problems that the public periodically rated as the "most important problem facing the nation today" (p. 301). To control for the influence of public opinion and real-world conditions on media coverage, Ader examined the length and prominence of news reports of pollution in the *Times* for three months before and three months after each Gallup poll was conducted as well as data from independent sources documenting real-world conditions for disposal of wastes, air quality, and water quality in these same periods.

Ader's findings affirmed the presence of a strong agenda setting effect, even when real-world conditions and prior public opinion both were taken into account. That is, even though objective measures showed that overall pollution had declined for the period studied, the *Times* increased its coverage of news stories about pollution; furthermore, the greater length and prominence of these stories correlated positively with a subsequent increase in readers' concerns about this issue. However, the opposite was not true; that is, the media did not appear to be mirroring public opinion. Ader concluded, "The findings suggest that *the amount of media attention devoted to pollution influenced the degree of public salience for the issue* [emphasis added]" (p. 309).

Finally, we should note that, while more individuals now find their news about the environment via digital technologies, traditional news media still exert an influence on decision makers. In a study of Canadian newspapers, for example, Soroka (2002) found that news media "show a significant impact of the media on both the public and policy agendas" when it comes to the environment (p. 279). Other research has also found that traditional news media agenda setting has an impact on decision-makers, including the effect of news photographs of environmental issues on policy-makers (though less so on the general public) (Hansen, 2015; Jenner, 2012). Overall, recent research is showing that news media "have strong agenda-setting impacts on environmental issues such as climate change" (Shanahan, McComas, & Deline, 2015).

While agenda setting theory may explain the salience or importance of an issue to the public, it doesn't claim to account for what people *think* about this issue. Therefore, it is important for us to look at other theories that focus on the role of the media in constructing *meaning* or ways of understanding environmental concerns.

Narrative Framing

As we noted earlier, journalists communicate not only facts about the environment but also media frames for making sense of these facts. Theories of media effects, therefore, have begun to focus on the ways in which such frames organize our perceptions or attitudes about the environment. To be clear, these theories do not argue that media reports, by themselves, *cause* public opinion; rather, they claim "media discourse is part of the process by which individuals construct meaning" (Gamson & Modigliani, 1989, p. 2). In this section, we describe the specific role of media framing in providing *narratives* or a story line about an event.

A narrative approach takes seriously the role of media frames that provide organizing themes that connect different elements of a news story into a coherent whole. **Narrative framing,** then, refers to the ways in which media organize the bits and facts of phenomena through stories to aid audiences' understanding and the potential for this organization to affect our relationships to the phenomena being represented. In other words, media frames can organize the facts in ways that provide a narrative structure—what is the problem, who is responsible, what is the solution, and so on. Principal proponents of a narrative framing approach, James Shanahan and Katherine McComas (1999), observed that environmental media coverage is "hardly ever the simple communication of a 'fact'"; instead, "journalists use narrative structures to build interesting environmental coverage. . . . [B]ecause journalists and media programmers must interest audiences, they must present their information in narrative packages" (pp. 34–35).

A case in point is Schlechtweg's (1992) early study of narrative framing in a Public Broadcasting Service (PBS) report on Earth First! protesters. In May 1990, Earth First! activists poured into California's northern redwood forests for "Redwood Summer," to "nonviolently blockade logging roads, [and] climb giant trees to prevent their being logged" (Cherney, 1990, p. 1). Although the organizers stressed that anyone who disagreed with nonviolence would be barred, tensions grew that summer among loggers, Earth First! activists, and rural communities.

On July 20, 1990, the PBS program *The MacNeil-Lehrer NewsHour* covered the Redwood Summer protests in a report titled, "Focus—Logjam." In his analysis, Schlechtweg found the *NewsHour* report established clear identities for protagonists and antagonists in a narrative frame of tense confrontation that suggested the real prospect of violence.

Protagonists in the broadcast were portrayed through key value terms: "workers," "timber people," and "regular people" who depended on "timber harvests" and "small-town economies" for "jobs," "livelihood," and their "way of life" (p. 273). Conversely, the report identified Earth First! protesters as "apocalyptic," "radical," "wrong people," "terrorists," and "violent" people who engaged in "confrontation," "tree spiking," "sabotage," and "civil disobedience" (p. 273). Schlechtweg argued that, as a result, "Focus—Logjam" implicitly constructed a narrative frame that pitted "regular people" against a "violent terrorist organization, willing to use sabotage . . . and tree spiking to save redwood forests" (pp. 273–274).

More recently, environmental communication scholars have explored the effect of various narrative frames on the public's responses to news of climate change. For example, many European newspapers are reframing climate change in terms of a narrative of "responsibility" (for solving the problem) and consequences, that is, how climate change will affect people (Dirikx & Gelders, 2009; Hansen, 2011, pp. 16). Others have suggested that "framing climate change in terms of public health and/or national security may make climate change more personally relevant and emotionally engaging to segments of the public who are currently disengaged or even dismissive of the issue" (Myers, Nisbet, Maibach, & Leiserowitz, 2012, p. 1105).

In a study testing the effects of different climate change frames, Myers et al. (2012) used an online survey of U.S. residents who were asked to read news articles about climate change that were framed variously as risks to (a) the environment, (b) public health, and (c) national security. Their results showed that across both sympathetic and doubtful audiences "the *public health* focus was the most likely to elicit emotional reactions consistent with support for climate change mitigation and adaptation." They also discovered that a national security frame could "possibly *boomerang* among audience segments already doubtful or dismissive of the issue, eliciting unintended feelings of anger [emphasis added]" (p. 1105).

In the analysis of both the Earth First! report and the study of climate change frames, we find that narrative frames sometimes may evoke deeper "interpretive packages" or ways of understanding the world that reflect cultural images, beliefs, or values that convey "particular ideological interpretations of nature and the environment" (Hansen, 2010, p. 105). This effect becomes more pronounced as people are exposed to particular frames over time and it is this theory of "cultivation" that we turn to next.

Cultivation Analysis

Akin to narrative theory is a cultivation model of media effects. Shanahan (1993) describes **cultivation analysis** as "a theory of story-telling, which assumes that repeated exposure to a set of messages is likely to produce agreement in an audience with opinions expressed in . . . those messages" (pp. 186–187). As its name implies, cultivation is not a claim about immediate or specific effects on an audience; instead, it is a process of gradual influence or cumulative effect. The model is associated with the work of media scholar George Gerbner (1990), who stated

> Cultivation is what a culture does. That is not simple causation, though culture is the basic medium in which humans live and learn. . . . Strictly speaking, cultivation means the specific independent (though not isolated) contribution that a particularly consistent and compelling symbolic stream [such as TV viewing] makes to the complex process of socialization and enculturation. (p. 249)

Gerbner's own research looked exclusively at the long-term effects of viewing violence on television—the cultivation of a worldview that he called the "**mean world syndrome**." This is a view of society as a dangerous place, peopled by others who want to harm us (Gerbner, Gross, Morgan, & Signorielli, 1986). Others who use cultivation theory to study the environment are interested in the longer-term effects of media on viewers' environmental attitudes and behavior.

Since news media report a stream of environmental threats or concerns, the easy hypothesis of cultivation might be that viewers who watch a lot of television are likely to develop more concern for the environment than light viewers. And some studies have, in fact, confirmed this (see, for example, Dahlstrom & Scheufele, 2010). But this is not always the case. Other analyses have found that heavy media exposure is sometimes correlated with lower levels of environmental concern (Ostman & Parker, 1987). In a study of college students' television viewing, for example, Shanahan and McComas (1999) reported that heavy exposure to television sometimes hampers the cultivation of pro-environmental attitudes. "This tends to go against the suggestion that media attention to the environment results in greater socio-environmental concerns. That television's heavy viewers tended to be less environmentally concerned suggests the opposite: Television's messages place a kind of 'brake' on the development of environmental concern, especially for heavy viewers" (p. 125).

Interestingly, Shanahan and McComas found that a decrease in environmental concern among heavy television viewers is stronger among politically active students. This finding appears to contradict what we said earlier about the effects of agenda setting; that is, the more frequent the coverage of a subject, the more salience it gained. How is this explained? Cultivation researchers explain this pattern as **mainstreaming,** or a narrowing of differences toward a cultural norm. Shanahan and McComas (1999) explained that television's consistent stream of messages may draw groups closer to the cultural mainstream, which, in many television programs, is "closer to the lower end of the environmental concern scale" (p. 130).

A second explanation for the decrease in environmental concern among heavy viewers of television is sometimes termed **cultivation in reverse** (Besley & Shanahan, 2004; Shanahan, 1993). In other words, media may cultivate an anti-environmental attitude through a persistent lack of environmental images or by directing viewers' attention to other, non-environmental stories. Thus, by ignoring or passively depicting the natural environment, television tends to marginalize its importance. Cultivation theorists (Shanahan & McComas, 1999) also call this phenomenon **symbolic annihilation**—the media's erasure of the importance of a theme by the indirect or passive de-emphasizing of that theme.

As a result, the effects of heavy media exposure on viewers' environmental attitudes may be more complex than first thought. For example, an important exception to *cultivation in reverse* is a study of TV viewers' perceptions of environment risks (Dahlstrom & Scheufele, 2010). This study introduced the variable of diversity of TV channels being viewed (from CNN, Discovery, and Fox News to Comedy Central), as well as the amount or frequency of TV viewing. The authors found that viewers' exposure to a wider diversity of TV channels led to increased concern about environmental risks "above and beyond . . . the effects of the amount of television watched" (p. 54).

Overall, the effects of news media on viewers' attitudes about the environment can be complex and difficult to pin down. Nevertheless, theories such as agenda setting, narrative framing, and the longer-term cultivation of viewers' outlooks do suggest broader influences. In the end, the media may provide, as Hansen (2011) observes, "an important cultural context from which various publics draw both vocabularies and frames of understanding for making sense of the environment generally, and of claims about environmental problems more specifically" (p. 20).

Digital Technologies and the Transformation of Environmental News

Although traditional news media continue to play an important role in the news coverage of the environment, "a tsunami-like impact" from social media and other digital technologies has emerged in the wake of the decline of traditional media (Friedman, 2015). Online services and social media have impacted not only reporters in their day-to-day work, but news sources themselves—from online environmental news services to citizens' use of social media to report eyewitness accounts of environmental events.

Digitizing Environmental Journalism

Changing Reporters' Routines

As newspapers, radio, and TV stations launched website versions of their news, the work of environmental reporters and the content of news itself changed dramatically. With their websites showcasing more (and longer) news stories, more (and crisper) photographs and video, and 24/7 coverage, environmental journalists are facing new demands for freshening content (and doing so via multiple feeds). The *Times Picayune* newspaper of New Orleans provides a vivid example of these changes.

In 2012, the *Times Picayune* cut its daily newspaper to three times a week, laying off some 200 employees. It also expanded its website. One reporter who stayed at the newspaper was the Pulitzer Prize-winning, environmental journalist Mark Schleifstein. As a result of the changes at the newspaper, Schleifstein said his "workday has already expanded in order to fill the daily web feed" (Sachsman & Valenti, 2015). He explained

> When I cover a meeting, I'll set up my laptop, using my cell phone as the hot spot, and sometimes I'll be tweeting, basically using that as my note-taking process . . . and also taking general notes, while at the same time trying to get up and take a picture of the speaker, or I may actually do some video. (Archibald, 2014, p. 21; quoted in Friedman, 2015)

Overall, as Sachsman and Valenti report, the *Time Picayune's* "expanded delivery via the Web now offers expanded stories, including aggregated input, some of which is . . . coming from Schleifstein's desk/computer/Droid operation."

Still, as traditional news media undergo changes, they're facing "increased multi-platform demands (video, audio, and text, along with blogs, Twitter, Facebook, Tumblr, Reddit, 4chan, and YouTube)," posing "significance challenges even to the most skilled and experienced reporters" (Boykoff & Yulsman, 2013, p, 7). One source for environmental news that has proved invaluable for many reporters in this new context has been the role of online environmental news organizations.

Online News Organizations

By the first decade of the 21st century, online platforms featured a wide range of news and information, including journalists' blogs, videos, news services, and websites of scientists, environmental groups, and governmental agencies such as the EPA. Among the earliest and most influential of these platforms were **environmental news services**. These are news aggregation sites—some with original reporting—offering access to journalists, scientists, government officials, and other readers looking for more in-depth environmental news and timely information.

One of the most prominent of these services is Environment & Energy Publishing (eenews.net). E&E offers a suite of news services, including a wire service, E&E TV, and special sites providing detailed coverage of energy, public lands issues, water, climate change, and other issues. Its Land Letter (eenews.net/11), for example, specializes in natural resources (e.g., wilderness, oil and gas drilling on public lands, etc.). More recently, E&E launched Climate Wire (eenews.net/cw), a premier source for daily news on climate-related stories.

While E&E is subscription-based, other news sites are freely available to the public as well as to reporters. For example, the leading online environmental news site, among 12 national news services, is *The Huffington Post,* with a daily Green section. HuffPost Green (huffingtonpost.com/green) also features 42 environmental blogs (Friedman, 2015).

Some of the best environmental reporting today is coming from nonprofit news sites such as Environmental Health News, ProPublica (which also covers environmental news), and smaller, start-up sites like InsideClimate News (insideclimate news.org). In fact, InsideClimate News won a Pulitzer Prize for national reporting for its news series, "The Dilbit Disaster: Inside the Biggest Oil Spill You've Never Heard Of." The series was a detailed investigation revealing the inept response to a 2010 pipeline rupture that spilled "a million gallons of bitumen, a thick, dirty oil from Canada's tar sands" into Michigan's Kalamazoo River (Brainard, 2015). Curtis Brainard, who covered the environment for the *Columbia Journalism Review* and is now at *Scientific American,* called InsideClimate News' "Dilbit Disaster" series "not only public service journalism at its best, it was new media at its best" (n.p.).

Social Media and Citizen Environmental Journalism

While the majority of original news reporting "still comes from the newspaper industry," social media and mobile applications are changing the ways that people

U.S. Environmental Protection Agency

Photo 5.3	Dilbit is a semi-solid form of petroleum containing oil-thinning chemicals that dissolve in water, which journalists covered in recent spills in Michigan and Arkansas. Research continues on whether or not it is more dangerous than other forms of oil and whether or not spill risks can be avoided near vital waterways. Continued journalist attention to this topic in the public sphere helps keep the topic on the agenda of environmental decision makers.

not only receive news, but are also helping to produce news (Mitchell, 2014, para. 5, 16). This is especially true when it comes to generating and receiving news about the environment.

Social Media and Eco-News

It is no accident that environmental reporters are tweeting and posting to blogs. As the latest Pew Research Center's Journalism Project reports, "half of Facebook and Twitter users get news on those sites as do 62% of reddit users" (Matsa & Mitchell, 2014, para. 2). And not only reporters, but environmental groups, government agencies, climate scientists, and a host of Twitter feeds are providing environmental news, from the EPA's Twitter feed @EPAGov and Global Warming (twitter.com/globalwarming), for example, to Green News Report (twitter.com/GreenNewsReport) and the National Oceanic and Atmospheric Agency's feed @NOAA.

While traditional news media assumed a kind of uni-directional, "broadcast" model for disseminating news, social networking users are also generating a cornucopia of news and information about the environment. Beyond Facebook, for example, social networking sites featuring news about the environment are proliferating—from familiar sites like Treehugger (treehugger.com) to MindBodyGreen (mind bodygreen.com) and Reddit: Environment (reddit.com/r/environment) to sites like Topix: Environmental Law (topix.com/forum/law/environmental), an active site for debate about environmental law and politics. (We explore the uses of new/social media in environmental activism further in Chapter 9.)

Still, some—journalism scholars especially—have posed some provocative questions about the popularity of new/social media for news and information about the environment, generally, and for handling "big" issues like climate change. (For some of these questions, see "Another Viewpoint: Some Questions About Social Media and Climate Change.")

Another Viewpoint: Some Questions About Social Media and Climate Change

How effective are new/social media in producing and circulating news and information about climate change? Climate change communication researchers Max Boykoff, Marisa McNatt, and Michael Goodman have posed several challenging questions about the influence of digital technologies in this important arena:

1. Does increased visibility of climate change in new/social media translate to improved communication or just more noise?

2. Do these spaces provide opportunities for new forms of deliberative community . . . and offline organizing and social movements? . . . Or has the content of this increased coverage shifted to polemics and arguments over measured analysis?

3. In this democratized space of content production, do new/social media provide more space for contrarian views to circulate?

4. And through its interactivity, does increased consumption through new/social media further fragment public discourse . . . through information silos where members of the public can stick to sources that help support their already held views?

SOURCE: Boykoff, M. T., McNatt, M. M., & Goodman, M. K. (2015).

Finally, more and more individuals are using their smartphones and other mobile devices to contribute to environmental reporting by taking photos or videos, and other actions (Matsa & Mitchell, 2014). Let's look a little further at this trend toward citizen journalism.

Citizen Environmental Journalism

With the widespread use of smartphones, iPads, and digital cameras, citizens are increasingly "playing important eyewitness roles around news events" (Mitchell, 2014, para. 16). Some are calling this trend **citizen environmental journalism,** or ordinary citizens who witness an environmental event sharing that information with others, including news organizations, scientists conducting studies, or with friends over Facebook or social media.

While ordinary citizens have often appeared as "sources" in traditional news stories—a mother concerned that her child's asthma is aggravated by pollution from a nearby coal plant, for example—many are increasingly assuming the role of reporters themselves. "Recent years have seen journalist-source relationships dramatically recast . . . with citizens making the most of the internet and digital technologies to engage in their own forms of reportage" (Allan & Ewart, 2015).

This reversal of journalist–citizen roles occurred, for example, during the disastrous oil spill in the Gulf of Mexico in 2010, when citizen journalists helped in the crowdsourcing of information:

> As journalists scrambled to gather and interpret official assertions in the face media blackouts, ordinary citizens . . . stepped in to provide valuable, near instant information via websites and social media platforms, such as Twitter. News organizations were quick to draw on these accounts, especially where the eyewitness accounts of those directly affected by events were concerned (Veil, Buehner, & Palenchar, 2011). (as quoted in Allan & Ewart, 2015)

Citizen eyewitness reports during the Deepwater Horizon disaster also involved the creative use of aerial digital cameras to map the spreading oil and its threats to marine life. One of the organizers explained

> For less than $100 in parts, we used helium balloons and kites to send cameras to over a thousand feet, and stitched the resulting images into high-resolution maps using our free, open-source software. Over a hundred volunteers hit the beaches to take tens of thousands of photos, depicting slicks, oiled wetlands, and the birds, fishes, and plants threatened by the disaster. (Griffith, Dosmagen, & Warren, 2012, quoted in Allan & Ewart, 2015)

In commenting on the role of citizen environmental journalists during the Gulf oil crisis, researchers Stuart Allan and Jacqueline Ewart (2015) observed, "there is little doubt that the value of citizen contributions was considerable, not least in helping to bridge the gaps between official expertise—both scientific and journalistic—and that of the lay publics committed to alternative forms of fact-finding."

* * *

Finally, we can't end this section without a demurral: As our journalist friends remind us, while digital technologies are immensely popular with many, traditional news media are still highly influential sources for most "elites," that is, for decision-makers and opinion leaders. Newspapers like the *New York Times*, *Wall Street Journal*, or the (London) *Times*, for example, and television news such as CNN and NBC or the BBC and Sky News (in the UK) remain pivotal news sources for many politicians, government officials, industry leaders, and others. That's one reason why we've spent time in this chapter on topics like *agenda setting* and *media effects*, since the attention given to environmental issues by traditional news media continues to exert an influence on those decision makers who often hold the power to affect the outcome of these issues.

SUMMARY

. Given the global reach of environmental crises and celebrations, we often rely on the reports of others, including TV reporters, social media tweets, newspaper articles, and stories on the radio. Yet news media are neither innocent nor neutral in their representations of the environment.

- In the first section of this chapter, we described the emergence, nature, and cycles of traditional news media.
- In the second section, we looked at some of the journalistic norms and constraints on news production, including newsworthiness, media frames, norms of objectivity and balance, political economy, and gatekeeping and newsroom routines.
- The third section examined media effects—the influence of news media on our attitudes toward the environment—and identified three theories: (1) agenda setting, (2) narrative framing, and (3) cultivation analysis.
- Finally, we described some of the changes that digital technologies are bringing to news organizations, including the impact on environmental reporters, the growth of online news services, the role of social media, and the rise of citizen environmental journalism.

SUGGESTED RESOURCES

- The Pew Research Center's *The State of the News Media* is an annual report that examines trends in the U.S. news media—its growth (or decline), role of digital technologies, revenue, and more. Available at http://www.journalism.org/pack ages/state-of-the-news-media
- Media Matters (report): "How Broadcast News Covered Climate Change in the Last Five Years," January 16, 2014; available at http://mediamatters.org/research/2014/01/16/study-how-broadcast-news-covered-climate-change/197612
- *Cosmos: A Spacetime Odyssey*, a visually stunning, science TV series: www.cosmosontv.com/
- Dale Slongwhite, Storytelling to Confront Injustice, Environmental Justice in Action Blog. U.S. EPA. Available at http://blog.epa.gov/ej/2014/06/storytelling/

KEY TERMS

Citizen environmental journalism   114

Cultivation analysis   109

Cultivation in reverse   110

Digital technologies   92

Environmental news services   112

Gatekeeping   105

DISCUSSION QUESTIONS

1. How do you feel about the journalistic norms of objectivity and balance in reporting environmental stories? What does it mean to be "objective" in reporting on climate scientists' forecasts? Should such reports be "balanced" with dissenting views?

2. Do news media affect your attitudes or behavior toward environmental issues? Nuclear power, vegetarianism, global warming, and more? Can you identify characteristics of a report that particularly influenced you?

3. Boykoff et al. caution about "information silos," where people visit only those news or information sources that agree with them. How often (if at all) have you checked out "opposing" news sources? For example, if you doubt humans' influence on the climate, have you ever visited websites featuring climate scientists? Or, vice versa? Is there any value in reading news sources with differing viewpoints?

4. Search "fossil fuels" (or another topic) via a digital news source you use; compare your results with someone sitting next to you who searched a different site or perspective. What do those results tell you about your news sources as well as the ways in which the "production" of news may affect your different understandings of fossil fuels?

NOTE

1. The decline in traditional news media's business model "does not seem to be affecting large Asian countries such as China and India, where newspaper reading is expanding. And no journalism crisis [has been] felt by science journalists—who often cover environmental issues—in Latin America, Asia, North and South Africa, in contrast to science journalists' concerns in the United States, Europe, and Canada" (Friedman, 2015).

PART III

Communicating in an Age of Ecological Crises

In recent years, environmental scientists have played a more visible role in public debates over the loss of biodiversity, global climate change, air pollution, species extinction, and other problems.

CHAPTER 6

Scientists, Technology, and Environmental Controversies

How will history judge us if we watch the threat unfold before our eyes, but fail to communicate the urgency of acting? . . . How would I explain to the future children of my 8-year-old daughter that her grandfather saw the threat, but didn't speak up in time?

—climate scientist Michael E. Mann (2014)

To advocate or not advocate? That question is one of the most basic ethical dilemmas facing environmental scientists today.

—Vucetich & Nelson (2010)

Should scientists speak up, publicly, about pollution, climate change, or a Sixth Extinction of species? Would scientists risk their objectivity by wading into controversies like these? This chapter explores the role of the environmental sciences and technology as sources of knowledge in a democratic society. We also look at moments when scientists' claims become a site of conflict, when the legitimacy of science itself is challenged by industry, climate change deniers, or other interests who seek to influence the public and government about the meaning and uses of science.

Chapter Preview

- The first section of this chapter describes how science became a source of symbolic legitimacy in a society increasingly defined by technical complexities and uncertainty.
- In the second section, we describe the *precautionary principle* as a way to manage the scientific and technological uncertainty basic to many environmental controversies.

(Continued)

(Continued)

- In the third section, we introduce the role of scientists as society's *early warners*, that is, the duty of scientists to "speak up," to alert and advise the public in moments of environmental danger.
- Section four looks at how anti-environmentalists draw on a rhetorical *trope of "uncertainty"* to challenge the credibility of international scientific consensus.
- Finally, we describe recent initiatives by scientists, journalists, and media producers to improve communication about the science and impacts of global climate change.

Upon finishing this chapter, we hope you will have gained an appreciation for the place of environmental sciences in a democratic society, as well as for the decision of some scientists to speak out in public controversies over climate change, "fracking," and other technologies with potential environmental consequences.

Science, Technology, and Symbolic Legitimacy

As we're writing, public officials, local communities, oil and gas publicists, and environmentalists are arguing over the possible dangers of **hydraulic fracturing**, or **"fracking,"** a method used to drill for natural gas by injecting large volumes of water and chemicals under high pressure into rock strata, creating fissures that release the trapped gas. Uses of this technology, however, have raised serious questions: Does fracking pollute local wells? Leak methane (a potent greenhouse gas) into the atmosphere? Disrupt local communities? Or cause earthquakes? Because fracking is a complex technology, its impacts on human health and the environment have been sources of public controversy for the past decade in the United States.

Similar problems of complexity appear in other major technologies with environmental and human health implications, including genetically modified organisms (GMOs), the storage of nuclear waste, the feasibility of various renewable energy sources, and the impacts of high-decibel Navy sonar on whales and dolphins (Passary, 2014). In the following sections, we describe some of these controversies, as well as the challenge of engaging the broader public in such issues.

Complexity and the "Eclipse" of the Public

In the early 20th century, the philosopher John Dewey wrote about the complexity of modern life as the American public experienced problems with urban sanitation, industrial safety, and dramatic changes in communication technologies. In his book *The Public and Its Problems,* Dewey (1927) warned of an "eclipse of the public" when citizens lacked the expertise to evaluate the complex issues before them. The decisions about these issues are so large, he wrote, "the technical matters involved are so specialized . . . that the public cannot for any length of time identify and hold itself [together]" (p. 137). With the need for technical expertise in making these decisions, Dewey feared the United States was moving from democracy to a form of government that he called **technocracy**, or rule by experts.

The solution to the "problem" of the public, favored by the Progressive movement of the 1920s and 1930s, was grounded in the reformers' faith in science and technology as sources of legitimacy for state and federal regulation of the new industries. In regulatory policy, the Progressives wanted to rely on trained experts working with government to discover an objective public interest (p. 12). Legitimacy for political decisions would come from these experts' use of "neutral, scientific criteria for judging public policy" (Williams & Matheny, 1995, p. 12).

Although the **Progressive ideal** of neutral, science-based policy would be challenged later, the legitimacy or credibility accorded to science by the public grew steadily stronger. Indeed, popular culture would give "tremendous prestige and power to our official, publicly validated knowledge system, namely science" (Harman, 1998, p. 116). As a result, the sciences acquired a kind of **symbolic legitimacy**, a perceived authority or credibility as a source of knowledge. This association of science as a source of legitimacy was particularly true for environmental matters. For example, when the Environmental Protection Agency (EPA) was criticized for taking too long in its study of the effects of hydraulic fracturing on drinking water, it replied that "sound science" was driving its research (Dlouhy, 2011, para. 13).

The appeals to "sound science" reflect the ideal that policy should be free of bias and grounded in reliable and valid knowledge. Still, the Progressive ideal of neutral, science-based policies often falls short of its promise. As we see later in this chapter, pressures from business interests, politics, budget cuts, ideology, and other factors sometimes restrict the degree to which scientific research guides public policies.

The Progressive ideal also falls short in one other important way: It does not envision the active engagement of the public. As a result, ordinary citizens who have become alarmed by the potential health or environmental impacts of technologies such as fracking, coal-burning power plants, or "factory" animal farming, have begun to insert themselves into the public debates and decisions affecting their communities and their lives.

Let's look at an example—the controversy over fracking technology and the different voices invoking the symbolic legitimacy of technical scientific expertise in such public policy disputes.

Fracking and the Environmental Sciences

There is perhaps no better illustration of the role of science—and the challenges that complex technologies pose for ordinary citizens' understanding and participation—than the ongoing controversy over hydraulic fracturing or fracking in the United States and, increasingly, in the UK, Europe, and China.

In the United States, fracking, combined with horizontal or "side-ways" drilling and other technologies, has "allowed domestic producers to quickly reverse a decades-long trend of declining natural gas production" (Wilmoth, 2014, para. 2). The boom in domestic energy production has been hailed by economists, politicians, industry, and local boosters for its economic benefits, greater energy security, a lessening of the need for coal in electricity generation, and other advantages (Hassett, 2013). Yet local communities, environmental activists, and documentary films like *Gasland* and *Gasland Part 2* (gaslandthemovie.com) also began to raise concerns: Reports of polluted drinking water, health problems, loss of home values, impacts from heavy truck traffic, and other negative impacts ("The Costs of Fracking," 2012; Food & Water Watch, 2013; SourceWatch, 2014).

As a result, battle lines have been drawn, particularly between anti-fracking activists in local communities and the oil and gas industry. At the center of this controversy is a rhetorical struggle over the science itself. The barriers to citizens' participation in the public discussion have been particularly challenging, fueled by the scarcity of research, complexity of the technology, and the public's limited access to information held by industry.

For many years, there was little research into the environmental and health effects of hydraulic fracturing. A provision in a 2005 energy law, dubbed the "Halliburton loophole," had stripped the EPA of its authority to regulate fracking and, therefore, prevented the agency from studying its effects ("The Halliburton Loophole," 2009, para. 1, 2). Although the EPA, as well as independent scholars, is now beginning to study fracking, the absence of definitive conclusions regarding fracking's safety and its impacts has opened space for a contentious public debate about what counts as credible knowledge claims of the effects of this complex technology.

While local residents and anti-fracking activists have aired stories of polluted drinking water, disrupted communities, and sicknesses from nearby fracking wells, the oil and gas industry has claimed the symbolic legitimacy of science, attempting to reassure the public and rebuke critics. A spokesperson for ExxonMobil, for example, accused fracking opponents of making "unsubstantiated statements . . . regardless of the scientific evidence undermining their position," urging, instead, that the public dialogue focus on "sound science" (Cohen, 2013, para.1, 8). Yet, the limited scientific research, itself, has produced different results,[1] leading one editorial writer to quip that such studies were "a kind of fracking Rorscharch test" (Nocera, 2013, p. A19).

Frustrating citizens' ability to challenge industry claims has been their struggle to get accurate information about the chemicals used in the fracking process. Between 2005 and 2009, for example, the oil and gas industry "used over 2,500 hydraulic fracturing products . . . containing 750 chemicals and components, some extremely toxic and carcinogenic like lead and benzene" (SourceWatch, 2013, para. 1). The industry has long insisted that information about many of these chemicals is proprietary, or trade secrets, and, therefore, unavailable for disclosure to the public. While the oil and gas industry does make a database of some fracking chemicals publicly available (see http://fracfocus.org), many engineers, environmental scientists, and energy experts, as well as anti-fracking activists, believe industry is "shielding too much information from public view" (Dlouhy, 2014, para.1).

The proprietary nature of many fracking chemicals illustrates a larger tension between what we've earlier called the "technical" and the "public" spheres of communication (Chapter 1). For the general public, environmentalists, and anti-fracking activists, this tension refers to the challenge of accessing critical information

©iStockphoto com/David Parsons

| Photo 6.1 | Hydraulic fracturing ("fracking") site, with drilling rig near homes. The environmental health and public safety of hydraulic fracturing and horizontal drilling are at the heart of contentious debates in many places in the United States and other nations as new sources of natural gas are exploited by these technologies. |

from more private or technical spheres—such as scientific or engineering reports, information designated as "trade secrets," or state regulators' databases—into the public sphere, where citizens can scrutinize or use this information themselves.

In summary, hydraulic fracturing remains a fiercely contested technology in the United States and other nations. With the continuing uncertainty about its impacts, many opponents have called for bans or moratoria on fracking. Yet, such controversy is not unusual for the environmental sciences where there is often not "conclusive" or "absolute proof" for a finding (something that is often beyond the reach of science). At the same time, such inconclusiveness provides the opponents of environmental rules with opportunities to contest the scientific claims, "to slow up application of scientific knowledge," as historian Samuel P. Hays (2000) has observed. Their rhetorical watchword, he said, is "insufficient proof" (p. 151).

To understand such conflicts over science, technology, and the environment, let's look at some of the ways the contending voices in society attempt to deal with uncertainty, as well as the ways in which science itself may become a site of controversy. As we see, the fault line in many of these conflicts occurs between those evoking a principle of *caution* and others whose interests are adversely affected by such caution and who seek, therefore, to contest the claims of science.

The Precautionary Principle

As we saw above, a paradox of environmental sciences is that knowledge about the effects of human behavior on the environment is always incomplete, yet the scale of our influence on the Earth often demands we take action. Although "certainty" evokes a powerful pull for social reformers, religious adherents, and popular radio commentators, "it is forever denied to scientists" (Ehrlich & Ehrlich, 1996, p. 27).

This absence of scientific certainty has spawned much debate in conflicts over such issues as the protection of wetlands, food safety, new technologies like hydraulic fracturing, and large-scale problems like climate change and the loss of biodiversity. In this section, we look at some of the ways that different voices in the public sphere have dealt with uncertainty in the face of environmental and public health threats and we introduce a key guide for taking action—the precautionary principle.

Uncertainty and Risk

The absence of scientific certainty has historically provided openings for delays in taking action on environmental and public health threats. For decades, the opponents of environmental regulation of industry have used the indeterminacy of science as a rationale for resisting new standards to regulate hazardous chemicals, such as lead, DDT, dioxin, and polychlorinated biphenyls (PCBs) in everything from gasoline and oil refineries to pesticides and spraying for mosquitoes.

An infamous case of industry's use of scientific uncertainty occurred in 1922 and involved the introduction of tetraethyl lead in gasoline for cars. Although public health officials thought lead posed a health risk and should be studied more carefully first, the industry argued that there was no scientific agreement on the danger and pushed ahead to market leaded gasoline for the next 50 years. Peter Montague (1999) of the Environmental Research Foundation described what happened: "The consequences of that . . . decision [to delay standards for leaded gasoline] are now a matter of record—tens of millions of Americans suffered brain damage, their IQs permanently diminished by exposure to lead dust" (para. 3).

Historically, the procedures for assessing risk have given the benefit of the doubt to new products and technologies, even though these may prove harmful later. For example, only a tiny fraction of the tens of thousands of chemicals commercially used in the United States today are "fully tested for their ability to cause harm to health and the environment" (Shabecoff, 2000, p. 149). As early as the 1960s, a number of scientists, environmentalists, and public health advocates in the United States had begun to warn of possible dangers to human health from new chemicals appearing in water, air, the food chain, and in mothers' breast milk. (We described Rachel Carson's 1962 warning of the dangers of DDT in *Silent Spring*, in Chapter 1.)

The Precautionary Principle

By the 1990s, other scientists and health advocates were arguing the burden of proof should be shifted to require use of a principle of precaution. In 1991, the National Research Council offered a compelling rationale for this new approach: "Until better evidence is developed, prudent public policy demands that a margin of safety be provided regarding potential health risks. . . . We do no less in designing bridges and buildings. . . . We must surely do no less when the health and quality of life of Americans are at stake" (p. 270).

An important step came in 1998, in Racine, Wisconsin, when scientists convened the Wingspread Conference on the Precautionary Principle. The 32 participants—scientists, researchers, philosophers, environmentalists, and labor leaders from the United States, Europe, and Canada—shared the belief that "compelling evidence that damage to humans and the worldwide environment is of such magnitude and seriousness that new principles for conducting human activities are necessary" (Science and Environmental Health Network, 1998, para. 3).

At the end of the three-day meeting, the participants issued the "Wingspread Statement on the Precautionary Principle," which called for government, business, and scientists to take into account the new principle in making decisions about environmental and human health (Raffensperger, 1998). The statement provided the following definition of the **precautionary principle**: "*When an activity raises threats of harm to human health or the environment, precautionary measures should be taken even if some cause and effect relationships are not fully established scientifically. In this context the proponent of an activity, rather than the public, should bear the burden of proof* [emphasis added]" (Science and Environmental Health Network, 1998, para. 5).

The new principle was intended to apply when an activity poses a combination of potential harm and scientific uncertainty. It requires (a) an ethic of prudence (avoidance of risk) and (b) an affirmative obligation to act to prevent harm. Importantly, the principle shifted the burden of proof to the proponents to show "their activity will not cause undue harm to human health or the ecosystem"; furthermore, it required agencies and corporations to take proactive measures to reduce or eliminate hazards, including "a duty to monitor, understand, investigate, inform, and act" when anything goes wrong (Montague, 1999, para. 13).

Invoking the precautionary principle can be rhetorically persuasive in public controversies. For example, public officials in Ohio recently ordered seven "fracking" drilling operations to be suspended, after a series of earthquakes rattled the area. In defending their action, officials said they acted "out of an abundance of caution" in suspending the work (Fountain, 2014, p. A20). After investigation, state geologists "linked earthquakes in a geologic formation deep under the Appalachians to gas drilling"; this, in turn, led state officials, to issue stricter conditions for hydraulic fracturing in some areas ("Ohio," 2014, p. A14). They explained, "While we can never be 100 percent sure that drilling activities are connected to a seismic event, caution dictates that we take these new steps to protect human health, safety, and the environment" (Soraghan, 2014, para. 4).

☞ **FYI**   **The Wingspread Statement**
on the Precautionary Principle

The release and use of toxic substances, the exploitation of resources, and physical alterations of the environment have had substantial unintended consequences affecting human health and the environment. Some of these concerns are high rates of learning deficiencies, asthma, cancer, birth defects, and species extinctions; along with global climate change, stratopheric ozone depletion, and worldwide contamination with toxic substances and nuclear materials.

We believe existing environmental regulations and other decisions, particularly those based on risk assessment, have failed to adequately protect human health and the environment—the larger system of which humans are but a part.

We believe there is compelling evidence that damage to humans and the worldwide environment is of such magnitude and seriousness that new principles for conducting human activities are necessary.

While we realize that human activities may involve hazards, people must proceed more carefully than has been the case in recent history. Corporations, government entities, organizations, communities, scientists, and other individuals must adopt a precautionary approach to all human endeavors.

Therefore, it is necessary to implement the Precautionary Principle: When an activity raises threats of harm to human health or the environment, precautionary measures should be taken even if some cause and effect relationships are not fully established scientifically.

In this context the proponent of an activity, rather than the public, should bear the burden of proof.

The process of applying the Precautionary Principle must be open, informed, and democratic and must include potentially affected parties. It must also involve an examination of the full range of alternatives, including no action.

Reprinted with permission from The Science and Environmental Health Network, downloaded from http://www.sehn.org/state.html

Not all parties rushed to embrace the precautionary principle. Some businesses, conservative policy centers, and politicians are concerned that the consequences of using the principle are at odds with assumptions in the dominant social paradigm (Chapter 3); that is, they object that the principle errs on the side of *too much* caution. Ronald Bailey (2002), at the libertarian Cato Institute, for example, argued that "the precautionary principle is an anti-science regulatory concept that allows regulators to ban new products on the barest suspicion that they might pose some unknown threat" (p. 5).

In many ways, the precautionary principle invites us to "look ahead," take caution or be aware. And it is to this "early warning" role of environmental scientists that we now turn.

Early Warners: Environmental
Scientists and the Public

When Elizabeth Kolbert published her riveting account of the looming loss of species, *The Sixth Extinction* (2014), she joined a growing cadre of scientists who believe they have a moral obligation to speak out or warn of threats from climate change, loss of biodiversity, species extinction, and other problems. Likewise, in the next few years, geologists will be deciding whether or not they want to officially declare the time period we are living in (since 1950) as a new geological epoch that is radically different than the Holocene (Johnstone, 2014). Others fear such public visibility hurts scientists' credibility or is incompatible with the ideal of science as an objective and neutral inquiry. Still, the role of scientists in warning of danger has gained greater urgency, as they warn of a *tipping point,* or a time when the catastrophic consequences of the warming of the planet will become irreversible.

In this section, we review the debate over the obligations of scientists as early warners. We also look at examples of political interference in scientific research, including an effort to censor a well-known National Aeronautics and Space Administration (NASA) scientist who was one of the earliest to speak out publicly about climate change.

Dilemmas of Neutrality and Scientists' Credibility

In the debate over whether they have a moral duty to speak publicly, environmental scientists find themselves choosing between two competing identities: Are they objective scientists whose duty is to remain neutral, limiting their communications to the technical sphere of academic journals? Or are they environmental physicians of a sort, guided by a medical ethic—the impulse to go beyond the diagnosis of problem to a prescription for its cure? And if not, do they, at least, have an ethical duty to warn or speak out when the public remains unaware of hidden or looming dangers?

This dilemma over scientists' obligations was heightened by the new field of conservation biology, emerging in the late 1980s. One of its founders, biologist Michael Soulé (1985), insisted that conservation biology was a **crisis discipline**— its emergence was necessitated by a worsening ecological disturbance with irreversible effects on species and ecosystems (p. 727). He believed that scientists, therefore, could not remain silent in the face of a "biodiversity crisis that will reach a crescendo in the first half of the twenty-first century" (Soulé, 1987, p. 4). Indeed, scientists, he argued, have an ethical duty to offer recommendations to address this worsening situation, even with imperfect knowledge, because "the risks of non-action may be greater than the risks of inappropriate action" (Soulé, 1986, p. 6).

Other scientists take a different view, assuming that the scientist, who is "free of preconceived values, seeks the truth and follows it wherever it leads. It is assumed that whatever the outcome of the search, it will benefit human welfare" (Shabecoff, 2000, p. 140). As a result, some scientists fear that abandoning this identity, by publicly advocating responses to problems, would violate an ethic of objectivity and undermine the credibility of scientists (Rykiel, 2001; Slobodkin, 2000).

The controversy over scientists' ideals and their ethical duties in the face of ecological and human challenges has its roots in earlier controversies, and it may be useful to briefly review this history.

Environmental Scientists as Early Warners

Scientists' role in developing the atomic bomb in 1945 prompted one of the first major debates over the ethical responsibilities of scientists. Molecular biologists also began to insist on a greater voice in informing the public of the consequences of new research emerging after World War II (Berg, Baltimore, Boyer, Cohen, & Davis, 1974). As a result of these debates, scientific associations like the Federation of American Scientists, along with journals such as the *Bulletin of the Atomic Scientists*, arose to represent scientists in the public realm.

Act Locally!

Should Scientists Be Society's Early Warners of Environmental Dangers?

To gain an appreciation for some of the ethical questions in debates over scientists' responsibility to speak publicly about possible dangers, consider inviting an environmental or climate scientist from your campus to speak with your class.

1. What is his or her view of the role of science and scientists in a free society?

2. How does he or she balance the duty as a scientist to remain neutral and an ethical concern about warnings to the public of possible health or environmental risks?

3. How far is he or she willing to go in speaking publicly in moments of environmental danger? Where should one draw a line?

Finally, ask the visiting scientist to describe the results of having spoken out, publicly (if he or she has ever done this). Did he or she receive pushback, criticism?

By 1969, the Union of Concerned Scientists (ucsusa.org) had formed to address problems of survival in the late 20th century, foremost among these, the threat of nuclear war. The UCS has since expanded its scope to problems of global warming and the potential dangers of genetic manipulation, and now includes more than 400,000 scientists and citizens. Echoing the mission of this group, theoretical physicist Stephen Hawking, author of *A Brief History of Time*, has spoken about the urgency for scientists to serve as early warners. In 2007, he told a gathering of scientists

> As citizens of the world, we have a duty to alert the public to the unnecessary risks that we live with every day, and to the perils we foresee if governments and societies do not take action now. . . . As we stand at the brink of . . . a period of unprecedented climate change, scientists have a special responsibility. (quoted in Connor, 2007, paras. 5, 6)

More recently, the Intergovernmental Panel on Climate Change (IPCC) has warned that, "*Warming of the climate system is unequivocal, human influence on the climate system is clear, and limiting climate change will require substantial and sustained reductions of greenhouse gas emissions* [emphasis in original]" (IPCC, 2014, para. 1). Along with the IPCC, NASA and other scientific bodies have issued warnings about the impacts of climate change. Among their regional forecasts are the following:

- **North America**: Decreasing snowpack . . . ; increased frequency, intensity and duration of heat waves
- **Latin America**: Gradual replacement of tropical forest by savannah in eastern Amazonia; risk of significant biodiversity loss through species extinction . . . ; significant changes in water availability for human consumption, [and] agriculture
- **Africa**: By 2020, between 75 and 250 million people . . . exposed to increased water stress; yields from rain-fed agriculture could be reduced by up to 50 percent in some regions by 2020; agricultural production, including access to food, may be severely compromised
- **Asia**: Freshwater availability projected to decrease . . . by the 2050s; coastal areas will be at risk due to increased flooding; death rate from disease associated with floods and droughts expected to rise (NASA, n.d., para. 8–12)

Such warnings have prompted extensive debate, actions by many governments, and public controversy over the science itself, in some cases. And while Stephen Hawking, the IPCC, and many other scientists have been free to speak publicly, some have faced interference and even censorship of their reports and web postings intended for the general public. Let's look at some cases of this push-back against scientists as early warners.

Censoring the Early Warnings of a NASA Scientist

On December 6, 2005, Dr. James Hansen, NASA's chief climate scientist, warned in a speech to the American Geophysical Union, "the Earth's climate is nearing, but has not passed, a tipping point, beyond which it will be impossible to avoid climate change with far-ranging undesirable consequences." Further warming of more than 1°C (or about 2°F), Hansen said, will leave "a different planet" (quoted in Bowen, 2008, p. 4). Hansen is former director of NASA's Goddard Institute for Space Studies and it was he who first introduced global warming into the public spotlight when he testified before the U.S. Senate nearly two decades earlier.

Hansen's warning set off more than just scientific alarm bells. In the days following his 2005 speech, NASA officials ordered its public affairs staff to monitor Hansen's lectures, scientific papers, postings on the Goddard website, and journalists' requests for interviews (Revkin, 2006a). When the *Washington Post* reported this story, NASA staff tried to discourage the *Post*'s reporter from interviewing him. They agreed, finally, that Hansen "could speak on the record only if an agency spokeswoman listened in on the conversation" (Eilperin, 2006, p. A1). Other NASA scientists complained to the *New York Times* about similar political pressures on them "to limit or flavor discussions of topics uncomfortable to the [George W.] Bush administration, particularly global warming" (Revkin, 2006b, A11).

Photo 6.2	Hubbard Glacier in Alaska calving. Further warming of more than 1°C will leave "a different planet" (Dr. James E. Hansen, Director of NASA's Goddard Institute for Space Studies).

NASA later assured reporters that its scientists could speak freely to the media, but the clumsy handling of the Hansen incident had set off a storm of criticism in the media and blogosphere. Dr. Hansen himself defended his actions: "Communicating with the public seems to be essential . . . because public concern is probably the only thing capable of overcoming the special interests that have obfuscated the topic" of global warming (Revkin, 2006a, p. A1).

Political Interference in Scientists' Communication With the Public

Similar reports of political interference in scientific reports and scientists' communication with the public also had appeared during the Bush administration, from 2001 to 2009 (Mooney, 2006; Revkin, 2005; Union of Concerned Scientists, 2004, 2008).

At times, the environmental sciences proved to be politically inconvenient. On March 7, 2001, Ian Thomas, a 33-year-old government cartographer, posted a map of caribou calving areas in the Arctic National Wildlife Refuge (ANWR) on a U.S. Geological Survey website. Thomas had been working on maps for all of the national wildlife refuges and national parks using new National Landcover Datasets (Thomas, 2001). Nevertheless, his timing in posting the map of caribou calving areas quickly landed him in the midst of a political controversy. At the time, the U.S. Congress was debating the administration's proposal to open parts of ANWR to oil and gas drilling, and the caribou calving grounds were in the way.

On his first day at work after posting his map, Thomas was fired and his website removed. A U.S. Geological Survey public affairs officer stated that Thomas had not had his maps "scientifically reviewed or approved" before posting them on the website (Harlow, 2001). Thomas, however, believed his dismissal was "a high-level political decision to set an example to other federal scientists" who might not support the administration's campaign to open the wildlife refuge for oil and gas drilling. "I thought that I was helping further public and scientific understanding and debate of the issues at ANWR by making some clearer maps," Thomas (2001) wrote in an e-mail to colleagues.

More evidence of political interference soon appeared. In 2004, the Union of Concerned Scientists released a report, *Scientific Integrity in Policymaking*, by more than 60 scientists (including 20 Nobel Prize winners) charging that White House officials had engaged in "a well-established pattern of suppression and distortion of scientific findings" (p. 2). Among its findings, the report claimed that officials had "misrepresented scientific consensus on global warming, censored at least one report on climate change, manipulated scientific findings on the emissions of mercury from power plants, and suppressed information on condom use" (Glanz, 2004, p. A21).

An especially egregious case of political interference in scientists' communication emerged the next year. In 2005, the *New York Times* obtained documents showing that Philip A. Cooney, Chief of Staff for the White House Council on Environmental Policy, had personally changed reports intended for public release by the federal Climate Change Science Program. In 2002 and 2003, Cooney "removed or adjusted descriptions of climate research that government scientists and their supervisors . . . had already approved" (Revkin, 2005, para. 2). *Times* reporter Andrew Revkin (2005) noted that the changes, "while sometimes as subtle as the insertion of the phrase 'significant and fundamental' before the word 'uncertainties,' tend to produce an air of doubt about findings that most climate experts say are robust" (para. 3). Nevertheless, it was this edited version of the federal report that the White House released to the public.

Recently, there has been a lessening of ideological battles over environmental science inside federal agencies in the U.S. government. Nevertheless, reports of restriction of scientists' communication continue elsewhere. The government of Canada, for example, recently placed roadblocks in the way of some government-funded scientists in communicating with the public about politically sensitive issues. The *New York Times* reported that the Canadian government has begun "to monitor and restrict the flow of scientific information, especially . . . research into climate change, fisheries, and anything to do with the Alberta tar sands" (Klinkenborg, 2013, p. 10). (Tar sands are a heavy, viscous type of oil whose mining and refining, many scientists believe, has a range of environmental impacts.)

Beyond such political interference, environmental sciences have frequently been the target of questioning or criticism by special interest groups, ideological bloggers, and corporations. Often, this criticism has taken the form of a potent rhetorical *trope*, suggesting the environmental sciences are "uncertain," and that, therefore, there is no need to act.

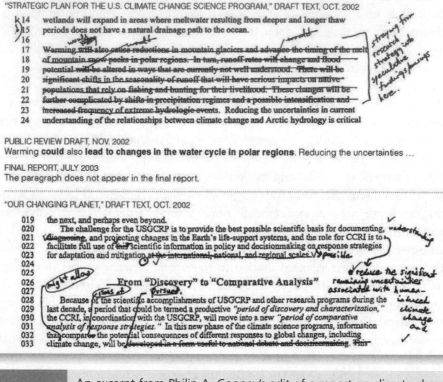

	An excerpt from Philip A. Cooney's edit of a report on climate change during the Bush administration. (Philip A. Cooney, Chief of Staff for the White House Council on Environmental Quality [and former lobbyist at the American Petroleum Institute] personally edited reports intended for public release by the federal Climate Science Program in 2002 and 2003.)
Photo 6.3	

Science and the Trope of Uncertainty

Although scientific advancements and newer technologies have produced "a cornucopia of material abundance for a substantial portion of the human race" (Shabecoff, 2000, p. 138), scientific evidence of the environmental impacts of modern industry often has been questioned. During much of the 20th century, for example, industry lobbyists, their political allies, and ideological opponents argued there must be "convincing proof of danger before policymakers had the right to intrude on the private reserve of industry in America" (Markowitz & Rosner, 2002, p. 287). Even as such proof emerged, some challenged the scientific consensus "at almost every step" when that knowledge might lead to new regulations (Hays, 2000, p. 138).

One result has been that the industries at risk of regulation by environmental science—especially petrochemicals, energy, real estate development, and utilities— and their supporters have perfected a powerful rhetorical response, a trope of uncertainty, in questioning the scientific rationale for new standards. (In Chapter 3, we described a *trope* as a "turn" or change in our perception that alters its meaning.) Let's look at this trope and several examples of its use to understand how some interests have used it to contest the symbolic legitimacy of science itself.

A Trope of Uncertainty

When opponents call for "further research," whether it's into the causes of climate change, the environmental links to breast cancer, or the effect of Pacific Northwest dams on salmon runs, they are using a familiar communication tactic in industry's challenge to the legitimacy of science. This **trope of uncertainty** functions in two ways: It nurtures doubt in the public's perception of a specific scientific claim, and this, in turn, serves to delay calls for action. In rhetorical terms, a trope of uncertainty *turns*, or alters, the public's understanding of what is at stake, suggesting there is a danger in acting prematurely or a risk of making the wrong decision. For this reason, "the call for more scientific evidence is often a stalling tactic" (Markowitz & Rosner, 2002, p. 10).

A classic source for the strategy for nurturing doubt about the legitimacy of a group or issue is public relations expert Philip Lesly's (1992) article, "Coping With Opposition Groups." Lesly advises corporate clients to design their communication to create uncertainty in the minds of the public:

> The weight of impressions on the public must be balanced so people *will have doubts and lack motivation to take action.* Accordingly, means are needed to get balancing information into the stream from sources that the public will find credible. There is no need for a clear-cut "victory." . . . Nurturing public doubts by demonstrating that this is not a clear-cut situation in support of the opponents usually is all that is necessary [emphasis added]. (p. 331)

Note that the aim is not necessarily "a clear-cut victory," but merely creating *doubt* in peoples' minds. Feeling uncertain about an issue, the advice goes, the public will be less motivated to demand action and the political will to solve a problem will weaken. For example, in opposing action to curb climate change, "Industry's PR strategy is not aimed at reversing the tide of public opinion. . . . Its goal is simply to stop people from mobilizing to do anything about the problem, *to create sufficient doubt in their minds about the seriousness of global warming that they will remain locked in debate and indecision* [emphasis added]" (Rampton & Stauber, 2002, p. 271).

Ironically, the trope of uncertainty reverses the precautionary principle, described earlier. This principle stresses that, when an activity raises threats of harm to human health or the environment, precautionary measures should be taken, even if some uncertainty about cause and effect remains. However, the call for "further research" turns the idea of "caution" against the principle itself, allowing the suspected activity to continue.

Challenging the Environmental Sciences

Let's look at several examples of this rhetorical trope and the deliberate uses of "uncertainty" to forestall action to slow or reverse climate change and other environmental harms.

Memo on Global Warming: Challenge the Science

Sometimes, the conflict over the legitimacy of scientific consensus may be fought on the terrain of language itself by engaging in what one political consultant to a major U.S. political party called the "environmental communications battle" (Luntz Research Companies, 2001, p. 136). In a memo titled "The Environment: A Cleaner, Safer Healthier America," consultant Frank Luntz (2001) warned Republican Party leaders in the United States that "the scientific debate [about global warming] is closing [against us] but not yet closed." Nevertheless, he advised, "There is still *a window of opportunity to challenge the science* [emphasis in original]" (p. 138).

Luntz's memo offers a rare look into a behind-the-scenes debate over rhetorical strategy in political circles. It is noteworthy especially for its explicit advice about the rhetorical use of uncertainty. "Should the public come to believe that the scientific issues are settled," he wrote, "their views about global warming will change accordingly." Advising party leaders, Luntz's memo advises, "'Therefore, *you need to continue to make the lack of scientific certainty a primary issue in the debate*'" (p. 137). It was a classic instance of the trope of uncertainty. The purpose, he explained, is to weaken voters' desire to call for action. Instead, Republican politicians should say, "Until we learn more, we should not commit America to any international [climate] document that handcuffs us either now or in the future" (p. 137).

Manufacturing Uncertainty: Industry and Conservative Think Tanks

Luntz's advice to "make the lack of scientific certainty a primary issue" also appears as a rhetorical strategy in other attempts by industry and ideological critics who have challenged scientific studies, ranging from chemical contamination to climate change. The objective of these efforts has been to instill public doubt of the legitimacy of scientific claims and scientific consensus about environmental problems, thereby discouraging any governmental responses to them. Let's look at some examples of this.

One of the earliest efforts by corporations to influence the public's views on climate change was the Global Climate Coalition (GCC), founded in 1989. Throughout the 1990s, the industry-funded GCC (Brown, 2000) drew heavily on the rhetorical trope of uncertainty in questioning climate science:

> Drawing upon a cadre of skeptic scientists, during the early and mid-1990s the GCC sought to emphasize the uncertainties of climate science and attack the mathematical models used to project future climate changes. The group and its proxies challenged the need for action on global warming, called the phenomenon natural rather than man-made. (Mooney, 2005, para. 21)

Although the GCC disbanded in 2002, other corporate-funded efforts had arisen to oppose U.S. governmental action on climate change. In 1998, *New York Times* reporter John Cushman (1998) uncovered a plan by the American Petroleum Institute, a powerful industry trade association, to spend millions of dollars to convince the American public that the then-new Kyoto treaty on global warming was based on "shaky science" (p. A1). The proposal included

> a campaign to recruit a cadre of scientists who share the industry's views of climate science and to train them in public relations so they can help convince journalists, politicians and the public that *the risk of global warming is too uncertain to justify controls on greenhouse gases* [emphasis added] like carbon dioxide that trap the sun's heat near Earth. (p. A1)

Beyond these corporate attempts, other efforts—skeptical bloggers, ideologues, and polemical documentary films like 2007 *The Great Global Warming Swindle*— have arisen, seeking to foster doubt in the public's mind about environmental science, in general, and climate change, in particular. In 2009, a media firestorm known as "Climategate" broke out, when skeptics claimed that hacked e-mails sent between well-known scientists proved that global warming was a hoax, "the worst scientific scandal of our generation" (Booker, 2009, para. 1). Although the scientists were vindicated in later inquiries,[1] the media controversy, nevertheless, had served to raise suspicion in the public's mind.

A more sustained effort to "manufacture uncertainty" about environmentalism has been the communication of conservative "think tanks" and other tax-exempt organizations (Jacques, Dunlap, & Freeman, 2008). Such think tanks are nonprofit, advocacy-based centers modeled on the image of neutral policy institutes (for example, the Heritage Foundation, the Cato Institute, and Heartland Institute). The role of seemingly "neutral" policy centers has been especially key. Many corporations learned that scientists who were funded by industry lacked the credibility of university scientists in debates on issues like cigarette smoking, "so providing political insulation for industry has become an essential role" for the conservative-leaning think tanks (Austin, 2002; Jacques et al., 2008, p. 362).

In their study of these think tanks, Jacques, Dunlap, and Freeman (2008) identified a key rhetorical strategy underlying their efforts: The promotion in books, press releases, and policy papers of an attitude of **environmental skepticism** that disputes the seriousness of environmental problems and questions the credibility of environmental science itself (p. 351). Such books, press releases, and papers suggested that the environmental sciences have been "corrupted by political agendas that lead it to unintentionally or maliciously fabricate or grossly exaggerate these global problems" (Jacques et al., 2008, p. 353). The result of this, Michaels and Monforton (2005) explained, was *"to 'manufacture uncertainty,' raising questions about the scientific basis for environmental problems* [emphasis added] and thereby undermining support for government regulations" (p. 362).

Another Viewpoint: Are Scientists at Fault for Poor Communication?

Recently, science journalist Nicole Heller (2011) asked why the public appears to be confused about climate change or believe that scientists are divided over the question of whether the earth is warming or if humans are a cause of warming . . . Are scientists confusing the public about the science of global warming?

An equally likely scenario . . . is that people do understand, but the range of their other interests and values prohibits them from "believing" it's true, or making the sacrifices necessary to adjust behaviorally.

Yale Professor Dan Kahan and colleagues' research on "cultural cognition" illustrates how the compatibility of empirical data with a person's cultural values influences his or her acceptance of that data. . . . For instance, people who hold an individualistic worldview may be more reluctant to accept scientific evidence of man-made climate change because addressing the problem would require restrictions on commerce and industry. In response to the communication disconnect, . . . it is not the scientists that are failing the public, rather the political and media environment is failing the scientist.

Nicole Heller, "Are Scientists Confusing the Public About Global Warming?" Climatecentral.org, 2011 (para. 38–40).

By the end of the 21st century's first decade, many scientists had become frustrated in their ability to inform the public of the risks from a warming planet. *New York Times* columnist Paul Krugman (2009) reported "Climate scientists have, en masse, become [like the Greek goddess] Cassandra—gifted with the ability to *prophesy* future disasters, but cursed with the *inability* to get anyone to believe them" (p. A21). It was not surprising, then, that U.S. news media, during this time, were "significantly more likely than media in other industrial nations to portray global warming as a controversial issue characterized by scientific uncertainty" (Jacques et al., 2008, p. 356).

More recently, many climate scientists have been stressing the need to be more vocal in public conversations. Dr. Michael Mann, director of the Earth System Science Center at Pennsylvania State University, for example, warned, "If [climate] scientists choose not to engage in the public debate, we leave a vacuum that will be filled by those whose agenda is one of short-term self-interest. There is a great cost to society," he wrote, "if scientists fail to participate in the larger conversation" (Mann, 2014, p. 8).

As the calls for more scientists to enter the public conversation grow, new ways and means for communicating with the public have arisen. In the final section, we survey some of the ways that scientists, filmmakers, news media, and others have begun to communicate, more effectively, the risks of global climate change.

Communicating Climate Science

Stung by charges that climate scientists were "losing the PR wars" and were "lousy communicators" (Begley, 2010, p. 20), many scientists have begun to reach out more proactively to the public. Dr. Gary W. Yohe of Wesleyan University, for example, noted, "A number of us, since 'Climategate' . . . have come to realize that the scientific community has an obligation to try to do better to communicate what it knows and what it doesn't know in an honest and clear way" to the public (quoted in Morello, 2011, para. 7).

There are, in fact, encouraging initiatives under way that go beyond "old media" (newspapers, TV broadcast news, etc.), using web-based media, film, innovative television programming (with robust online platforms), social media, and other digital technologies. Many are engaging the public, not simply by providing more and clearer information about climate science (although they are doing that), but by adapting to different audiences and their ways of understanding the world. Let's look at some of these initiatives.

Climate Scientists Go Digital

Increasingly, climate scientists are initiating new media for interacting with the public, journalists, policy makers, and television producers. Among recent initiatives are the following:

- The Union of Concerned Scientists has launched a comprehensive online platform (www.ucsusa.org/global_warming) explaining not only the basics of climate science, but big picture "solutions" to climate change. The site is rich with infographics (for example, "Climate Risks of Natural Gas"), blogs, videos, and recommendations for ways that regions can adapt to the impacts from climate change.
- More than 100 scientists also have launched the Climate Science Rapid Response Team (http://climaterapidresponse.org), "a match-making service to connect climate scientists with lawmakers and the media." The "rapid response" service has sought to narrow the "wide gap between what scientists know about climate change and what the public knows" (Climate Science, 2014, para. 3).
- Many science and popular blogs are also providing responses to common misconceptions and arguments by climate change deniers (see, for example, www.grist.org/article/series/skeptics and www.skepticalscience.com/argument.php). The Skeptical Science site also offers breaking news, videos, graphs, and articles examining the deniers' objections. Other sites, such as www.realclimate.org, offer analyses from scientists of more technical claims questioning climate science.

While these efforts enable climate scientists to reach new audiences, some caution that merely providing more or clearer information about climate change fails to move away from an older, **information deficit model of science communication**. This is the assumption that—because the public lacks scientific knowledge—providing more information or facts will change their attitudes. And changing attitudes—and behavior—is for many an even more urgent goal of climate change communicators.

Importantly, the relationship between *facts* and changes in the public's *attitudes* about climate change is much more complex. As Susanne Priest, editor of the journal *Science Communication*, explains, "Audience members bring their own attitudes, expectations, values, and beliefs about both science and environment to their interpretation of media stories" (2015). In seeking to adapt to such diverse audiences, science communicators are realizing the need to use different approaches and messages. As a result, scientists, educators, artists, filmmakers, and others have also begun to team up with television, film, and even weather forecasters to communicate in different ways with popular audiences.

Media and Popular Culture

Such initiatives come at a timely moment for science in popular media. With "TED talks, YouTube channels like MinutePhysics and Vsauce, and TV programs such as PBS' *Nova*," one media critic observed, "it's a golden age for communicating science to popular audiences" (Price, 2014, p. A1). And, as we write, *Cosmos: A Space-Time Odyssey* a highly acclaimed science series hosted by astrophysicist Neil deGrasse Tyson, is bringing science to mainstream television viewers in brilliant images and with compelling narratives. Film, television, and other popular media are also becoming sites for new ventures in communication climate science.

This media trend began several years ago. Groups like the American Association for the Advancement of Science and the American Geophysical Union, for example, began sponsoring workshops to demystify climate change for journalists, TV weather forecasters, talk radio hosts, and other media sources. The American Geophysical Union "offered a veritable smorgasbord of communications-oriented workshops" for more than 19,000 registrants at its annual meeting (Yale Forum on Climate Change and the Media, 2011, para. 1). And the American Meteorological Society, working with the National Communication Association, brought together meteorologists, climate researchers, and communication practitioners in 2011 to aid TV meteorologists' understanding and communication of climate science.

☞ **FYI** **Climate Fiction or "Cli-Fi"**

Literary scholars and novelists have begun to describe an emerging genre of science fiction dedicated to the climate crisis, called "cli-fi." Here are some definitions in their words:

Margaret Atwood (2013):

There's a new term, cli-fi (for climate fiction, a play on sci-fi), that's being used to describe books in which an altered climate is part of the plot. Dystopic novels used to concentrate only on hideous political regimes, as in George Orwell's *Nineteen Eighty-Four*. Now, however, they're more likely to take place in a challenging landscape that no longer resembles the hospitable planet we've taken for granted. (para. 3)

Richard Pérez-Peña (2014):

The mushrooming subgenre of speculative fiction known as climate fiction or cli fi[:] Novels like *Odds Against Tomorrow*, by Nathanial Rich, and *Solar*, by Ian McEwan. Novels set against a backdrop of ruinous climate change have rapidly gained in number, popularity, and critical acclaim over the last few years, works like *The Wind-Up Girl*, by Paolo Bacigalupi; *Finitude*, by Hamish MacDonald; *From Here*, by Daniel Kramb; and *The Carbon Diaries 2015*, by Saci Lloyd. (pp. A1, A15)

Dan Bloom (2014) [first coined the term "cli-fi" in 2007]:

What makes cli-fi, cli-fi? Novelists, screenwriters, literary critics, and academics will determine what makes cli-fi in an organic way over the next 100 years. This is just the beginning of a whole new world of literary and cinematic expression. (quoted in Holmes, 2014, para. 6, 11)

Meanwhile, scientists have joined with journalists and media producers to develop other online and popular media platforms, not merely to convey timely information, but adapting to the ways in which people understand and relate to their worlds. One initiative that has received praise is multimedia ClimateCentral.org. Sponsored by a nonprofit journalism and research organization, the site is dedicated to "helping mainstream Americans understand how climate change connects to them." ClimateCentral.org features dramatic videos, news, and interactive graphics, and visualizations of local areas impacted—or soon to be—by climate change that help to localize the impacts of climate change. For example, rising sea levels along the U.S. coast and declining global wheat yields.

And in 2013, climate scientists and Hollywood stars teamed to produce such visually stunning documentaries as Showtime's *Years of Living Dangerously* (http://yearsofliv ingdangerously.com), aimed "to humanize global warming by examining the effects on everyday individuals in the United States and beyond" (Schor, 2013, para. 5). Called by one scientist as "perhaps the most important climate change multimedia communication endeavor" in recent history (Abraham, 2014, para. 1), *Years of Living Dangerously* came with a full-spectrum, online platform, enabling viewers to take action and providing resources for journalists. And as we're writing this chapter, Leonardo DiCaprio's *Carbon*, an innovative short documentary, has come out (www.greenworldrising.org). "We cannot sit idly by and watch the fossil fuel industry make billions at our collective expense. We must put a price on carbon—now," DiCaprio says in his narration of the film (quoted in McCalmont, 2014, para. 2).

The efforts of ClimateCentral.org, the workshops for weather forecasters, the *Years of Living Dangerously*, *Carbon*, and other media initiatives reflect a shift from the information "deficit" model of communicating climate change. Along with NBC television's "deep-dive" special on climate change in 2014, they have been likened to "a 'vanguard' for a changing media landscape" (Robbins, 2014, para. 3, 8). Let's conclude, then, by looking at an important rationale for this new approach to engaging popular audiences and the diverse ways they understand the world.

Inventing New Climate Change Messages

If you've talked about global warming to friends, family, other students, or with strangers, you've undoubtedly encountered widely differing viewpoints. As we noted earlier, different people bring their own personal attitudes, expectations, values, and beliefs to their understanding of science and technology, especially subjects like climate change policy.

Scholars at George Mason University's Center for Climate Change Communication report a variety of perspectives among Americans about this subject, "with some dismissing the danger, some entirely unaware of its significance, and still others highly concerned and motivated to take action." As a result, they stress that, "Understanding the sources of these diverse perspectives is key to effective audience engagement" (Roser-Renouf, Stenhouse, Rolfe-Redding, Maibach, & Leiserowitz, 2014, p. 1). (See Figure 6.1, "Global Warming's Six Americas.")

Researchers at the center have been exploring the cultural, psychological, and political underpinnings of people's views on climate change and have recently identified messaging strategies that are more likely to succeed with these diverse audiences. (We describe "messages" more in Chapter 8.) For example, the center found that individuals who are less interested, "cautious," or "disengaged" with climate change "are unlikely to pay attention if understanding the content requires cognitive effort," and therefore, in order to engage their attention and involvement with this issue, scientists and other communicators "must turn to methods that are not effortful" (Roser-Renouf et al., 2014, p. 18). These "low-involvement communication strategies"

- require only peripheral/heuristic information processing, e.g., visual imagery, humor, and attractive or highly credible sources;
- promote positive social norms by demonstrating that climate-friendly behaviors are popular, respected, and common;
- show rather than tell what is happening, thereby triggering automatic information processing;
- personalize the threat by showing impacts on places that are physically close or emotionally significant; and
- generate involvement through the use of narratives. (pp. 18–19)

"Showing," rather than telling, narratives, and highly credible individuals may be the idea behind production of the popular documentary film *Chasing Ice* (www.chasingice.com). In filming *Chasing Ice*, for example, photographer James Balog used time-lapse cameras "to capture images to help tell the story of the Earth's changing climate" by showing the dramatic retreat of Arctic glaciers (Synopsis, 2014, para. 1).

The point is, climate scientists are turning not only to a variety of communication media, but also by engaging audiences with a range of rhetorical appeals that respect their ways of understanding the environment. Moving away from imagining communication as a one-way model from experts to the public toward more engaged approaches will continue to be a vital lesson for scientists to learn (Lindenfeld, Hall, McGreavy, Silka, & Hart, 2010).

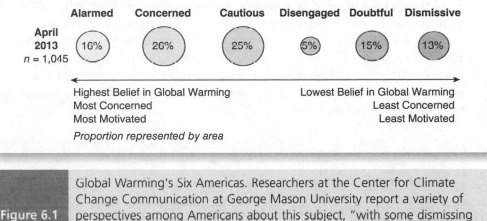

	Alarmed	Concerned	Cautious	Disengaged	Doubtful	Dismissive

April 2013 $n = 1,045$

16% 26% 25% 5% 15% 13%

Highest Belief in Global Warming   Lowest Belief in Global Warming
Most Concerned   Least Concerned
Most Motivated   Least Motivated

Proportion represented by area

Figure 6.1 Global Warming's Six Americas. Researchers at the Center for Climate Change Communication at George Mason University report a variety of perspectives among Americans about this subject, "with some dismissing the danger, some entirely unaware of its significance, and still others highly concerned and motivated to take action."

Roser-Renouf, Stenhouse, Rolfe-Redding, Maibach, & Leiserowitz (2014), used with permission of the George Mason University Center for Climate Change Communication.

SUMMARY

In this chapter, we've considered several provocative questions about the role of science in environmental controversies. How should the public interpret the meaning of scientific claims when the research is characterized by uncertainty? When, if at all, should scientists speak up, raise awareness publicly, or warn the public about environmental or climate risks? How can scientists effectively communicate with diverse publics about global climate change?

- The first section described science as a source of symbolic legitimacy in a society beset with complexity and technical knowledge.
- In the second section, we introduced the *precautionary principle* as a way to manage the problem of scientific uncertainty. The precautionary principle states that, when an activity threatens human health or the environment, even if some cause-and-effect relationships are not fully established scientifically, caution should be taken.
- In the third section, we described the debate over scientists' role as *early warners*, the duty of scientists to inform, advise, or alert the public in moments of environmental danger.
- Section four explored a trope of "uncertainty" used to challenge the symbolic legitimacy of the environmental sciences.
- Finally, we described recent initiatives by scientists, filmmakers, journalists, novelists, and media producers to improve communication about the science and impacts of climate change.

The challenges involved in scientists' communication with the public about climate change and other environmental dangers can be serious, particularly when opponents launch campaigns to question or undermine the public's confidence in scientific research. At stake is the public's perception of the symbolic legitimacy of science itself as well as scientists' pleas for action to address environmental risks. We explore this sense of urgency in more detail in the next chapter, "Environmental Risk Communication and the Public."

Act Locally!

Communication Strategies About Climate Change

What communication strategies about climate change are most likely to succeed in engaging

- your friends?
- your family?
- students at your campus who disagree with you?
- strangers?

Read the Center for Climate Change Communication's 2014 report, Engaging Diverse Audiences With Climate Change: Message Strategies for Global Warming's Six Americas, at www.climatechangecommunication.org/report/engaging-diverse-audiences-climate-change-message-strategies-global-warmings-six-americas.

Then, working in groups, sketch the basic idea for a TV program, website, documentary film, advertising campaign, etc. that engages one of the "Six Americas" in ways that create awareness and/or willingness to act on climate change. For example, what approaches would most likely succeed in engaging your friends, family, other classmates, and others who are "Cautious" or "Dismissive" about this issue?

1. Which recommendations from the Center's report would apply? How would you implement these strategies with your targeted audience?

2. Can you think of other strategies? For example, what *rhetorical genre* or *communication frame* (Chapter 3) would you use to engage your audience about climate change?

Finally, you might ask classmates or your professor to observe and comment on your choices or suggest differing approaches or strategies.

SUGGESTED RESOURCES

- *Years of Living Dangerously*, a nine-episode television documentary (Showtime) on "the crippling effects of climate change-related weather events and the ways individuals, communities, companies, and governments are struggling to find solutions to the biggest threat our world has ever faced" (http://yearsoflivingdangerously.com/). Available on DVD and also streaming on iTunes, Amazon, and Vimeo on Demand.

- *Climate Central:* "Sound science and vibrant media" (www.climatecentral.org). One of the best websites devoted to independent, multimedia journalism, and research, bringing "the immediacy and relevance" of climate change and its impacts to visitors to its website.
- Fred Pierce, *The Climate Files: The Battle for the Truth About Global Warming.* London: Guardian Books, 2010 (an independent journalist's look into the infamous Climategate e-mail controversy).
- *Gasland Part II* (www.gaslandthemovie.com) is the follow-up to filmmaker Josh Fox's provocative film about the effects of fracking on local communities. (2010 *Gasland* [Part I] is also available at the film's website.)
- Elizabeth Kolbert, *The Sixth Extinction: An Unnatural History.* New York, NY: Henry Holt. 2014. Takes readers to the Amazon, Great Barrier Reef, and other places where the signs of species extension are already visible.
- *Chasing Ice* (www.chasingice.com): time-lapse film showing the impact of global warming on the dramatic retreat of glaciers in the Artic.

KEY TERMS

Crisis discipline   129

Environmental skepticism   137

Hydraulic fracturing   122

Information deficit model of science communication   139

Precautionary principle   127

Progressive ideal   123

Symbolic legitimacy   123

Technocracy   122

Trope of uncertainty   135

DISCUSSION QUESTIONS

1. "What does today owe tomorrow?" As Justin Gillis (2014), a blogger at the *New York Times,* recently observed, "That simple question goes to the heart of figuring out what we should spend now on efforts to deal with climate change" (p. D3A). A vibrant public sphere requires challenging questions and honest discussion—and debate—in answering them. How would you answer this question?

2. Is the precautionary principle a clear guide in dealing with uncertainty? How should we go about weighing the uncertainty or risks of environmental danger versus the economic and other benefits that come from an activity like fracking?

3. Should environmental scientists ever serve as advocates in the public sphere? Where do you draw the line—if at all—in how far scientists should go in speaking, warning, or recommending actions to the public?

4. Climate deniers have charged that scientists faked their data and that global warming is a "hoax." What do you believe? What sources of information do you use in forming a judgment? How accessible is climate science itself to the public?

5. Climate scientists, filmmakers, and others are warning of catastrophic impacts of climate change—sea level rise, droughts, crop failures, regional conflicts, and starvation. How effective do you think are such fear appeals in motivating the public to care about global warming? What approaches would be effective?

NOTE

1. In late 2009, hackers broke into computers at the University of East Anglia, a major climate research center in the UK, downloading more than 1,000 personal e-mails from climate scientists. Although bloggers claimed these showed a conspiracy to lie about the science of climate change, later independent inquiries cleared the scientists of charges that they had falsified their research and exonerated the basic findings of climate science itself (Gulledge, 2011).

On January 9, 2014, national headlines in the United States reported a toxic chemical leak into the water supply of over 300,000 residents in West Virginia. This disaster prompted public outrage about outdated toxic chemical legislation: "Last night our twitter chat with Leah Segedie of Bookieboo LLC. and owner of Mamavation.com was a success as we discussed the need for real reform of our toxic chemical laws. We reached over 14 million people in just 90 minutes!" (Safer Chemicals, Healthy Families Blog, January 24, 2014).

CHAPTER 7

Environmental Risk Communication and the Public

We view the emergence of the downwinder community as a response not only to perceived environmental hazards, but also to constraints on democratic public discourse.

—William J. Kinsella and Jay Mullen, on people living downwind from the Hanford Nuclear Facility, 2008 (p. 76)

H umans always have faced danger from weather and geological events such as storms, earthquakes, famine, and crop failure. Today, however, we face ever-increasing dangers from human sources as well—toxic chemical contamination, oil spills, nuclear accidents, climate disasters, and more. Spurred by disasters, predictable crises, and preventable dangers, experts have tried to evaluate and to share information about the public health risks of environmental hazards in the public sphere. But who counts as an "expert"? What should everyday citizens expect from formally trained experts? And what might formally trained experts learn from everyday citizens? These are the types of questions negotiated through *risk communication* in the public sphere.

This chapter explores some of the ways we understand risk through communication and what these differing perspectives assume about risk, expertise, and the possibilities of the public sphere.

Chapter Preview

- This chapter begins by describing why some call our current era a *risk society*. We also describe two ways of assessing risk and its meaning: (1) a technical approach and (2) a cultural theory model.
- The second section introduces the practice of *risk communication* through both the technical model and cultural approach.
- The third section briefly notes how *citizen scientists* have established their own experiential expertise as vital to risk assessment in their communities.
- The final section explores the ways media shape our perceptions of risks. We describe two dimensions of media reports:

  1. Debate over the accuracy of news reports of risk

  2. Different voices quoted in media reports of risk; first, society's *legitimizers*—e.g., public officials—and, second, the voices of the *side effects*—residents, parents of sick children, and others who are most affected by environmental dangers

In sharing information and beliefs about environmental and public health risks, government officials, news media, corporate spokespeople, scientists, lawyers, public relations coordinators, and the general public engage in an important and sometimes controversial communication practice called **risk communication**, the symbolic mode of interaction that we use in identifying, defining, assessing, and negotiating environmental and public health dangers.

When you've finished the chapter, you should be able to recognize the broad cultural context of risk we live in today and how communication serves instrumental and constitutive functions for those who publically address risk.

Dangerous Environments: Assessing Risk

The popular film *Erin Brockovich* (2000) dramatized the real-life experiences of residents in the small, rural, working-class town of Hinkley, California, who discovered that toxic contamination of their drinking water from Pacific Gas and Electric Company (PG&E) may have caused increased incidences of Hodgkin's disease, breast cancer, and other diseases in their community. The film portrayed residents' anger, suspicions, and worries after they learned that PG&E's assurances that their water was safe were untrue. Ultimately, the company settled a multimillion-dollar lawsuit with 634 residents and pledged to clean up the contaminated groundwater (Pezzullo, 2006).

Erin Brockovich calls popular attention to the all too real dangers to human health and the environment that exist today. The filmmakers also ask us to look closely at communication about risk expressed by technical experts, business owners, health officials, legal counsel, and the media, as well as at attempts to improve communication by those who may be affected most directly by such dangers.

Risk Society

In his book *Risk Society* and other writings, sociologist Ulrich Beck (1992, 2000, 2009) argued social life has changed fundamentally in its ability to manage the consequences of its successful technological and economic development. Beck explained, "The gain in power from techno-economic 'progress' is being increasingly overshadowed by the production of risks" (1992, p. 13). This shift, he claims, is fundamental to **modern society**, which is a historical period that can be defined by the rise of industrial capitalism, the increased significance of the nation-state, and the scientific revolution. Unlike risks that are considered natural disasters (like earthquakes, tsunamis, and volcanic eruptions) or industrial disasters that affect individuals or relatively small groups of workers (like fire hazards, rat infestations, and work injuries), he claims today's **risk society** is defined by the introduction of large-scale nature of risks and the potential for irreversible threats to human life from modernization itself. These risks include such far-reaching and consequential hazards as nuclear power plant accidents, global climate change, toxic chemical pollution of our reproductive systems, and the alteration of genetic strains from bioengineering in the global food supply.

In a risk society, rapid scientific and technological changes may bring unintended consequences. What are most disturbing are what are called **black swan events,** a concept developed by the mathematical financier Nassim Nicholas Taleb (2010). According to Taleb, *black swan events* refer to unexpected, high-magnitude events that are beyond what modern society can usually predict. Recent examples are the unexpected and devastating impact of the earthquake and tsunami on the Fukushima nuclear power plants in Japan in 2011 and the Freedom Industries toxic spill in the water supply of West Virginia in 2014. Some argue climate change also poses a high risk of black swan events, such as massive releases of methane from thawing tundra, complete melting of Greenland's ice sheet, or the weakening or shutdown of the Gulf Stream (Pope, 2011).

In addition, exposure to risk in modern society is unevenly distributed across the population. That is because the burden of coping with the hazards of new technologies and environmental pollutants often falls on the most vulnerable—elderly people, children with respiratory problems, pregnant women, and, as we discuss further in Chapter 10, residents of working-class and people of color neighborhoods. As a result, serious conflicts occurred in the 1980s between the residents of at-risk communities and the technical experts who assured them that polluting factories or buried toxic wastes posed no harm. Lois Gibbs (1994), a former resident of Love Canal, New York, expressed the feelings of many: "Communities perceive many flaws in risk assessment. The first is who is being asked to take the risk and who is getting the benefit?" (pp. 328–329).

Since defining risk is a judgment call based on the best information available, it is important to distinguish its different meanings and what constitutes *acceptable risk* for different parties. Therefore, we look at both technical and cultural meanings of *risk*.

Risk Assessment

By the 1980s, the public's fear of environmental hazards had led to pressure on the U.S. government to evaluate risk accurately and to do a better job in communicating with affected communities. In 1984, the EPA Administrator William Ruckelshaus proposed the term *risk assessment* "as a common language for justifying regulatory proposals across the agency" (Andrews, 2006, p. 266). In general, **risk assessment** is defined as the evaluation of the degree of harm or danger from some condition such as exposure to a toxic chemical. In the years following, the EPA dramatically increased its analysis of risks from nuclear power, pollution of drinking water, pesticides, and other chemicals in order to justify new health and safety standards. Environmental policy expert Richard N. Andrews (2006) observed that, by the end of the 1980s, "the rhetoric of risk had become the agency's primary language for justifying its decisions" (p. 266).

In the United States, the EPA and the Food and Drug Administration (FDA) are two of the agencies responsible for evaluating health and environmental risks. For example, the FDA routinely evaluates the risks of food contamination and issues recalls of tainted products like peanut butter, eggs, ground beef, or pet foods. Understanding the technical model of risk used by these agencies is important in its own right, but this also allows us to appreciate its limitations and why such an approach to risk communication has generated controversy among some affected publics.

Technical Model of Risk Assessment

In everyday terms, risk is a rough estimate of the chances of something negative happening to us, such falling while rock climbing or getting electrocuted if playing with an outlet. However, for many agencies, risk is a solely quantitative concept. Such a **technical approach to risk assessment** is the expected annual mortality (or other severity) that results from some exposure. For example, the EPA defines risk as "the chance of harmful effects to human health or to ecological systems resulting from exposure to an environmental stressor" (United States Environmental Protection Agency, 2010b, para. 2). *Technical risk* is, then, a calculation of the likelihood that a certain number of people (or an ecological system) will suffer some harm over time (usually one year) from exposure to a hazard or environmental stressor; in other words, Risk = Severity × Likelihood.

How does an agency such as the EPA know the severity of a hazard and its likelihood of occurrence? In answering these questions, technical risk assessment uses the findings of research labs and experts such as toxicologists, epidemiologists, and other scientists. This process typically involves a **four-step procedure for risk assessment**: (1) identify the hazard, (2) define the pathways of human exposure to it, (3) determine humans' response to different levels of exposure, and (4) characterize the risk; that is, is the risk safe or unacceptable? (Office of Environmental Health Hazard Assessment, 2001).

Let's take an example from Chapter 6—fracking for natural gas—to see how this four-step procedure might work in an ongoing controversy. In 2010, the EPA announced it would investigate the potential adverse impacts of hydraulic fracturing (or fracking) in drilling for natural gas on human health and the environment. As a result of public complaints, the EPA has begun to assess the human and environmental risks.

1. *What is the hazard or the potential source of danger?* In our example, the EPA is identifying multiple, potential sources of danger, including "the chemicals and fluids used in the fracturing process, biogeochemical and physical-chemical reactions triggered by [hydraulic fracturing], leakage from gas-bearing formations," and runoff of chemicals from the drilling site (United States Environmental Protection Agency, 2010a, p. 6).

2. *What are the pathways of human exposure to these sources?* In its study of fracking, the EPA is looking at activities that may "introduce contaminants into water, food, air, soils, and other materials over the [hydraulic fracturing] lifecycle"; specifically, the exposure pathways being explored by EPA include the "ingestion, inhalation, dermal exposure [of these contaminants] through water, air, food, and environmental exposures" (United States Environmental Protection Agency, 2010a, p. 8). The EPA will identify the different levels of exposure that populations may be receiving from the chemicals entering these pathways. (These first two steps are important to assess the potential *severity* of exposure.)

3. *What is the relationship between the level or dosage that is received and any harmful responses or illnesses in the exposed population?* That is, what is the likelihood, given these levels of exposure, to this particular hazard, of specific health effects to X number of individuals?

4. *Is the risk safe or unacceptable?* A progress report of the EPA's study of fracking was released in 2012, and a draft was released in 2014 for peer review and public comment. This fourth step draws on the prior three steps to estimate the mortality or other severities that can be expected from an exposure to the hazard (for example, an estimate of 0.22 cases of cancer over 10 years). Technical models of risk use such numerical values as the basis for judgments of **acceptable risk**. Acceptable or unacceptable risk, however, involves values other than just numerical estimates. Ultimately, an acceptable risk is a judgment of the harms or dangers that society or specific populations are willing to accept (or not). Many communities desire such studies be conducted before their bodies and the environments of which they are a part are placed at risk. While the EPA currently does not find the risks from fracking unacceptable, more and more cities and states are finding that they are.

Technical judgments of risk may also involve a consideration of who is subject to the danger, the costs that are required to reduce the danger, and the benefits that might be accrued from accepting the danger. **Cost-benefit analysis** is a type of economic evaluation of technical risk in which both the disadvantages (costs) and advantages (benefits) of an investment, purchase, or process are weighed in financial, supposedly neutral, mathematical ways.

Limitations of the Technical Approach

Disagreements over what is safe or an acceptable risk are a major challenge to technical risk assessments. There are several reasons for this: there may be insufficient lab results; different studies may differ in their estimates of danger; the studies may not be able to detect which of several sources may be causing a problem; and so on. One of the main limitations of this model is the **androcentric bias** of the data collected; that is, much of technical risk assessment is based on the assumption that the person being exposed to a danger is an average adult male.

The status quo in toxicology also involves **acute single-dose tests** in a lab, which means risks often are assessed by considering an individual hazard in isolation with a short-term exposure at a relatively high level, despite the fact that we are exposed to thousands over a lifetime in various unsterile environments where chemicals interact with each other and the specificity of our bodies. Without knowing any chemistry, many of us realize how problematic an androcentric, acute single-dose risk assessment is, based on what we know of alcohol. When you consider how much one beer might affect someone's body, what do you take into consideration? Weight? How much someone ate earlier in the day? If one consumed other alcoholic drinks or drugs earlier? Over what period of time one drinks the beer? Personal tolerance? And so forth. The impact of one beer varies depending on one's answer, just as it would if we were assessing the impact of a toxic chemical in the body.

An extended example of a technical risk communication controversy is the public debate over the chemical known as bisphenol A, or BPA. BPA is used in thousands of products, including plastic bottles, baby food containers, and "the lining of nearly every soft drink and canned food product" (Parker-Pope, 2008, p. A21). We've been exposed to BPA so routinely that it has been detected in more than 90% of the U.S. population (Stein, 2010). Recent research, however, has raised concerns about the possible health effects from exposure to BPA products, including cancer (Grady, 2010), lowered sperm counts (Stein, 2010), and neural and behavioral effects in infants and babies from low-level, long-term exposures (Grady, 2010; Szabo, 2011). Nevertheless, the FDA had insisted in 2008 that BPA was safe. The director of the FDA's Office of Food Additive Safety told the *Washington Post,* "We have confidence in the data we've looked at to say that the margin of safety is adequate" (quoted in Layton, 2008, p. A3).

Nevertheless, the FDA asked an independent scientific panel to review the research on which the agency, at that time, had based its characterization of the risk from BPA. The panel held a public hearing (see Chapter 12) to receive testimony from the public, other scientists, and health groups. Public comments revealed sharp disagreements

Photo 7.1	**Technical risk communication** is the process of informing the public or local communities of a potential environmental or health hazard. In 2013, the Environmental Working Group (EWG) launched a Skin Deep® mobile app, so that consumers may scan a barcode and access data on toxic risks posed by everyday product ingredients.

with the FDA's risk characterization. For example, a member of the Union of Concerned Scientists (UCS) complained to the panel that the FDA had based its characterization of BPA's risk largely on two studies funded by industry "while downplaying the results of hundreds of other studies" (quoted in Layton, 2008, p. A3).

Photo 7.2	In May 2013, people from six continents and over 40 countries protested the lack of government oversight of the corporation Monsanto's role in global food supplies. Despite the company reassuring the public that its experts use risk assessment to identify the best choices for the greatest number of people, many express doubt that genetically modified crops are fit for consumption.

In a "blistering report," the panel found that the FDA had "ignored important evidence in reassuring consumers about the safety" of BPA (Parker-Pope, 2008, p. A21).

Even today, the controversy continues, with "mountains of data" in technical risk studies showing "conflicting results as to whether BPA is dangerous" (Grady, 2010, p. D4). As a result of these limitations, new, federally financed research has been authorized, particularly for effects of BPA on infants. For now, the U.S. government recommends reducing BPA exposure (see, for example, http://www.hhs.gov/safety/bpa/). Meanwhile, the European Union and Canada already have banned BPA.

Despite all of these limitations, most of us rely on technical risk assessments to make judgments in our everyday lives and lobby our governments to conduct more studies when we believe there is cause for alarm.

A Cultural Theory of Risk Assessment

Despite the value of technical experts, everyday citizens repeatedly have voiced their disappointment in not being included in the decision making that directly impacts their lives. As Beck (1998) explained, "There is a big difference between those who take risks and *those who are victimized by risks others take* [emphasis added]" (p. 10). As a result, many government agencies and corporations have been lobbied to solicit the views of affected communities in their assessment of risks and their judgments about what is an acceptable risk through interviews, public hearings, surveys, and other communication practices.

Anthropologist Mary Douglas and political scientist Aaron Wildavsky (1982) are credited with first establishing a cultural theory of risk. A **cultural theory of risk assessment** rejects an individualist, rationalist notion of risk in favor of one that believes our perceptions of risk are informed by cultural values. Take, for example, our cultural perception of an adult's death versus a child's. Both count as a death of one quantitatively and, financially, the adult usually earns more than the child so that the adult's death would have a greater financial impact; nevertheless, most cultures appear to find it more tragic that a child might die from a risk than an adult. A cultural theory of risk does not assume the data themselves are neutral, since the impact of a danger may vary across bodies being exposed due to age, gender, occupation, belief systems, and more. Let us illustrate with examples.

Environmental Hazards Versus Outrage

Public responses to an environmental danger sometimes differ from a technical risk assessment. While some who study risk perceptions remind us that most individuals are "irrational" in judging risks such as the odds of a shark attack or from dying in a plane crash (Ropeik, 2010; Slovic, 1987; Sunstein, 2004), others point to well-grounded reasons for our concerns. Political scientist Frank Fischer (2000) explains that the context in which an environmental risk is embedded raises questions that may affect one's judgment of whether a risk is acceptable or not: "Is the risk imposed by distant or unknown officials? Is it engaged in voluntarily? Is it reversible?" (p. 65).

Peter Sandman (1987, 2011), a well-known risk communication consultant, makes a similar point. Sandman proposed that risk be defined as a combination of technical risk and social factors that people often consider in judging risk. He suggested that what technical analysts call a *risk* instead be called a *hazard* and that other social and experiential concerns be called *outrage*. **Hazard** is what most experts mean by risk (expected annual mortality or danger to human health) and **outrage** refers collectively to those factors that the public considers in assessing whether their exposure to a hazard is an *acceptable risk*. According to Sandman (1987), "risk, then, is the sum of hazard and outrage" (p. 21).

Consider, for example, the 2014 Freedom Industries toxic spill in West Virginia that we noted at the opening of the chapter. Although everyday people (14 million on Twitter alone) seemed to express a sense of outrage that 300,000 residents were without water, the U.S. government has yet to update its national toxic regulation laws. In a comedy skit on the spill, *The Daily Show* (January 13, 2014, "Coal Miner's Water—A Terrorist Plot?") showed an outraged reporter asking for an investigation into the military intelligence failure and the launching of a war to exact revenge on this "act of terrorism" on the American way of life; when the reporter was informed that it was a corporation and not "foreign terrorists" that caused the toxic spill, he joked that there is no story because it is business as usual. The humor of the skit homed in on how risk perceptions of acceptability are not just of the health impacts or the people affected, but also our cultural perceptions of the *source* of the risk.

Here are some of the main factors of outrage that Sandman believes people consider in judging an environmental risk:

1. *Voluntariness:* Do people assume a risk voluntarily, or is it coerced or imposed on them?

2. *Control:* Can individuals prevent or control the risk themselves?

3. *Fairness:* Are people asked to endure greater risks than their neighbors or others, especially without access to greater benefits?

4. *Diffusion in time and space:* Is the risk spread over a large population or concentrated in one's own community?

Sandman's model of Hazard + Outrage is particularly useful in calling our attention to the experiences of those left out of technical risk calculations. Sandman's definition is not without its critics, however. Some have suggested that this definition subtly characterizes scientific or technical assessments of hazards as rational and the emotional outrage of communities as irrational. They fear such characterizations can be used to trivialize community voices in debates about risk. So, let's look at this cultural side of risk more closely.

Cultural Rationality and Risk

Judgments about social context or the experience of exposed communities to risk raises the question of whether there exists a kind of cultural rationality about such dangers. Pioneering scholars in risk communication Alonzo Plough and Sheldon Krimsky (1987) define **cultural rationality** as a type of knowledge that includes personal, familiar, and social concerns in evaluating a real risk event. As distinct from technical analysis of risk, *cultural rationality* "is shaped by the circumstances under which the risk is identified and publicized, the standing or place of the individual in his or her community, and the social values of the community as a whole" (Fischer, 2000, pp. 132–133). Unlike technical rationality—the valorizing of scientific methods and expertise—*cultural rationality* includes folk wisdom, the insights of peer groups, an understanding of how risk impacts one's family and community, and sensitivity to particular events.

Harvard University professors Phil Brown and Edwin J. Mikkelsen (1990) provide a disturbing example of the differences between *risk assessments* made in a technical sphere and those made in a wider public sphere. In their classic study *No Safe Place: Toxic Waste, Leukemia, and Community Action,* they cite the experience of residents in Friendly Hills, a suburb of Denver, Colorado. In the early 1980s, mothers from the neighborhood wondered why so many of their children were sick or dying. After EPA and state health officials refused to study the problem, the women decided to canvass door to door to document the extent of the problem. They found that 15 children who lived in the neighborhood from 1976 to 1984 had died from cancer, severe birth defects, and other immunological diseases. The mothers suspected that the cause of these deaths was toxic waste discharge from a nearby industrial facility owned by the corporation Martin Marietta.

State health officials dismissed the mothers' findings and denied there was any environmental or other unusual cause of the children's deaths. They insisted that all the illnesses were within "expected limits," and although there were more childhood cancer cases than expected, the officials said that "they might be due to chance" (Brown & Mikkelsen, 1990, p. 143). One week after the officials declared the waste discharges safe, "the Air Force, which runs a test facility on the Martin Marietta site, admitted that the groundwater was contaminated by toxic chemicals" (p. 143).

Brown and Mikkelsen describe what happened next: "Residents found that Martin Marietta had a record of toxic spills. . . . Several months later, EPA scientists found serious contamination . . . in a plume, or underground wave, stretching from the Martin Marietta site toward the water plant" (pp. 143–144). The Harvard professors concluded, "The belated discovery of what residents knew long before is eerie and infuriating—and, sadly, it is common to many toxic waste sites" (p. 144). In cases such as Friendly Hills, a cultural theory of risk assessment challenged the credibility of technical agencies and their methods.

The differences between technical and cultural models of risk assessment also raise important questions about how we communicate risk in the public sphere. Let's look at the different ways these approaches to assessment shape how environmental risks are communicated.

Communicating Environmental Risks in the Public Sphere

Although the study of risk communication barely existed before the late 1970s, the field has grown dramatically. In this section, we look at two different contemporary models of risk communication, each reflecting one of the meanings of risk we described in the first section of this chapter: technical and cultural.

Technical Risk Communication

Early experiences with risk communication grew out of the need of federal managers of environmental projects to gain the public's acceptance of their risk assessments (for example, declaring a cleanup completed). Other experiences grew from health agencies' need to communicate about risk to target populations (for example, messages to underage smokers). This early approach was influenced heavily by *technical risk assessment*. This **technical model of risk communication** is defined as the translation of technical data about environmental or human health risks for public understanding, with the goal of educating a target audience. Communication from this perspective usually is imagined as a one-way model with a sender or source attempting to share information with a receiver. An example of this approach was the EPA's public announcement that, while small amounts of pollutants were created by the controlled burning of oil from the BP oil spill in the Gulf of Mexico, "the levels that workers and residents would have been exposed to were below EPA's levels of concern" (United States Environmental Protection Agency, 2010e, para. 1). This approach often is called the "decide-announce-defend" model within the U.S. EPA (2009).

Inform, Change, and Assure

The primary goal of the technical model of risk communication is to share quantitative risk assessments with a broader public. As used by environmental and health agencies, this goal traditionally has three objectives: to inform, to change behavior, and to assure the public.

1. *To inform:* The U.S. EPA's (2007) guide for managers, *Risk Communication in Action*, defines risk communication explicitly this way: "The process of informing people about potential hazards to their person, property, or community" (Reckelhoff-Dangel & Petersen, 2007, p. 1). The agency explains that risk communication is a "science-based approach" whose purpose is to help affected communities understand risk assessment and management by forming "scientifically valid perceptions of the likely hazards" (p. 1). Given the scale of the risks we face in modern society—usually invisible to the naked eye and on a scale previously unimaginable—there is a growing need for scientists willing to study risks with cutting edge technology in ways that are impossible for most of us to establish in our everyday lives.

In the United States, pregnant women often are targeted for a good deal of information about risks, including not smoking or drinking alcohol, avoiding cat litter, reducing fish consumption, and avoiding soft cheese. Of course, if something is risky for a pregnant woman, it probably is not healthy for anyone. After the 2014 Freedom Industries toxic spill in West Virginia, pregnant women were cautioned by the U.S. Centers for Disease Control and Prevention (CDC) not to drink the water after the rest of the population was assured it was *acceptable*. West Virginia Democratic Senator Joe Manchin and Republican Representative Shelley Moor Capito wrote a joint letter stating the CDC's action "resulted in confusion, fear and mistrust" and that they believe "people must have confidence that their water supply is safe for use" (Drajem, 2014).

2. *To change behaviors:* Government agencies have long had the goal of warning the public about unsafe food products, environmental dangers, and risky personal behaviors in order to avoid risks. By changing individual behaviors (for example, quitting smoking) or by avoiding certain products or exposures (such as not drinking from BPA containers), the risks are presumably reduced.

Recently, for example, the National Cancer Institute issued a warning of a link between our exposure to everyday chemicals—in drinking water, vehicle exhaust, canned foods, and more—and a risk of cancer and that such risks previously have been "grossly underestimated" (quoted in "U.S. Cancer Institute Issues Stark Warning," 2010, para. 2). The institute reported that there are nearly 80,000 "unregulated chemicals" in use, including "a variety of carcinogenic compounds that many people are regularly exposed to in their daily lives" (paras. 4, 5). To change behaviors, the National Cancer Institute urged the U.S. government to "identify and eliminate [the] environmental carcinogens" that it had warned about, from workplaces, schools and homes" ("U.S. Cancer Institute Issues Stark Warning," 2010, para. 4).

3. *To assure:* Most institutional experts' initial impulse in a technical communication model is to downplay potential or ongoing risks. During the Gulf of Mexico oil spill in 2010, for example, many voiced their concerns through the media about food safety and other environmental dangers from the spill. With images of oil sheens and oil-soaked pelicans in the news daily, the U.S. FDA (2010) assured Gulf Coast residents that "there is no reason to believe that any contaminated product has made its way to the market" (para. 1). The Louisiana Seafood and Marketing Board also launched an "ad campaign in major newspapers boasting that Louisiana's seafood is indeed safe to eat because most of the state's waterways remain untouched by oil" (Gray, 2010, para. 4).

Despite reassurances, many communication scholars, as well as community activists, argue that the technical model of risk communication, by itself, fails to acknowledge the concerns of those individuals who are most intimately affected by environmental dangers. One result has been that, in recent years, some government agencies have turned to a more culturally sensitive model of risk communication.

Cultural Approaches to Risk Communication

In an early, influential essay "Technical and Democratic Values in Risk Analysis," former EPA advisor Daniel Fiorino (1989) argued "the lay public are not fools" in judging environmental risks (p. 294). He identified key areas in which the public's intuitive and experiential judgments differed most from technical risk analysis, including:

- *"Concern about low-probability but high-consequence events"*: For example, there may be a 1% chance that an accident will occur, but it will take a terrible death toll if it does occur.
- *"Desire for consent and control in social management of risks"*: The public's feeling that they have a say in decisions about risk is the opposite of a coerced or involuntary imposition of risk.
- *"The relationship of judgments about risk to judgments about social institutions"*: In other words, the acceptability of risk may depend on citizens' confidence in the institution that is conducting a study, managing a facility, or monitoring its safety. (p. 294)

Research since Fiorino's essay confirms that public input into risk assessment often increases the likelihood that decisions about risks will be perceived as legitimate and, therefore, viable in a democratic culture. In its study of decisions about science, the National Research Council found success depends on the ability of affected parties to participate in the risk decision process (Dietz & Stern, 2008). Let's look at what this more **dialogic** (or two-way) **communication** about risk requires.

Citizen Participation in Risk Communication

One sign of a shift in risk communication has been the trend of some agencies to use an approach that draws upon the cultural rationality of communities. This **cultural model of risk communication** involves the affected public in assessing risk and designing risk communication campaigns and fosters democratic dialogue in the public sphere about risk. A major step forward was taken in 1996 when the National Research Council (1996) released its report, *Understanding Risk: Informing Decisions in a Democratic Society*. The report acknowledged that *technical risk assessment* was no longer sufficient to cope with the public's concerns about environmental dangers. It called for greater public participation and use of local knowledge in risk assessment studies, and it pointedly noted that understanding risk requires "a broad understanding of the relevant losses, harms, or consequences to the interested and affected parties, including what the affected parties believe the risks to be in particular situations" (p. 2).

Some health and environmental risk agencies have taken this principle further to develop new practices in risk communication that recognize such cultural knowledge and experience of local communities. For example, the U.S. EPA (2007) has acknowledged "an ideal risk communication tool would put a risk in context, make comparisons with other risks, and *encourage a dialogue between the sender and the receiver of the message* [emphasis added]" (p. 3).

Similar interest in a more democratic, culturally informed, two-way process have begun to appear in the risk communication campaigns by agencies that work with vulnerable populations, such as children, expectant mothers, the elderly, or individuals suffering from asthma or other respiratory illnesses. In each case, agencies often interview affected groups or identify community partners to collaborate in designing communication campaigns that warn of specific risks. Let us look at one extended example of how this process of negotiation might occur, first through a technical perspective and then through a cultural one.

Mercury Poisoning and Fish Advisories: A Technical Model of Risk Communication

Mercury is a highly dangerous **neurotoxin**—a chemical substance that can cause adverse effects in the brain and nervous system. It is present in everyday products such as thermometers and compact fluorescent light bulbs (CFLs) and is dispersed widely in air pollution from coal-fired power plants. The EPA reports that people are most commonly exposed to mercury by eating fish or shellfish that accumulate methylmercury in fatty tissues. Although adults can be affected, risks become more significant for the young, due to smaller bodies and more sensitive developmental vulnerabilities. Numerous studies have found that "for fetuses, infants, and children, the primary health effect of methylmercury is impaired neurological development. Methylmercury exposure in the womb . . . can adversely affect a baby's growing brain and nervous system" (United States Environmental Protection Agency, 2010c, para. 5).

Due to significant health risks, the U.S. EPA provides extensive, technical risk information in advisories and on its website for mercury (www.epa.gov/hg). However, the challenge of risk communication in this case goes beyond technical information. Health officials also hope to change the eating practices of populations who consume fish. The Union of Concerned Scientists (2009) has estimated that, just 1/70th of a teaspoon of mercury in a 25-acre lake can make fish unsafe to eat. As a result, the EPA and other agencies attempt to provide culturally sensitive risk communication materials online (www.epa.gov/waterscience/fish) and to fund local health agencies to conduct public outreach to target populations.

Among the materials that the EPA and FDA make available is a brochure, *What You Need to Know About Mercury in Fish and Shellfish*. It is a guide for pregnant or nursing women and young children on selecting and eating fish in order to reduce their exposure to mercury. The brochure poses a series of questions and answers. For example, it states "I'm a woman who could have children but I'm not pregnant—so why should I be concerned about methylmercury?" In response, it advises:

If you regularly eat types of fish that are high in methylmercury, it can accumulate in your bloodstream over time. Methylmercury is removed from the body naturally, but it may take over a year for the levels to drop significantly. Thus, it may be present in a woman even before she becomes pregnant. This is the reason why women who are trying to become pregnant should also avoid eating certain types of fish. (United States Food and Drug Administration & United States Environmental Protection Agency, 2004, p. 3)

The brochure can be downloaded in English, Spanish, Chinese, Cambodian, Portuguese, Hmong, Korean, and Vietnamese—for different communities in the United States that rely heavily on fish in their diets. Similar guides and fish advisories are available as posters, illustrated brochures, key chain tags, and magnets for the kitchen. This campaign arguably is a successful example of a *technical model of risk communication*, which has taken into account the cultures of its target audiences.

Mercury Poisoning and Fish Advisories: A Cultural Model of Risk Communication

In association with the U.S. National Institute of Environmental Health Sciences (NIEHS), the University of Wisconsin's Marine and Freshwater Biomedical

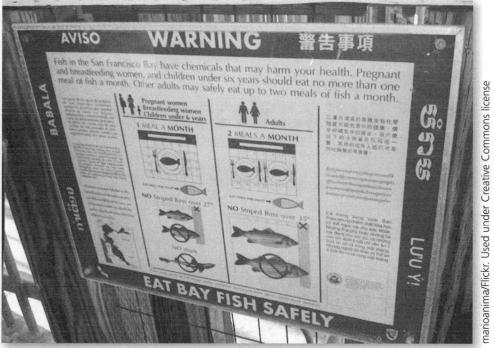

Photo 7.3 The challenge of risk communication goes beyond technical information into cultural values about our everyday attitudes, behaviors, and values. The reasons some people catch their own fish, eat it, and how often are bound in a range of cultural factors, such as ethnic heritage, income, and dietary preferences.

Sciences Center also has initiated a public education campaign about mercury poisoning and fish advisories. One of the center's objectives was "to increase the knowledge and involvement of minority communities in environmental health issues, including awareness of the risks and benefits of fish consumption by the Hmong community in Milwaukee" (National Institute of Environmental Health Sciences, 2003, para. 1). Let us look at this example as a successful cultural model of risk communication campaign.

The Hmong are immigrants from Southeast Asia. In Vietnam, they depended heavily upon fish for their diets. In fishing in Wisconsin waters and in the Great Lakes as immigrants to the United States, however, many initially had little understanding of existing pollution and contamination. As a result, the Marine and Freshwater Biomedical Sciences Center departed from the technical model of risk communication and proposed, instead, to work in collaboration with the community to design its campaign. Collaboration between the center's scientists and the community included the Hmong American Friendship Association and the Sixteenth Street Community Health Center, the major health care provider for Hmong residents.

The cultural model of risk communication plan that grew from this dialogic approach relied not only on the Hmong language, but also a broader awareness of Hmong cultural traditions. The main product of the process included a bilingual Hmong–English video titled *Nyob Paug Hauv Qab Thu* (*Beneath the Surface*). The video, which features local Hmong residents, "communicates in a simple,

Table 7.1 Models of Risk Communication		
	Technical Model	**Cultural Model**
Type of communication:	Usually one way (experts to laypeople)	Collaborative (citizens, experts, agencies)
Source of knowledge of risk:	Science/technology	Science plus local, cultural knowledge and experience
Objectives:	1. To translate and inform	1. To inform by recognizing social contexts of meaning
	2. To change risky behavior	2. To change risky behavior when in the interests of affected groups
	3. To assure concerned groups	3. To involve affected groups in judgments of acceptable and unacceptable risks

SOURCE: Adapted from table located at: http://oehha.ca.gov/pdf/HRSguide2001.pdf, page 5. Office of Environmental Health and Hazard Assessment.

understandable, and culturally sensitive way the risks of eating contaminated fish and teaches methods of catching and preparing fish that can reduce these risks" (Thigpen & Petering, 2004). Since pregnant women and children were at higher risk, the center and its partners also decided to develop approaches relating specifically to those populations. In addition, they relied upon Hmong focus groups to discuss the framing of the content of the communication campaign.

The center also relied upon volunteers in the Hmong community to distribute the video to households and to show it at Hmong festivals. Finally, the center and its partners worked with the local middle school to develop a module for its life science class that would educate inner-city students about eating contaminated fish. The project is ongoing among the Hmong community, largely as a result of the support and participation of local leaders, residents, and professionals in the community itself.

These examples reflect many of the differences between technical and cultural models of risk communication (see Table 7.1). The success of both risk communication initiatives reflected the concerns and viewpoints of those who were most affected by mercury poisoning. Government agencies increasingly are recognizing that these democratic values must be part of any communication among scientists, agency officials, spokespeople, and members of the most impacted communities.

Citizens Becoming Scientists

While some environmental risks are recognized in the public sphere, many remain contested. When a particular institution benefits from not addressing a risk, heated debates can occur. Political theorist Michael R. Reich (1991) studied toxic chemical disasters in the United States, Japan, and Italy to attempt to identify cross-cultural patterns in responses. Overall, he found people most directly impacted by the health risks of a toxic disaster attempted to publicize their grievances and people who benefitted the most from business-as-usual tried to privatize the grievances.

In response to environmental disasters and ongoing public health risks, some everyday people have spent countless hours creating their own assessment tools of risk and ways of sharing them. People who never formerly studied toxicology, biochemistry, or other related disciplines find themselves seeking out technical reports, experts willing to share their knowledge, and creating their own hypotheses about environmental dangers, health-related concerns, and solutions to the problems their communities face. Lois Gibbs's organization, the Center for Health and Environmental Justice (CHEJ), provides a clearinghouse of technical and firsthand knowledge to those seeking help in assessing risks (see http://chej.org/). In addition to formally trained scientists and public health experts, CHEJ and other concerned citizens draw on **experiential knowledge**, or that which has been learned through direct experience. Examples of experiential knowledge include tallies of hazardous material trucks driving through an intersection, stories of health complications, and water or air samples collected by everyday people and sent to labs for analysis.

Sometimes, those who gather and interpret knowledge from their own experiences and observation, as well as collecting relevant secondhand technical data, call themselves **citizen scientists**. Jason Corburn (2005), a professor of City and Regional Planning, has detailed how citizen scientists are transforming urban planning departments and finding compelling ways to map their experiential knowledge. For example, he shows how a group of high school students calling themselves El Puente Toxic Avengers (after the comic book) worked with university researchers to collaboratively draw maps identifying multiple pollution sources overburdening their community based on their experiential knowledge.

In addition to these less formal modes of communication, risk remains communicated through news media as well, of course. It is to these voices in the public sphere that we now turn.

Mainstream News Media and Environmental Risk

News reports, TV ads, marketing campaigns, websites, blogs, and more media provide public forums for discussing environmental and public health risks ranging from climate change crises to toxic chemicals used in baby toys and equipment. And, although media can be invaluable sources about risk and debates surrounding risk, most mainstream news media still labor under many of the same requirements for newsworthiness that we described in Chapter 5.

In this section, we describe some of the ways in which risk is represented and the factors that contribute to coverage of environmental dangers in mainstream news media. We also point to some of the challenges people most impacted by the costs of our *risk society* have in gaining recognition in mainstream news media.

News Media Reports of Risk: Accurate Information or Sensational Stories?

As environmental media scholar Libby Lester (2010) reminds us, knowledge of risks—"how we become aware of them, [and] assess their dangers"—relies upon "public" claims made by a variety of actors, who compete to have the legitimacy of their claims recognized in the public sphere" (p. 102). This, in turn, is influenced by any number of challenges—for example, how well reporters understand a technical risk report, if dramatic visuals of an event overwhelm a more sober analysis, or whether the public views scientific claims themselves as sites of disagreement.

A common criticism from scientists is that journalists often report inaccurate information rather than substantive coverage of an environmental danger. A classic illustration of such misinformation occurred after the damage to the reactor core and release of radioactive materials at Pennsylvania's Three Mile Island nuclear plant on March 28, 1979. Rhetorical scholars Thomas Farrell and Thomas Goodnight (1981) described communication about the accident by mainstream news media, government spokespersons, and industry as a "conspicuous confusion and failure" (p. 283). In their study, they found that "reporters were unable to judge the validity of technical statements.... Government sources, frequently at odds with one another, could not decide what information to release.... [And] some representatives of the nuclear power industry made misleading statements" (Farrell & Goodnight, 1981, p. 273). And, in 2010, news reports of the amount of oil spilled into the Gulf of Mexico by the blowout of the BP oil well reflected a similar confusion among scientists, the EPA, and oil company spokespersons.

News media also face other constraints in covering environmental risks, particularly the more complex news stories about biodiversity, climate change, or energy shortages. Thus, while attempting to provide information about serious hazards, reporters and editors also must negotiate a thicket of journalistic norms; among them, Is the story newsworthy (Chapter 5)? Dr. John Graham, former director of the Harvard Center for Risk Analysis, has complained, for example, that "what constitutes news is not necessarily what constitutes a significant health problem" and that reporters often stress "the bizarre, the mysterious," rather than more realistic risks (quoted in Murray, Schwartz, & Lichter, 2001, p. 116).

Mainstream news media, for example, often over-report the risks generated by acute crises, such as oil spills or tornados, while underreporting slower, less visible threats such as the loss of biodiversity (Allan, Adam, & Carter, 2000). A study of network TV found similar results. Airplane accidents received 29 times more coverage than the danger of asbestos, which scientists believe is 41 times more likely to kill Americans (Greenberg, Sachsman, Sandman, & Salomone, 1989, p. 272). As such, news media can foster public anxiety about issues that scientists find less worrisome.

In fairness, journalists face difficult constraints in reporting risk issues. For one thing, scientific research itself may be ambiguous and hence difficult for reporters to interpret fairly or accurately for laypeople. And risk stories—like everything else—compete with other breaking news. Speaking at a 2008 meeting of the American Association for the Advancement of Science, former *New York Times* science reporter Andrew Revkin said that despite interest in the environment, global warming remains a "fourth-tier" story in the press. Among the reasons, he explained, is "the 'tyranny of the news peg,' a dearth of print space, and different learning curves for complex stories like climate science.... You don't get extra room in a newspaper just because the story's harder" (quoted in Brainard, 2008a, para. 2). Not helping the situation, the *New York Times* decided to close its environmental desk in 2013.

Luckily, investigative journalists remain on the environmental beat. In 2014, *New York Times* reporter Nicholas Kristof listened to reader feedback and covered climate change in his column as the most important underreported story of 2013. Other newspapers have continued coverage of environmental risk and largely have overshadowed the *New York Times'* efforts. In particular, the *Chicago Tribune's* 2012 multi-award winning investigative series "Playing With Fire" about toxic flame retardants and their risk to public health is an exemplar of what journalists can offer public debate about risk communication and assessment.

☞ **FYI**   **News Media Coverage of Environmental Risk**

A survey of U.S. mainstream news media coverage of risk by Lundgren and McMakin (2009) echoed many of the criticisms of news media by scientists and others. Their chief findings included:

1. "Scientific risk had little to do with the environmental coverage presented on the nightly news. Instead the coverage appeared driven by the traditional journalistic news values of timeliness, geographic proximity, prominence, consequence, and human interest, along with the television criterion of visual impact" (p. 210).

2. "Mass media disproportionately focus on hazards that are catastrophic and violent in nature, new, and associated with the United States.... Drama, symbolism, and identifiable victims, particularly children or celebrities, make risks more memorable" (p. 211).

3. "Concepts important to technical professionals, such as probabilities, uncertainties, risk ranges, acute versus chronic risks, and risk tradeoffs, do not translate well in many mass media formats" (p. 212).

4. "To humanize and personalize the risk story, news organizations often use the plight of an individual affected by a hazard, regardless of how representative the person's situation is" (p. 212).

Lundgren and McMakin (2009) stressed that these ways of simplifying or personalizing information about risk "may make information more accessible to the public, but may result in incomplete and sometimes unbalanced information for making personal risk decisions" (p. 212).

SOURCE: R. E. Lundgren & A. H. McMakin. (2009). *Risk Communication: A Handbook for Communicating Environmental, Safety, and Health Risks* (4th ed.). Hoboken, NJ: John Wiley.

Whose Voices Speak of Risk?

Public understanding of environmental risk depends not simply on information about risks but also on who speaks about or interprets the information about risk. Let us now turn to two main sources of risk information used by journalists in mainstream news media: *legitimizers* and *voices of the "side effects."*

Legitimizers as Sources for Risk

Not surprisingly, mainstream news media stories about risk rely upon **legitimizers,** or sources such as official spokespersons and experts. In fact, mainstream news media often establish their credibility through citing officially sanctioned sources. In a study of this tendency, public relations scholar Donalyn Pompper (2004) surveyed 15 years of environmental risk stories in three U.S. newspapers that target different social groups: the *New York Times, USA Today,* and the *National Enquirer.* Her conclusion was that mainstream news media such as the *New York Times* and *USA Today* relied heavily upon government and industry sources, while the *National Enquirer* relied mostly on individuals, community members, and so forth. Pompper not only found that cited sources of risk distinguished mainstream news media from other media outlets, but also that the sources communicated about risk quite differently.

As a result, stories about environmental risk may quote an EPA spokesperson or a scientist to provide "content" for the coverage while using local residents or environmental groups for nontechnical aspects of the story such as "color, emotion, and human elements" (Pompper, 2004, p. 106). These patterns reflect the contrast between technical and cultural models of risk assessment. That is, government, science, and industry voices were more likely to characterize risk in terms of official studies and assurances of safety, suggesting that risk could be controlled, responsible agencies were providing oversight, and so on. On the other hand, everyday people quoted spoke of personal concerns about dangers, such as cancer and industrial accidents.

Pompper's (2004) conclusion is stark:

> Voices of common people who live with environmental risks every day and voices of groups organized to save the environment . . . are drowned out by elites cited most often in environmental risk stories. For non-elites . . . this study's major finding has grim implications, indeed. The news essentially ignore them. (p. 128)

Classifying news sources as *elite* versus *non-elite*, however, may be unfair. Sometimes, legitimizers themselves differ about environmental or health risks: Is BPA in plastic bottles safe? Does air pollution from coal-burning power plants harm infants and young children? In cases of controversy among legitimizers, news media will often revert to conflict frames in covering the story. Earlier in this chapter, for example, we saw scientists challenging the FDA for relying on risk studies funded by industry (Parker-Pope, 2008). Further, we all make judgments about risk informed at some level by legitimizers. Perhaps our understanding of how risk is reported in the public sphere might be informed more fairly by considering what happens when sources conflict with each other.

Let's look briefly at an example of when mainstream news media quote industry legitimizers who challenge government legitimizers. For years, the U.S. Food and Drug Administration (FDA) has been reviewing public comment on proposed guidelines that limit the use of antibiotics in farm animals. For years, large agribusinesses and small farmers alike have "routinely fed antibiotics to their cattle, pigs, and chickens to protect them from infectious diseases but also to spur growth and weight gain" ("Antibiotics and Agriculture," 2010, p. A24). Medical experts, however, believe the overuse of antibiotics in these animals has led to "the emergence of antibiotic-resistant bacteria, including dangerous E-coli strains that account for millions of bladder infections each year, as well as resistant types of salmonella and other microbes" (Eckholm, 2010, p. A13).

Mainstream news media coverage of the FDA's proposed guidelines featured conflicting risk assessments, quoting farmers and livestock producers as well as medical experts. For example, the *New York Times* framed a 2010 news story about antibiotics in farm animals as a conflict between industry and science. The story quotes the National Pork Producers Council, an industry source, that insisted the risks were remote: "There is no conclusive . . . evidence that antibiotics used in food animals have a significant impact on the effectiveness of antibiotics in people"; then the *Times* story stated, "But leading medical experts say the threat is real and growing" (Eckholm, 2010, A13).

As the antibiotics story illustrates, the coverage of *risk assessment* in mainstream news media faces many of the same characteristics of news production related to environmental matters more broadly, including the tension between norms of objectivity and balance that we reviewed in Chapter 5.

Voices of the "Side Effects"

The dominance of government, scientific, and industry sources in news media certainly affects public perceptions of risk. This dominance also raises an important question about the opportunities for cultural rationality in media outlets. Indeed, an important debate over media reporting of risk concerns what Beck (1992) called the **voices of the "side effects"** (p. 61). Beck is referring to those individuals (and their children) who suffer the side effects of living in a risk society, such as asthma and other illnesses from air pollutants, chemical contamination, and the like. Reflecting the tension between technical and cultural models of risk communication that we discussed earlier, these voices of the "side effects" seek news media recognition of a very different understanding of environmental dangers and the burdens of risk. Beck explains in the following way:

> What scientists call "latent side effects" and "unproven connections" are for them their "coughing children" who turn blue in foggy weather and gasp for air, with a rattle in their throat. On their side of the fence, "side effects" have voices, fears, eyes, and tears. And yet they must soon learn that their own statements and experiences are worth nothing so long as they collide with the established scientific [views]. (p. 61)

As a result, voices of these "side effects" too often are not given the journalistic space to offer alternative, cultural rationality in news about environmental risks. A rather extreme example of this occurred in a news story about the toll that the Gulf of Mexico oil spill in 2010 had on local citizens. Medical experts at the time observed that "the impact of the disaster on human health and well-being has not even begun to be quantified" (Walsh, 2010, para. 1). The story on *TIME's Health & Science* blog included interviews with experts from various universities, the National Institute for Occupational Safety and Health, and the Children's Health Fund, commenting on the impacts of the oil spill on individuals. Typical was Dr. Irwin Redlener, director of the National Center for Disaster Preparedness at Columbia University. "These are people in a serious crisis," he said. "They're at ground zero of a catastrophe" (quoted in Walsh, 2010, para. 3).

Curiously, the news account never interviewed a single person who might have been affected directly by this catastrophe—unemployed workers, residents, or cleanup crews working under hazardous conditions—for insight into their well-being or health. The story did, however, open dramatically with this account:

> [A]n Alabama fisherman who reluctantly took an oil-spill cleanup job with BP, was found dead from a self-inflicted gunshot wound. He left no suicide note, and we'll likely never know why he took his own life, but friends told the media that he had been deeply troubled by the spill and the destruction it caused on the coast. In this instance, the voice of the "side effects," the Alabama fisherman, was literally silent, but used, nonetheless, for "emotion, and [a] human element." (Pompper, 2004, p. 106)

Media communication scholar Simon Cottle (2000) considered examples such as this in a study of environmental news on British television. Cottle found that "ordinary voices" were more frequently (37%) cited in TV news than either government or scientific sources. However, he cautioned that ordinary voices and expressions of "lived experience" are used mainly to provide human interest or a "human face" for stories rather than for substantive analysis. The inclusion of such personal interviews is, of course, important for newsworthiness in stories. Nevertheless, these experiences are "sought out and positioned to play a symbolic role, not to elaborate a form of 'social rationality'" (Cottle, 2000, pp. 37–38). That is, they are used to provide color, variety, and human interest rather than to offer insight into a problem.

Act Locally!

How Do Local News Media Report Environmental Risks?

How do your local news media cover possible environmental or health risks? Follow a story-line across time or read across headlines for one week.

1. Do the news stories provide enough information, background, or insight about the issue?

2. Is the tone sensational, opinionated, or factual?

3. What sources are included in the stories? Do they include the voices of the "side effects"?

4. What function do these voices serve? Legitimizing official policies? Offering color or personal interest for readers or viewers?

An important caveat to the limitations of mainstream news media, of course, is that online channels are now available for individuals (including the voices of the "side effects") to publicize their concerns. As risk communication scholar Sheldon Krimsky (2007) points out, "The Internet . . . has . . . expanded the breadth and channels of risk communication, while also providing new opportunities for stakeholders to influence the message" (p. 157). Numerous blogs, clearinghouses for different diseases, chat rooms, and advocacy groups' sites now populate social media and online sites and make possible discursive spaces for sharing of views about risks and other environmental dangers. Consider, for example, the blog West Virginia Water Crisis launched by Krista Bryson, a rhetorics graduate student, after the toxic spill, which provides video testimony of people most directly impacted, as well as "answers" to questions regarding water safety and health risk concerns: http://westvirgin iawatercrisis.wordpress.com/. Such media forums are redefining risk communication through broadening public access to information and to impacts of risk.

SUMMARY

This chapter introduced the idea of *risk communication,* including different philoso-phies of the relationship between risk and communication, as well as how risk should be addressed in the public sphere.

- The first section described the idea of a *risk society* or the dangers from modern society itself. We also introduced two approaches for assessing risk: (1) technical and (2) cultural.
- The second section introduced the practice of *risk communication* through two models, a traditional one-way technical model of risk communication and an interactive, dialogic cultural model of risk communication.
- The third section illustrated how *citizen scientists* have established their own experiential expertise as vital to risk assessment impacting their communities.
- The final section explored the ways that mainstream news media shape our perceptions of risks and two dimensions of journalist reports:
  - Debate over the accuracy of news reports of risk; that is, are they factual or sensational?
  - Different voices who speak in news media coverage of risk—society's legitimizers (e.g., public officials), and the voices of the "side effects": residents, parents of sick children, and others who are most affected by environmental dangers.

At the beginning of this chapter, you read that the ways in which environmental risks are assessed and publicized help to constitute the public's perceptions and judg-ments of risk. We hope, then, that this discussion of technical and cultural models of risk communication has helped you appreciate some of the reasons our risk society seems conflicted in the public sphere about what we should and should not worry about. Mainstream news media continues to play a significant and telling role in risk communication in the public sphere, despite growing new media forums. We have shown that the ways various actors in the public sphere—government officials, news media, corporate spokespeople, scientists, lawyers, public relations coordinators, and the general public—judge risks as acceptable or unacceptable are constituted, judged, and negotiated through communication.

SUGGESTED RESOURCES

- Compare and contrast risk communication in *Erin Brockovich* (2000), a film starring Julia Roberts as an unemployed mother who, as a paralegal, exposes a powerful corporation's toxic pollution of a rural California town, with *A Civil Action* (1998), a film starring John Travolta, who, as a personal injury lawyer, becomes bankrupt in helping a community sue a powerful corporation for the toxic pollution of their small Massachusetts town.
- Tribune Watchdog: Play With Fire. This *Chicago Tribune* website is worth exploring in depth. It provides primary documents, videos, and gripping analysis about how babies born in the United States now have more toxic flame retardants in their bodies than a child anywhere else on Earth—and why this is an unacceptable risk.
- Peter Sandman has a detailed website with handouts elaborating on risk communication, including handouts on topics such as Activists and Media; Explaining Risk Data; and Safety and Health Communication: http://www.psandman.com/handouts/topical.htm#three.
- For more advanced discussion of dialogic approaches to risk, see Jennifer Duffield Hamilton, Exploring Technical and Cultural Appeals in Strategic Risk Communication: The Fernald Radium Case. *Risk Analysis, 23*(2) (April 2003), 291–302; and Jennifer Duffield Hamilton and Caitlin Wills-Toker, Reconceptualizing Dialogue in Environmental Public Participation. *Policy Studies Journal 34*(4) (November 2006), 755–775.
- "Waste Land: One Town's Atomic Legacy" is the story of one woman's relentless efforts to clean up the toxic waste from a factory producing nuclear fuel for submarines, which was causing sickness in her small Florida town. See John R. Emshwiller, *New York Times*, November 22, 2013, at http://online.wsj.com/news/articles/SB10001424052702304868404579194231922830904

KEY TERMS

Acceptable risk   153

Acute single-dose tests   154

Androcentric bias   154

Black swan events   151

Citizen scientists   166

Cost-benefit analysis   154

Cultural model of risk communication   161

Cultural rationality   158

Cultural theory of risk assessment   156

Dialogic communication   161

Experiential knowledge   165

Four-step procedure for risk assessment   152

Hazard   157

Legitimizers   168

Modern society   151

DISCUSSION QUESTIONS

1. Do you trust government warnings about the health risks of smoking, texting while driving, tanning salons, or consuming alcohol while pregnant? Why or why not? Are these warnings effective? Do they affect your own behavior?

2. Is the public's outrage over environmental hazards rational? Although a toxic waste landfill may inconvenience those living near it, doesn't it have to go somewhere? Or does it? Does society manage fairly the risks associated with our chemical culture?

3. Recently, the Environmental Working Group (2010) found concentrations of hexavalent chromium, or chromium 6 (also featured in the movie *Erin Brockovich*), in the drinking water of 31 cities in the United States. If you lived in one of these cities, what would you want in a risk communication plan about your drinking water? What would persuade you that your water was safe or not? What would cause you to change your behavior, for example, to use bottled water instead?

4. What do the voices of the "side effects" (such as parents of sick children) contribute to risk communication? Are these voices merely emotional, or do they have relevant insight into health or environmental dangers in their communities?

5. Imagine a 2-minute Public Service Announcement about various topics of risk (for example, coal plants polluting the air, flame retardants in kids' pajamas, etc. What are the biggest challenges to communicating risks? How did you try to visualize risk? Which words did you think were most persuasive for a message?

PART IV

Environmental Campaigns and Movements

Advocacy campaigns are a uniquely democratic form of communication, offering a collective voice for those who wish to influence decision makers. The Sierra Student Coalition's Coal Free campaign is mobilizing students across the United States to persuade their boards of trustees and university presidents to move away from coal as a source of energy on their campuses and toward renewable energy.

CHAPTER 8

Advocacy Campaigns and Message Construction

I am now on my first International Speaking Tour. . . . I talk about how kids can . . . get involved in projects . . . important to them, because this planet is not a place kids will inherit at some point far off in the distant future, we live here now, we share this planet already.

—Milo Cress, age 12 (2013, para. 1–4)

The Be Straw Free campaign illustrates a form of environmental communication called **advocacy**, the act of persuading or arguing in support of a specific cause, policy, idea, or set of values. The campaign urges us to stop using plastic straws in order to reduce unnecessary waste and pollution. Milo Cress started it in 2011 when he was 11 years old. Now, it's an international campaign, changing the behaviors of thousands of people across the world.[1]

Campaigns are just one form of advocacy used by businesses, candidates for public office, public relations firms, environmental groups, and more. Advocacy takes many forms, including advertising, political campaigning, community organizing, marches and demonstrations, legal argument, and so forth. In this chapter, however, we focus on one significant form of advocacy—the *environmental advocacy campaign.*

Chapter Preview

- In the first section of this chapter, we describe *advocacy* in general and distinguish *advocacy campaigns* and *critical rhetorics,* the questioning or criticism of the status quo.
- In the second section, we explore the basic elements of advocacy campaigns, focusing on (a) objectives, (b) audiences, and (c) strategies.

(Continued)

(Continued)

- We go behind the scenes in the third section by describing the design of a successful advocacy campaign—the Zuni Salt Lake Coalition's campaign against plans to strip-mine for coal near sacred Native American lands.
- In the final section, we describe the role of *message construction* in campaigns and identify two challenges: (1) overcoming the *attitude-behavior gap* and (2) the use of values in the construction of messages that seek to motivate supporters.

Our hope is that, when you have finished this chapter, you'll be more aware of the wide range of communication practices that environmental advocates engage in and that you'll also appreciate the rhetorical constraints they face in building public demand for, and winning, environmental protection.

Environmental Advocacy

Advocacy is a powerful tool for a wide range of social change organizations. Groups—whose goals range from educational awareness about the significance of recycling to demands for compensation for small island nations most immediately and dramatically affected by climate change—constitute public forums and avenues of action for diverse voices and concerns. Such collectives of people seek to hold more powerful institutions accountable to democratic and humane principles and often achieve changes that protect more vulnerable populations and public interests.

Advocacy groups often act as intermediaries between grassroots individuals and the large, often impersonal institutions of public life. This position has been particularly true of environmental groups. Former *New York Times* writer Philip Shabecoff (2000) argues that a chief role of environmental groups is to act as "intermediaries between science and the public, the media, and lawmakers" (p. 152). For example, the Sierra Student Coalition's Beyond Coal campaign (opening photo) has successfully served this intermediary role on many campuses. On campuses where we have worked—Indiana University and the University of North Carolina—student groups drew on the knowledge of faculty, physical plant workers, and alumni to emphasize the public health, environmental, and economic costs of relying on campus coal-burning power plants, as well as the option of renewable energy sources. As intermediaries, these small numbers of students enabled other students to gain expertise and to express their concerns. In both cases, they succeeded in persuading university officials to commit to retiring our campuses' reliance on our on-campus coal plants.[2]

Modes of Environmental Advocacy

Environmentalists engage in a wide variety of advocacy. Advocates may differ in their goals, the media they use, and the audiences they target. Types of advocacy include public education, community organizing, lobbying campaigns, boycotts, direct action protests such as sit-ins, and more. (See Table 8.1.) Here and in following chapters, we

Table 8.1 Modes of Environmental Advocacy Campaigns

Mode of Advocacy	Objective
Official Political and Legal Channels:	
1. Legislation	To influence laws or regulations
2. Litigation	To seek compliance with environmental standards by agencies and businesses
3. Electoral politics	To mobilize voters for candidates and referenda
Public Awareness Campaigns and Actions:	
4. Critical rhetoric	Questioning or criticism of a behavior, policy, societal value, or ideology
5. Public education	To influence societal attitudes and behavior
6. Direct action	To influence specific behaviors through acts of protest, including civil disobedience
7. Media events	To create publicity or news coverage to broaden advocacy effects
8. Community organizing	To mobilize citizens or residents to act
Consumer Appeals and the Market:	
9. Green consumerism	To use consumers' purchasing power to influence corporate behavior
10. Corporate accountability	To influence corporate behavior through consumer boycotts and shareholder actions

describe some of these modes of advocacy in more detail. For now, we describe two broad forms of advocacy: advocacy campaigns and critical rhetoric.

Campaigns Differ From Critical Rhetoric

Before an environmental advocacy campaign starts, there is often a period in which the status quo is questioned and a desire to find a better way is expressed. This is the role of *critical rhetoric*. Although they are different in some ways, campaigns and critical rhetorics can function in complementary ways, and it is therefore important to understand each of these modes of advocacy in more detail.

Critical Rhetoric

Critical rhetoric is the questioning or criticism of a behavior, policy, societal value, or ideology. Such rhetoric may also include the articulation of an alternate policy, vision, or ideology. Throughout the modern environmental movement, many voices have questioned or criticized the taken-for-granted views of culture. For example, Rachel Carson's (1962) classic book *Silent Spring* sharply criticized the practices of the pesticide industry and government agencies that exposed the public to harmful chemicals. Other voices have urged an alternative vision of society. A report by the UK World Wildlife Fund envisions that "by 2050, power, transport, industrial and domestic energy needs could be met overwhelmingly from renewable sources" (WWF, 2011, para. 3).

Many consider the global Occupy Movement, which reached its initial peak in 2011, an exemplar of critical rhetoric. Occupy named the majority of the world as a united 99% that needs to demand accountability from the 1% of elites. Although this critique is not a concrete success (such as campaign finance reform or an end to predatory loans for higher education), many believe the fact that we can say "99%" and that most people know what we are referencing when we do so is a symbolic achievement of collective imagination, through which new political alliances and actions might emerge. As a result, critical rhetorics frequently serve not only to question taken-for-granted realities but also expand the range of visions that are eclipsed in the day-to-day political struggles of a campaign.

Advocacy Campaigns

Although campaigns also advocate social change, they differ from critical rhetorics. Most important, campaigns are organized around concrete actions that move us closer to larger goals. In this sense, an **advocacy campaign** can be defined broadly as

Courtesy of Greenpeace

| Photo 8.1 | Greenpeace International declared a campaign victory in 2013 when Japan's top casual wear clothing brand, Uniqlo, committed "to eliminate all releases of hazardous chemicals throughout its entire global supply chain and products by 2020." Although other brands such as Levi's and Esprit have made similar commitments, the campaign continues to focus on brands such as GAP and Calvin Klein. |

a strategic course of action, involving communication, which is undertaken for a specific purpose or objective. A campaign is waged to win a victory or secure a specific outcome; it, therefore, goes beyond the questioning or criticism of a policy or societal value. This is an important point. For example, local citizens might criticize plans to build a toxic waste landfill in their neighborhood. A campaign, however, will go further; it might pursue the objective of blocking a permit for the construction of the landfill by organizing local residents, businesses, and church and synagogue leaders to call, visit, and lobby city council members who have the power to decide for or against the landfill's permit.

Advocacy campaigns draw upon several of the advocacy modes listed in Table 8.1. For example, many environmental advocacy campaigns will involve legislative and electoral politics; they also might engage in public education, community organizing, and corporate accountability campaigns. Overall, a campaign may rely on multiple forms of advocacy as part of a strategic and time-limited course of action for a specific purpose.

In contemporary society, advocacy campaigns are used by many groups for a range of purposes. In information and behavior-change campaigns, for example, the campaign form is used, among other purposes, to reduce underage smoking and to encourage drivers to conserve energy by properly inflating their tires. Environmental advocacy campaigns share some characteristics of these information and behavior-change campaigns, and it is important to recognize these similarities before looking at their differences.

☞ FYI   Features of Campaigns

In their classic study of campaigns, Everett Rogers and Douglas Storey (1987) identified four features shared by most campaigns:

1. *A campaign is purposeful.* That is, "specific outcomes are intended to result from the communication efforts of a campaign" (p. 818).

2. *A campaign is aimed at a large audience.* A campaign's purpose usually requires an organized effort going beyond communication with just one or a few people; it aims to persuade enough people to make a difference.

3. *A campaign often has a specifically defined time limit.* The desired outcome of a campaign—a vote or passage of a law, for example—must be achieved by some deadline, when the window for any further action will close.

4. *A campaign involves an organized set of communication activities.* These activities are particularly evident in construction of the campaign's message and in efforts to educate and/or mobilize different constituencies.

SOURCE: Everett M. Rogers & Douglas D. Storey, "Communication Campaigns," in C. R. Berger & S. H. Chaffee (Eds.), *Handbook of Communication Science* (pp. 817–846). Newbury Park, CA: Sage, 1987.

Although they share features with information and behavior-change campaigns, environmental advocacy campaigns differ from them in basic ways. Two differences in particular stand out.

First, access to decision-making power and resources differ due to those who lead the campaigns. Most information and behavior-change campaigns aim to reduce risk or influence individual behavior and are *institutionally sponsored by a legally recognized, centralized entity;* that is, they are initiated by a government health agency, a university, or a company targeting individuals (like students or employees). Environmental advocacy campaigns, on the other hand, usually are waged by people and/or organizations that exist outside of state or corporate institutions—concerned individuals, environmental organizations, or small community action groups, for example.

Second, the type of change encouraged tends to differ. That is, most public health information and behavior-change campaigns seek *to modify individuals' choices about their personal lifestyles,* such as drug use or car maintenance practices. These changes usually do not require the sponsoring entity to undergo broader changes in laws or corporate practices, other than perhaps adding signage to designate smoking areas or remind people not to idle in their cars or to turn off the lights when they leave a room. Most environmental advocacy campaigns, on the other hand, seek to change something beyond one's personal lifestyle through *external conditions*—for example, the closing of a coal plant or the cleanup of an abandoned toxic waste site—or more *systemic change,* that is, alter the policies or practices of a governmental or corporate body to provide health care for all or build affordable public transportation.

These distinctions are not to say there is no overlap. The Be Straw Free Campaign mentioned at the beginning of this chapter highlights how personal behavior choice campaigns can be launched by individuals who then push for systemic changes in the restaurant industry and beyond. Communication scholars and practitioners, however, must be aware of who is launching any given campaign, and for what ends, in order to provide useful analysis or ideas.

From our own experiences in the U.S. environmental movement, we've been convinced that while information and behavior-change campaigns can make a difference, advocacy campaigns also are important in shaping public debate and civic decisions about environmental policy and practices that exceed the impact of individual choices. Therefore, in the following sections, we describe, in more detail, the basic design that many advocacy campaigns use and also provide more examples of successful campaigns.

Environmental Advocacy Campaigns

By the time of the first Earth Day in 1970, the ecology movement had begun to change the way citizens communicated with public officials about the environment. Not content to rely simply on magazine articles or nature programs on television to educate the public, many environmental groups began to design advocacy campaigns to achieve specific changes. One architect of this new strategy was Michael McCloskey, the former executive director of the Sierra Club. In an interview, McCloskey reflected on his role in the environmental movement's turn to campaigns:

> What I have emphasized has been a serious approach toward achieving our ends. I thought that we were not here just to bear witness or to pledge allegiance to the faith, but in fact we were here to bring that faith into reality. . . . That means we could not rest content with having said the right things . . . but we also had to plan to achieve them. We had to know how the political system worked, how to identify the decision makers. . . . We had to have people concerned with all the practical details of getting our programs accomplished. (Gendlin, 1982, p. 41)

The shift described by McCloskey echoes the basic difference between critical rhetoric and campaigns—between "saying the right things" and having a "plan to achieve them." A *plan* means that advocates must ask, "What do we need to do to implement a strategic course of action, involving communication, to achieve our purpose?" From our observations of successful campaigns,[1] we've found that environmental leaders usually ask themselves three basic questions:

1. What exactly does the campaign want to accomplish?

2. Which decision makers have the ability to respond?

3. What will persuade these decision makers to act on the campaign's objectives?

These three questions ask, respectively, about a campaign's (1) objectives, (2) audiences, and (3) strategies. (We discuss each of these questions below.)

In answering these questions, campaigns also pursue important **communication tasks**. Campaigns, for example, must compose persuasive messages to win public support for their objectives; in doing this, campaigns strive to mobilize this support from constituencies (audiences) relevant to their strategy for influencing key decision makers. Of course, it is important to be aware that campaigns take place in the context of other, competing voices and countercampaigns. Successful campaigns adapt to these challenges in what is often an ever-changing communication environment.

In the remainder of this section, we describe the basic questions and corresponding communication tasks that advocacy campaigns confront. (See Figure 8.1 for a model of the advocacy campaign.)

Campaigns' Objectives

Successful advocacy campaigns require a clear-eyed focus on a concrete objective. For example, John Muir's preservation campaign to protect Yosemite Valley (Chapter 2) focused on the passage of a single piece of legislation in the U.S. Congress in 1890 that designated the mountains around Yosemite Valley as a National Park. Therefore, the first question environmental advocates face in designing a campaign is about the group's objectives. It asks, "*What exactly does the campaign want to accomplish?*"

Goals Versus Objectives

Campaigns flounder when their objectives are unclear or when they confuse a broad goal or vision with near-term, achievable, and specific actions or decisions. It is one thing to declare, "The world should protect old-growth forests," and quite another to mobilize relevant constituencies to persuade the relevant government agency in, for example, the United States, Canada, Brazil, Russia, or elsewhere, to issue an official ruling to halt the building of roads into these specific native forests. (Roads give access to commercial logging operations.) While halting roads in a specific forest contributes to the broader *goal* of protecting global old-growth forests, it is important to distinguish a campaign's need to focus on this specific *objective* from the broader effort that would presumably be needed to protect the remaining old-growth areas.

What, then, does it mean to answer the first question, "What exactly does the campaign want to accomplish?" First, it is important to distinguish between a campaign's long-term *goals* and its specific *objectives*. As it is used here, the term **goal** refers to a long-term vision or value, such as the desire to protect old-growth forests, reduce arsenic in drinking water, or reduce the levels of greenhouse gases entering the atmosphere. Critical rhetorics are often important in articulating these broader visions, but they are not campaigns.

On the other hand, the term **objective** refers to a specific action, event, or decision that moves a campaign closer to its broader goal. An objective is a concrete and

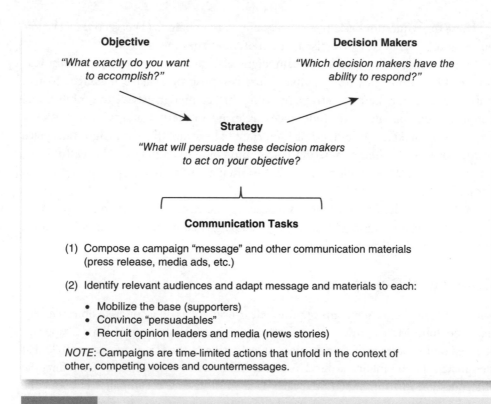

Objective

"What exactly do you want to accomplish?"

Decision Makers

"Which decision makers have the ability to respond?"

Strategy

"What will persuade these decision makers to act on your objective?

Communication Tasks

(1) Compose a campaign "message" and other communication materials (press release, media ads, etc.)

(2) Identify relevant audiences and adapt message and materials to each:

- Mobilize the base (supporters)
- Convince "persuadables"
- Recruit opinion leaders and media (news stories)

NOTE: Campaigns are time-limited actions that unfold in the context of other, competing voices and countermessages.

Figure 8.1   Design of the Environmental Advocacy Campaign

time-limited decision or action. For example, the U.S. Environmental Protection Agency (EPA) can issue a regulation imposing stricter limits on the number of parts per billion of arsenic allowed in drinking water. That's why the emphasis in this first question is to ask, "What *exactly* do you want?"

Most successful campaigns answer this question by identifying an objective that is a concrete, specific, and time-limited action or decision. Typical objectives from past campaigns have included the passage of a referendum in support of clean water bonds, a city council's vote for a zoning ordinance that banned hazardous waste facilities within 10 miles of a school, and a state utility commission's decision to deny a permit for a coal-burning power plant. Each of these objectives—once achieved—furthered a broader goal but, in themselves, were concrete, achievable decisions or actions.

Identifying Key Decision Makers

Once an advocacy campaign decides what exactly it wants to achieve, it must ask, "Which decision makers have the ability to respond?" In answering this, campaign organizers succeed when they identify (and develop a strategy to influence) relevant decision makers, for example, a governmental agency, legislative body, university's Board of Trustees, business leaders, or other responsible parties who have the authority to act on the campaign's objectives.

Primary Versus Secondary Audiences

In identifying a relevant decision maker, successful campaigns seek to distinguish two types of audiences: (1) the **primary audience** is the decision maker who has the authority to act or implement the objectives of a campaign and (2) **secondary audiences** (also called *public audiences*) are the various segments of the public, supporters, coalition partners, opinion leaders, and news media; the support of these constituencies is often pivotal in holding a primary audience or decision maker accountable for the campaign's objectives.

A campaign cannot achieve an objective until someone with the ability or authority to decide on the objective responds favorably. This decision maker is a campaign's primary audience. For example, if the campaign's objective is to prohibit flashing (digital) billboards along state highways, then the primary audience is most likely to be the members of the state legislature's budget or commerce committee. On the other hand, if a campaign wants tighter regulation of emissions of mercury from coal-burning power plants, then the primary audience will be the EPA, which administers the Clean Air Act.

Developing a Strategy to Influence Decision Makers

The third question a campaign asks, therefore, is, "What will persuade these decision makers to act on the group's objective?" This is a quintessential question about strategy. Strategy can be a surprisingly slippery concept and it is often confused with a campaign's tactics, so let's look more closely at this term.

Strategy Versus Tactics

Environmental educator David Orr (1992) once said that questions about strategy land us squarely in the realm of **praxis**, the study of efficient action or the best means to achieve an objective. In communication studies, we further emphasize the importance of critical theory to this study and define *praxis* as an ongoing process of critical theoretical reflection and embodied action in which the two inform each other in order to provide further insight and practice that may improve the world within specific contexts. Whereas critical rhetorics may help us to imagine a desired future, an advocacy campaign goes further and asks, "How do we actually get to this future?" "And how can reflection on previous action and theory help us achieve our goals?" Answering this is the heart of strategy.

The idea of strategy can be confusing, so it might be best to begin with a definition and an example. Simply defined, a **strategy** is a critical source of influence or *leverage* that, if fully implemented, is able to persuade a primary decision maker to act on a campaign's objective. **Leverage** often is said to arise from Archimedes' famous claim, "Given me a place to stand and a lever long enough, and I will move the world," that is, "the application of a certain kind of action (assuming 'a place to stand') produces a dynamic that can move—or leverage—*a much larger force* [emphasis added]" (Cox, 2010, p. 128). Sometimes, specific actions—signing a petition, protests, and so forth—are mistaken for strategy. These are usually what are called *tactics*, not the wider leverage or influence that a campaign may be exerting by using these tactics. **Tactics** are concrete acts that carry out or implement the broader strategy.

A much-studied example of a strategy is the leverage used by environmental and health groups to persuade the multinational fast-food chain McDonald's to change its food purchasing policies. In 2003, McDonald's acknowledged that the use of growth-stimulating antibiotics by large factory farms that raise and sell poultry and beef threatened human health. Many scientists believe that the "indiscriminate use of antibiotics in cows, pigs and chickens raised for meat . . . has endangered human health by fueling the growing epidemic of antibiotic resistance" (Tavernise, 2013, p. A1). In its announcement, McDonald's agreed to phase out its purchase of chickens injected with such growth hormones; this, in turn, added pressure on the poultry industry to begin to change its practices[3] (Greider, 2003; see also Strom, 2014). What, then, influenced McDonald's to institute this change?

Strategy as Leverage: Influencing McDonald's

In the McDonald's case, a coalition of 13 environmental, religious, and public health organizations, including Environmental Defense, the Humane Society, and National Catholic Rural Life Conference, decided a creative use of market forces might be a source of influence or leverage on McDonald's purchasing practices. This strategy drew on the power of consumers to change industry behavior "not by one purchase at a time, but on a grand scale *by targeting large brands in the middleman position* [emphasis added]" (Greider, 2003, p. 8). That is, their campaign chose to influence the behavior of the meat industry by targeting one of the largest purchasers of its products: McDonald's.

Journalist William Greider closely studied the innovative strategy used in the McDonald's campaign. "What has changed [in the case of McDonald's] is an essential strategic insight" (p. 10). Grieder explained,

> Consumers are in a weak position and have very little actual leverage over the content of what they buy or how it is produced. . . . Instead of browbeating individual consumers, new reform campaigns focus on the structure of industry itself and attempt to leverage entire sectors. The activists identify and target the larger corporate "consumers" who buy an industrial sector's output and sell it at retail under popular brand names. They can't stand the heat so easily, since they regularly proclaim that the customer is king. When one of these big names folds to consumer pressure, it sends a tremor through the supplier base, much as McDonald's has. (2003, p. 10)

In the McDonald's campaign, the strategy sought to use the purchasing power of the fast-food giant itself rather than individual customers to influence the poultry industry. By linking the familiar brand and logo of this global icon in the public's mind with health risks from growth hormones in its food, the campaign was able to leverage the buying power of McDonald's to influence the behavior of its suppliers. If the factory farms that sold meat products to McDonald's wanted to continue to do business, they would have to reduce their use of growth hormones in poultry and perhaps in other animals as well.

The campaign to influence McDonald's, and thereby the poultry industry, illustrates the difference between strategy and tactics. In this case, strategy, as a source of influence or leverage, was the use of a powerful corporation's brand and buying power to affect the poultry industry's use of growth hormones in chickens. The tactics that carried out this strategy included the materials distributed to McDonald's, meetings with company officials, organizing of protests outside McDonald's restaurants, and so forth. Each of these was important, but their critical function was to implement the wider strategy—using the vulnerability of McDonald's brand to public pressure and, subsequently, its purchasing power to affect changes in the meat industry. (We define and discuss boycott campaigns and corporate accountability more in Chapter 11.)

For now, we want to emphasize that strategy is easy to overlook or not be clearly understood in designing a campaign. When it is unclear, a campaign usually suffers as a result. In his discussion of this problem, Orr (1992) recalled the cartoon shown in Photo 8.2, which appeared in the journal *American Scientist*. The cartoon shows a scientist who, in balancing an equation, has inserted this curious step: "then a miracle occurs." Orr observed, "Most strategies of social change have similar dependence on the miraculous" (p. 61).

Political theorist Douglas Torgerson (1999) has claimed that a dependence on the miraculous is particularly true of environmental strategies. He argued that a simple,

Reprinted with permission from Sidney Harris, © www.ScienceCartoonsPlus.com

"I think you should be more explicit here in step two."

Photo 8.2	"A miracle needed?" Campaigns often fail when they overlook the important step of identifying a viable strategy for influencing key decision makers.

although cynical assumption sometimes underlies green strategic thought: "Environmental problems are sure to get worse . . . and when they do, more and more people will be moved to join the green cause, thus enhancing its power and its chance of making a real difference" (p. 22). Torgerson believed that such an assumption borders on belief in the miraculous. We agree. Environmental campaigns more often succeed when they identify a source of influence that is able to affect a larger power or decision maker. Exercising that leverage, rather than waiting for a miracle, is the meaning of strategy.

A Campaign's Communication Tasks

Earlier, we defined an advocacy campaign as a strategic course of action, *involving communication*, that is undertaken for a specific purpose. It's good to remember that, in answering the questions about objectives, decision makers, and strategy, a campaign is also engaged in several important communication decisions and tasks. Campaigns not only must articulate a clear objective—*what exactly does it wish to achieve*—but compose persuasive messages and content, carefully target which medium to use, adapt to key audiences, and more. We explore examples of these communication choices later in this and subsequent chapters. For now, let's look at two critical communication tasks for advocacy campaigns: mobilizing key audiences or constituencies and constructing a "message."

Mobilizing Constituencies

In designing a strategy, campaigns often need the support of others—media professionals, opinion leaders, community members, and more—who can exert influence on the primary audience to act on the campaign's objective. The ability to fulfill this task assumes that decision makers, in fact, are ultimately accountable to voters, the media, or other groups. (This assumption goes to the heart of democratic politics and is, itself, a subject of much debate.) Let's look at this task more closely.

In mobilizing the support of others, it is useful, first, to distinguish between the media and opinion leaders, on one hand, and members of the general public, on the other. **Opinion leaders** are those persons whose statements are influential with the media and members of the primary audience. For example, the Natural Resources Defense Council relies upon well-known figures like environmentalist Robert F. Kennedy Jr. and actors Robert Redford or Leonardo DiCaprio as spokespersons, while a smaller, local environmental group may turn to a respected member of that community—a well-respected business owner or local sports star—to speak for the group publicly. Reporters are more likely to quote such opinion leaders in covering a story than ordinary citizens. And, as we saw in our discussion of "agenda setting" (Chapter 5), media coverage of a group's campaign may help to raise the salience or importance of an issue in the eyes of potential decision makers.

When environmental advocacy groups turn to the general public sphere, they confront three very different types of audiences: (1) the campaign's **base** (its core supporters and potential coalition partners), (2) **opponents** (those who strongly disagree and are unlikely to be persuaded), and (3) **persuadables** (members of the public who are undecided but potentially sympathetic to a campaign's objectives). Normally, a campaign does not try to persuade its opponents, since they are committed already to their own objectives. Instead, persuadables often constitute the heart of a campaign's communication because their support often makes the difference in the outcome of the campaign.

It's important to note that, although persuadables may be potential supporters of a campaign, they are usually undecided or unaware, at first, of a campaign's objectives. This was true with the Beyond Coal campaign. As noted earlier, this campaign aimed to close coal-burning power plants, due to their harmful air pollution as well as emissions of carbon dioxide (CO_2), a leading cause of climate change. Campaign persuadables (students, parents, and others living in communities near the power plants) were initially unaware of the impacts of coal plant emissions on either climate change or their families' health, including mercury poisoning and childhood asthma. Nevertheless, the Beyond Coal campaign has viewed them as potentially open to information about these impacts; as a result, it has succeeded in many cases in mobilizing area residents to attend and speak at public hearings about the risks from these power plants.

A Campaign's Message

Both in mobilizing supporters and communicating its objective to decision makers, an advocacy campaign is challenged to create appropriate educational and persuasive content. A particularly important element of a campaign's strategic communication is its message. As developed by many advocacy groups, a **message** is usually a phrase or sentence that concisely expresses a campaign's objective and the values at stake in the goals it seeks. Although campaigns have much information and develop a range of arguments, the message itself is usually short, compelling, and memorable. It accompanies all of a campaign's communication materials, from posters to radio ads to websites. Let's look at two examples of compelling advocacy campaign messages.

First, a number of civil society groups in the United States, Europe, Asia, South America, and Africa have joined together in a "Water Is Life" campaign (http://water islife.com). The advocacy campaign brings clean drinking water and sanitation to those in need in developing countries. By explicitly identifying water with the core value of *life,* the campaign's message signals the urgency of the campaign's advocacy work for many people, animals, and habitats in drought-prone areas. And a second, classic message is "Extinction Is Forever." The message was composed many years ago by the Center for Environmental Education (n.d.) for its campaign for a ban on the commercial hunting of whales—many species were nearing extinction—and it has since been adopted by other groups that campaign to protect endangered species. The message appeals to a sense of the **irreparable**, that is, a warning to act before it is too late or a feeling that, once something of value is lost, it cannot be recovered (Cox, 2001a). "Extinction Is Forever," therefore, evokes powerful feelings about mortality and life itself.

Messages are only one part of a campaign's communication, but they serve an important purpose. They summarize a campaign's objective, state its core values, and provide a frame for understanding and reception of the details of its other informational materials. In developing messages, campaigns attempt to identify values and language that resonate with their base and persuadables—those who may be sympathetic to the campaign's objectives, but are undecided. Because messages are so important to environmental advocacy campaigns, we return to this topic and the role of values in mobilizing supporters later in this chapter.

Evaluating an Environmental Advocacy Campaign

How do we know if a campaign is likely to succeed, or later, why it succeeded? Or, for campaigns that failed, what went wrong? Here are some of the basic questions that may explain why a campaign succeeded or failed to achieve its objective:

Questions about a campaign's basic design:

- Did the campaign identify a clear *objective*? What specific action or decision was required?
- Did the campaign identify a relevant *decision maker*, i.e., someone or group that had the ability to act on the campaign's objective?
- Did the campaign develop a realistic *strategy*, i.e., a source of influence or leverage, able to persuade the decision maker?

Questions about a campaign's communication tasks:

- Did the campaign compose and consistently use a compelling *message* in all of its communication?
- Did this message clearly convey the campaign's core value or objective?
- Did the campaign mobilize its *base* of supporters? Relevant opinion leaders and media?
- Did the campaign identify key *values* in its message and appeals, able to convince "persuadable" audiences and motivate them to act?

In summary, advocacy campaigns usually succeed by identifying a concrete, achievable objective, a decision maker able to respond, and a strategy for influencing this decision maker, along with constructing a compelling message and other content to mobilize the support of relevant constituencies. When environmental advocacy campaigns are designed well, they have several advantages over isolated protests or critical rhetorics:

- By planning a strategic course of action over time, advocacy campaigns increase the chances of reaching more people in the public sphere to help achieve their objectives.
- Advocacy campaigns draw on the collective strength of people and resources for both planning and implementing a course of action, which tends to have a greater impact on public life than what one person can do alone.
- Advocacy campaigns serve as intermediaries between individuals in their private lives and the large, often impersonal, institutions of public life, which forms the basis for democratic social change.

Each of these strengths of an advocacy campaign can be illustrated by looking at a successful campaign in more depth.

The Campaign to Protect Zuni Salt Lake

On August 4, 2003, the Salt River Project (SRP), third-largest electric power company in the United States, cancelled its plans for a coal strip-mine near the Zuni Salt Lake in western New Mexico. The company's announcement was a victory for a coalition of Native American tribes, environmental and religious groups, and the Zuni people themselves who waged a multiyear campaign to protect the sacred Zuni Salt Lake and surrounding lands from mining and other environmental threats.

We use this example because it clearly illustrates the three core elements of design that advocacy campaigns must consider: (1) a clear *objective,* (2) a clearly identified *decision maker,* and (3) a *strategy* to persuade the primary decision maker to act on this objective. The Zuni Salt Lake campaign also illustrates the ability of a small group, working with allies and coalition partners, to use the principles of an advocacy campaign to achieve an important objective—safeguarding a sacred tribal site.

Zuni Salt Lake and a Coal Mine

SRP company's plans called for strip-mining more than 80 million tons of coal from 18,000 acres of federal, state, and private lands. (**Strip-mining** is the removal of surface land to expose the underlying mineral seams.) To settle the coal dust from such mining, SRP planned to pump 85 gallons of water per minute from underground aquifers (Valtin, 2003). The New Mexico Department of Energy, Minerals, and Natural Resources had granted permits for the company to begin construction of the mine in 1996, although work did not immediately begin. By June 22, 2001, opposition to the mine had grown.

To the Zunis and area tribes, the Salt Lake is sacred. It is home to the Zunis' deity *Ma'l Oyattsik'i,* the Salt Mother, who, Zunis believe, has provided salt for centuries for tribal religious ceremonies. (In dry season, the water evaporates, leaving behind salt flats, the source of salt for Zunis and neighboring tribes.)

The region surrounding Zuni Salt Lake is known as the Sanctuary or *A:shiwi A:wan Ma'k'yay'a dap an'ullapna Dek'ohannan Dehyakya Dehwanne.* It has burial grounds and other sacred sites and is laced with trails that are used by the Zunis, Navajos, Acomas, Hopis, Lagunas, Apaches, and other Southwestern tribes to reach the Zuni Salt Lake. By tradition, the Sanctuary is a neutral zone where warring tribes put their weapons down and share in the gathering of "the salt which embodies the flesh of the Salt Mother herself" (Sacred Land Film Project, 2003, p. 1).

The strip-mine would have been located in the heart of the Sanctuary, 10 miles from Zuni Salt Lake. Although the mine itself would not be on Zuni land, tribal leaders feared that the company's plans to pump large volumes of groundwater from the same desert aquifer that feeds the Salt Lake would dry up the lake. Malcolm Bowekaty, former Zuni Pueblo governor, told reporters, "If they vent a lot of pressure that's forcing the water up, we will no longer have the salt" (Valtin, 2003, p. 3).

A Coalition's Campaign

By 2001, Zuni leaders had assembled a coalition that began working together to protect Zuni Salt Lake and the Sanctuary. For two days, the group met informally in the kitchen of a Zuni leader to design a two-year advocacy campaign plan.[4] On November 30, 2001, leaders from the Zuni tribe, Water Information Network, Center for Biological Diversity, Citizens Coal Council, Tonatierra (an indigenous group), Friends of the Earth, Sierra Club, and Seventh Generation Fund for Indian

Development publicly announced the formation of the Zuni Salt Lake Coalition. In what follows, we describe how this campaign embodied the core elements of an advocacy campaign.

Campaign Objectives

From the beginning, the Zuni Salt Lake Coalition saw its goal to "get SRP to drop its plans for the Fence Lake Coal Mine [and] protect Zuni Salt Lake for the long-term" (Zuni Salt Lake Coalition, 2001). More immediately, coalition planners faced the prospect of SRP's imminent preparation of the mine site, including plans to drill into the Dakota Aquifer, which fed Zuni Salt Lake.

Therefore, the coalition identified two immediate, more specific *objectives*: (1) "make sure that SRP does not tap Dakota Aquifer" and (2) persuade the State of New Mexico and the Department of the Interior to deny the permits needed to open the coal mine. If these permits were granted, then the objective would be to appeal these decisions in order to delay actual construction of the mine (Zuni Salt Lake Coalition, 2001). Coalition members believed that, if they could achieve either one of these objectives, they would succeed in persuading SRP to cancel its plans for the project.

Identifying Key Decision Makers

The Zuni Salt Lake Coalition identified two sets of primary decision makers. Ultimately, they sought to persuade SRP officials to withdraw plans for the coal mine. Related to this goal and the campaign's two more concrete objectives, the coalition targeted the U.S. Department of the Interior and the officials in New Mexico who oversaw the state's permitting process.

Strategy: Influencing the Primary Decision Makers

Given its goal to persuade SRP officials to withdraw their plans for the mine, the Zuni Salt Lake Coalition decided the best strategy would be to raise the costs to the company as it pursued permits for the mine. At its very first meeting, the coalition pledged to hold SRP accountable by making "it so hard for them [SRP officials] that they want to drop it. Make them feel that the Fence Lake project is a fruitless effort" (Zuni Salt Lake Coalition, 2001). This core strategy—raising the costs (in time and money) to SRP—would guide subsequent decisions and activities of the coalition.

Specifically, the coalition sought ways to influence SRP and federal officials responsible for issuing the mine's permits (a) by introducing scientific evidence of the ecological effects on Zuni Salt Lake of pumping water from the aquifer and (b) by launching an aggressive outreach to opinion leaders, news media, and New Mexico public officials. (We return to this second, outreach strategy in a moment.) By organizing around these actions, the coalition intended to place continual roadblocks in SRP's path and thereby raise the costs to SRP, increasing pressure on the company to cancel its plans for the coal mine.

The first element of the coalition's strategy was to introduce evidence of environmental damage to Zuni Salt Lake as a basis for challenging the state and federal permits that had been issued.[2] New research was a critical part of the effort to hold the Department of Interior accountable under the National Environmental Policy Act (NEPA) requirement for an environmental impact statement. (We describe the importance of NEPA in Chapter 12.) For example, the coalition argued that "every hydrological study, except SRP's own, shows that this pumping will detrimentally affect the lake" (Zuni Salt Lake Coalition, 2003). Based on its hydrological information (pumping tests), the coalition requested that Interior conduct a supplemental environmental impact study. Similarly, it appealed the state's water permit pending completion of further pumping tests on the aquifer.

Finally, the coalition's threat to file a challenge in the event that Interior failed to consider possible impacts of pumping water from the underground aquifers promised to add delay, and therefore more costs, to SRP's plans to start construction of the coal mine.

Mobilizing Support From Key Constituencies

Also pivotal to the Zuni Salt Lake Coalition's strategy was its plan for an aggressive outreach to opinion leaders, news media, and New Mexico public officials in an effort to influence the primary decision makers. Influencing a powerful utility like SRP may seem unrealistic. SRP officials were not publicly elected; therefore, they were unaffected by voters. Nevertheless, the coalition believed that the company's credibility and ability to secure cooperation (including its permits) depended on a number of key constituencies, including public officials, opinion leaders, and the media; they believed further that some of these groups could be mobilized.

In implementing this outreach strategy, the coalition began with its base—the Zuni people themselves and their allies among area tribes. In addition, it sought to mobilize support from persuadable groups—area churches, environmental groups, and people of faith generally (Zuni Salt Lake Coalition, 2001). In turn, support from these groups ultimately attracted support from opinion leaders, the media, and, in the end, from elected officials.

An important rationale for the outreach strategy was also to respond to SRP's own public relations campaign. (Remember, campaigns unfold in the context of other, competing voices and countermessages.) This strategy supplemented the coalition's work on the permits by keeping the threat to Zuni Salt Lake before the wider public.

At the heart of the outreach strategy were efforts to generate "lots of publicity" (Zuni Salt Lake Coalition, 2001). From 2001 to 2003, the Zuni Salt Lake campaign generated thousands of letters to allied groups, newspapers, and public officials. The coalition also publicized resolutions of support from tribal councils and from the New Mexico Conference of Churches, and it mounted two *fax attacks*—deluges of fax messages—on the Department of the Interior to urge delays in its approval of the permits.

In seeking "lots of publicity," the campaign also employed creative approaches for generating media coverage. Coalition organizer Andy Bessler explained,

> Tribal members have a different approach, which made us think "outside the box." Where the Sierra Club might air a radio spot to convey our message, the Zuni suggested sending runners [from Zuni Pueblo to SRP's Phoenix headquarters]. And where we did run radio ads, we had scripts in English, Spanish, Zuni, Navajo, Hopi, and Apache so the spots could run on tribal radio stations as well as on mainstream stations in Phoenix and Albuquerque. (quoted in Valtin, 2003, p. 3)

Along with the use of traditional runners to generate media coverage, the coalition also scheduled a people's hearing on Zuni Salt Lake in Zuni Pueblo. The event included the showing of a video, updates on the campaign, and a public forum. Over 500 people attended and offered their own testimony. "At the conclusion of the hearing, the sky opened up and let loose a torrential downpour, which the Zuni took as a blessing from heaven" (Valtin, 2003, p. 1). Finally, the coalition won the National Trust for Historic Preservation's listing of the Zuni Salt Lake area as one of America's most endangered places ("Victory," 2003, p. 6).

In mobilizing these supporters, the coalition drew on several important sources for persuasion. Especially relevant to mobilizing its base and other key supporters, the coalition highlighted the spiritual and cultural values associated with the Zuni Salt Lake, the Zuni tribe's history, and the indigenous cultures of this region. Related to these values, the coalition spoke of the *irreparable* nature of the threats to the Zuni Salt Lake. (Earlier, we defined the irreparable as a warning to act before it is too late.) Such a warning implicitly invites an audience to feel that (a) something [the Zuni Salt Lake] is unique or rare and therefore of great value; (b) its existence is threatened or precarious; (c) its loss or destruction cannot be reversed; and, therefore, (d) action to protect it is timely or urgent.

The coalition's media materials reflected these persuasive appeals. For Southwest indigenous peoples, the Zuni Salt Lake and Sanctuary are very powerful places. Zuni Council Member Arden Kucate reminded the coalitions' supporters of these values when he warned of the challenge before them: "We have to start thinking in the traditional way. It is not the earth, it is Mother Earth. Zuni people will not sacrifice our Salt Woman for cheap coal to serve Arizona or California, because she is irreplaceable" (LaDuke, 2002).

Using its support from tribal councils, churches, opinion leaders, and other public constituencies, the Zuni Salt Lake Coalition succeeded in gaining the attention of news media and, ultimately, in enlisting the support of public officials in New Mexico. (We return to these developments.)

Communication Message

Finally, at the heart of all its communication, the coalition's campaign reiterated a memorable and compelling message: "*SRP Is Targeting Our Sacred Lands. Save Zuni Salt Lake.*" This message combined a statement of threat—"Targeting Our Sacred

Lands"—and a call to action—"Save Zuni Salt Lake." As a result, the message reinforced the sources for persuasion used by the coalition—the values associated with the Salt Lake and Zuni heritage, and the harm from strip-mining. For example, religious leaders sent postcards to the SRP president, emphasizing respect for sacred sites and the potentially irreparable danger to Zuni Salt Lake. The postcard read "People of faith don't want any sacred areas to be desecrated by a strip mine . . . for cheap electricity from dirty coal: Not the Vatican, not Mecca, not Temple Square in Salt Lake City . . . and not Zuni Salt Lake."

One of the creative ways the campaign kept its message before the public was a panel truck with the words, "SRP Is Targeting Our Sacred Lands. Save Zuni Salt Lake" prominently displayed on its side. (See Photo 8.2) The panel displayed a large photo of Zuni Salt Lake with a rifle's crosshairs on it and, in large letters, the campaign's message. Bessler recalled, "We drove the truck . . . all over Arizona and New Mexico to tribal pueblos, and we got a lot of people to sign petitions" (quoted in Valtin, 2003, p. 3).

The coalition's effort to frame the news (Chapter 5) of the controversy as a struggle over spiritual and ecological values began to pay dividends. In July 2003, the entire U.S. Congressional delegation from New Mexico sent a letter to the Secretary of the Interior, asking her to stop the mining permit until new studies of the aquifer could be completed. Their letter also made clear that they were planning to bring a lawsuit under NEPA if Interior officials refused to prepare a supplemental environmental impact statement. The prospect of a lengthy lawsuit threatened to delay SRP's plans even further, continuing the campaign's overall strategy of making it "so hard for them that they want to drop" plans for the strip-mine.

Courtesy of the Zuni Salt Lake Coalition

| Photo 8.3 | Members of the Zuni Salt Lake Coalition pose beside the mobile billboard truck used in their campaign. |

Success for Zuni Salt Lake

On August 4, 2003, SRP announced that the company had canceled its plans for the coal mine and would also relinquish its permits and the coal leases it had acquired for the mine. This was a rare victory for indigenous peoples and environmental groups, since such energy projects usually proceed. For this reason, we believed that study of this environmental justice campaign is worthwhile.

After the announcement by SRP, Zuni Tribal Councilman Arden Kucate led a delegation to the edge of Zuni Salt Lake to pray and to make an offering of turquoise and bread to *Ma'l Oyattsik'i*, the Salt Mother. Back at Zuni Pueblo, the tribe's Head Councilman Carlton Albert expressed his feelings of relief and appreciation to coalition partners who had worked with the Zuni Salt Lake campaign: "It has been a long . . . struggle . . . but we have had our voices heard. . . . If there is a lesson to be learned it is to never give up and [to] stay focused on what you want to accomplish" (Seciwa, 2003, p. 2).

Act Locally!

Design an Environmental Advocacy Campaign for Your Campus or Community

What is one important step that your campus or community can take to support environmental values? Convert the university's fleet of cars and trucks to biofuels? Reduce the use of paper? Divest portions of your school's portfolio that is invested in fossil fuels?

Work with (or start) a campus or community group to design a campaign that pursues a specific, achievable objective (for example, persuade campus officials to phase out the use of coal in generating energy by a deadline: 2025? or convince a business or community center to start using solar energy within five years?).

In designing this campaign, how would you answer these questions?

1. What *exactly* do you want to accomplish?

2. Who has the ability to respond?

3. What will influence this person or authority to respond?

In designing your strategy, ask, "What groups are likely to be our base of support?" "Who will be our coalition partners?" "Who are our persuadables?" "What message and other communication materials will be required to perform the related communication tasks of mobilizing support to influence decision makers?"

As an exercise, you might answer these questions in the form of a proposal to submit to a campus or community group interested in pursuing such a campaign.

Message Construction

The Zuni Salt Lake advocacy campaign succeeded in mobilizing area tribes, churches, elected officials, and others whose support was critical to their success. Yet, this is not always possible. In some cases, advocates may succeed in changing audiences' beliefs or attitudes but fail to mobilize them or change their behaviors. This disconnect between people's attitudes and their behaviors is called the *attitude–behavior gap* and is a major challenge facing advocacy campaigns.

In this final section, therefore, we describe this attitude–behavior gap and the question that environmental campaigns must address: How can advocates construct messages or persuasive appeals that will mobilize or influence their audiences *to act* in support of the campaign?

The Attitude–Behavior Gap and the Importance of Values

The Attitude–Behavior Gap

People generally have high regard for environmental amenities such as clean air and water, chemical-free food, parks, and open space. Yet, these attitudes do not always predict what people actually will do. Kollmuss and Agyeman (2002), for example, found that we tend to engage in those environmental behaviors that demand the least cost, not only in money, but in "the time and effort needed to undertake a pro-environmental behavior"; that is, while many of us recycle (low-cost), we may "not necessarily engage in activities that are more costly and inconvenient such as driving or flying less" (p. 252).

Social scientists call this disconnect the **attitude–behavior gap**. The gap refers to the fact that, although individuals may have favorable attitudes or beliefs about environmental issues, they may not take any action; their *behavior*, therefore, is disconnected from their *attitudes*. Literary critic Stanley Fish (2008) provides a self-confessed example of this gap:

> Now don't get me wrong. I am wholly persuaded by the arguments in support of the practices I resist. . . . But it is possible to believe something and still resist taking the actions your belief seems to require. . . . I know that in the great Book of Environmentalism my name will be on the page reserved for serial polluters. But I just can't get too worked up about it. (para. 9)

Like Stanley Fish, we may, for example, be convinced that using disposable paper cups is bad for the environment, but resist doing anything about it (e.g., bringing a reusable mug to the coffee shop or carrying a reusable water bottle). More troubling, scholars have found that, while many individuals believe global climate change is real and happening now, they may not feel any urgency to change their own behaviors or speak out (Moser, 2010). This gap is also seen in consumer behavior. Research by OgilvyEarth (2011) found a "green gap" in Americans' buying behavior: Although "82% of Americans have good green intentions . . . only 16% are dedicated to fulfilling these intentions" (para. 3).

The difficulty in changing people's behavior has been a concern, especially, in public information campaigns encouraging consumers to save energy or install energy-efficient appliances in their homes. In recent years, utility companies have expended enormous time and energy trying to improve energy efficiency in homes and businesses, often with limited success. Merrian Fuller, a researcher at the Lawrence Berkeley National Laboratory, explained, "Convincing millions of Americans to divert their time and resources into upgrading their homes to eliminate energy waste, avoid high utility bills and help stimulate the economy is one of the great challenges facing energy efficiency programs around the country" (quoted in Mandel, 2010, para. 8).

One of the reasons that behavior-change campaigns often fail is that they assume that providing information—educating people—is enough. Simply knowing that better insulation in our attics will save us money on our energy bills, for example, is usually not enough to persuade us to purchase (and install) higher R-rated (energy-efficient) insulation. The reason, Fuller explained, is that, when information campaigns "address the issue of energy efficiency benefits, they . . . neglect the issue of *how to motivate consumers* [emphasis added]" to actually take action (quoted in Mandel, 2010, para. 9). The results of the Lawrence Berkeley study point to the importance of emotional, or affective, as well as educational elements, in designing messages for a campaign that expects people to take an action as a result of the campaign's communication. (We return to an example of such a campaign below.)

Campaigns do succeed (sometimes) in persuading people to change their behaviors. Although many factors are involved, an important component in successful campaigns is the construction of messages that are framed in terms of values that are important to those the campaign is aiming to reach.

Values and Pro-Environment Behavior

While our beliefs often don't directly influence our behaviors, our values and cultural norms do play a role. Indeed, there is a great deal of evidence that pro-environmental behaviors are related to certain values (Crompton, 2008; Schultz & Zelezny 2003). In an earlier, classic study of the environmental movement, Stern, Dietz, Abel, Guagnano, and Kalof (1999) found that "individuals who accept a movement's basic values, believe that valued objects are threatened, and believe that their actions can help restore those values" are more likely to feel an obligation to act or provide support for the movement (p. 81).

Let's look at the different values that environmental advocacy campaigns sometimes consider in their messages.

Recent research suggests that there are three broad categories of values associated with environmental behaviors (Farrior, 2005, p. 11):

1. Egoistic concerns focusing on the self (health, quality of life, prosperity, convenience)

2. Social–altruistic concerns focusing on other people (children, family, community, humanity)

3. Biospheric concerns focusing on the well-being of living things (plants, animals)

Some people may be concerned about water pollution because of the dangers to themselves (for example, "I don't want to drink polluted water"). Others may be motivated by social–altruistic concerns about their children or communities ("I don't want my children to drink polluted water"). Finally, others may be concerned about the effects of polluted water on marine animals, living coral reefs, that is, they are motivated by wider, biospheric concerns.

What category of values is predominant? The answer may differ among different groups, regions, or cultures. For example, an international survey among college students found that social–altruistic values rated the highest. The survey, however, found that students in different countries differed about the importance of other values. In the United States, a majority rated egoistic concerns higher than biospheric, while students in Latin American countries placed biospheric concerns higher than egoistic (Schultz & Zelezny, 2003, pp. 129–130).

This finding presents an interesting dilemma for some advocates in choosing the values they'll use in their campaign's messages. For example, in arguing for the value of wilderness, the radical group Earth First! (2014) rejects all self-interested rationales for wilderness, such as recreation or medicines from native plants. Instead, the group voices a clear, biospheric value in its messaging. In stating there should be "No Compromise in Defense of Mother Earth," the group explains,

> Guided by a philosophy of deep ecology, Earth First! does not accept a human-centered worldview of "nature for people's sake." Instead, we believe that life exists for its own sake, that industrial civilization and its philosophy are anti-Earth, anti-woman and anti-liberty. . . . To put it simply, the Earth must come first. (paras. 5–6)

Earth First!, therefore, potentially faces a dilemma: Can appeals to biospheric values still gain a hearing from those motivated principally by social-altruistic or self-enhancement values? Or must wilderness advocates appeal to individuals' egoistic concerns or their social–altruistic values to mobilize support from a wider public?

Responses to the dilemma about values have varied. The Biodiversity Project (now Bluestem Communications) in the U.S. Midwest, for example, has recommended that campaigns to protect natural resources be based on "messages that address socio–altruistic concerns or make biodiversity relevant to everyday life" (Farrior, 2005, p. 11). The key is using a "diversity of messages that will appeal to people with a different range of value orientation" (p. 11, quoting Schultz & Zelezny, 2003, p. 134). Similarly, a survey in the UK of values motivating people to adopt lower-carbon lifestyles found that, although biospheric values were important, more participants in the survey rated altruistic values significantly higher; for example, they expressed concern "about the plight of poorer people who will suffer from climate change" (Howell, 2013, p. 281).

As the Biodiversity Project and UK survey make clear, a campaign must choose values that potentially motivate or influence an audience, and it does this in the construction of its message.

Message Construction: Values and Framing

As we noted earlier in this chapter, a campaign's strategy has an important communication task—the identification of the appropriate educational and persuasive messages, spokespersons, and media for communicating with the campaign's supporters and primary audience. We described a *message* as a phrase or sentence that succinctly expresses the campaign's objective and, sometimes, the values that are at stake. It is usually compelling, memorable, and is used in all of a campaign's communication materials.

A campaign's message, therefore, can play a pivotal role in addressing the attitude–behavior gap that occurs for many audiences. This is particularly true when a campaign refers, in its message, to an important value that its audience perceives as threatened, such as their health or a natural area that has special meaning. Let's look more closely, therefore, at the role of values in a campaign's messaging and the ways in which campaigns frame these messages.

Framing and a Campaign's Values

The role of values in environment campaigns can be illustrated in the messages used by many U.S. environmental groups in their fight against pollution from coal-burning power plants. The *New York Times* reported that many of these groups have been "stepping up their transition toward a new health-centric message" (Schor, 2011, para. 1). The Sierra Club's online campaign, for example, led with this statement:

> Not only is coal burning responsible for one third of US carbon emissions—the main contributor to climate disruption—but it is also *making us sick*, leading to as many as *13,000 premature deaths* every year and *more than $100 billion in annual health costs* [emphasis added]. (Sierra Club, 2014, para. 1)

Similarly, the Natural Resources Defense Council (NRDC) and other groups commissioned a dramatic photo-ad showing a young girl wearing a respirator and the message "It's our air, but big polluters treat it like they own it. They dump millions of tons of dangerous pollution into our air, threatening the health of all Americans" (NRDC, 2014, para 1).

Words and phrases such as "*making us sick*," "*premature deaths*," "*health costs*," and "*dangerous pollutants*" and a photo of a child with a respirator are intended to evoke a powerful frame in the minds of a campaign's target audience. In Chapter 3, we described a *frame* as a cognitive map or pattern of interpretation that we use to organize our understanding of reality. It is important to understand that frames are not just words but are deeper, often unconscious, mental structures. Cognitive linguist George Lakoff (2010) claims that "All of our knowledge makes use of frames, and every word is defined through the frames it neutrally activates" (p. 71). Moreover, he argues, many of our "frame-circuits have direct connections to the emotional regions of the brain" (p. 72).

A campaign's message, therefore, must evoke an existing and emotionally relevant frame to be successful. Environmentalists' campaigns against coal-burning power plants, as we saw, used words that were designed to evoke an existing, powerful frame for many Americans—a concern for their health and, especially, the health of their children.

Framing an Energy Savings Message

Let's illustrate the use of emotionally relevant frames in one of the campaigns to convince "millions of Americans" to save energy, which we mentioned earlier.

After surveying the best practices of 14 home energy-efficiency programs in the United States, researchers at Lawrence Berkeley National Laboratory concluded that, in urging consumers to save energy, it is not enough to provide information. An information campaign's communication must address something people value—"health benefits, improved comfort, community pride, or other benefits that consumers tend to care about" (Fuller et al., 2010, p. 2). Researchers, therefore, recommended that the energy campaigns spend time

© Survival International

Photo 8.4 Some advocacy campaigns take years or decades to achieve their goal. The Dongria Kondh tribe in India declared victory in 2014 after almost a decade-long campaign to thwart Vedanta Resources from pushing them off their land in order to mine bauxite. International groups such as Survival and Amnesty International worked in a coalition with local organizations and people to amplify Dongria's campaign.

studying their audience—residential customers—and "tailor messages to this audience" (Fuller et al., p. 2). In constructing their messages, they advised the programs to "avoid meaningless or negatively associated words like 'retrofit' and 'audit.' [Instead,] use words and ways of communicating that *tap into customer's existing mental frames* [emphasis added]" (Fuller et al., p. 2).

One of the successful programs studied by the Berkeley Lab was an experiment in rural, conservative Kansas, urging people to save energy. "Don't mention global warming," Nancy Jackson warned. "And don't mention Al Gore. People out here just hate him" (quoted in Kaufman, 2010, p. A1). Jackson heads the Climate Action and Energy Project, a nonprofit group in Kansas whose goal is to persuade people to reduce fossil fuel emissions that contribute to climate change. But any talk about climate change or global warming is very unpopular in rural Kansas. So, how could the project go about constructing a message to persuade people to cut their use of oil or coal-generated electricity?

Jackson felt that saving energy was a different matter. The project's message could separate this from appeals about stopping global warming. She explained, "If the goal was to persuade people to reduce their use of fossil fuels, why not identify issues that motivated them instead of getting stuck on something that did not?" (Kaufman, 2010, p. A4). The project, therefore, commissioned a study of independent voters and Republicans in the area around Wichita and Kansas City to identify what people cared about, what worried them, or what values motivated them.

Based on its study, the project ran an experiment "to see if by focusing [its messages] on thrift, patriotism, spiritual conviction, and economic prosperity, it could rally residents of six Kansas towns to take meaningful steps to conserve energy and consider renewable fuels" (Kaufman, 2010, p. A4). It adapted its message in a number of ways as it worked with civic leaders, churches, and schools. For example, Jackson talked with civic leaders about jobs in renewable energy, such as wind power, as a way of boosting local economies.

Jackson also spoke to Kansas ministers about "Creation Care," the duty of Christians "to act as stewards of the world that God gave them" (Kaufman, 2010, p. A4). And, importantly, Jackson used the appeal of thrift to persuade the six towns to compete to see which could save the most energy and money. As part of the competition, for example, schoolchildren "searched for 'vampire' electric loads, or appliances that sap energy even when they seem to be off," and towns' restaurants served meals by candlelight for Valentine's Day. The project discovered, while many of the towns' residents believed global warming was a "hoax," they cared about "saving money"; as one man explained, "That's what really motivated them" (quoted in Kaufman, p. A4).

By the end of the first year of the experiment, the project saw signs of success. Overall, the six towns experienced energy savings of more than 6 million kilowatt-hours (Fuller et al., 2010, p. 13). This amounted to a decline in energy use in the towns "by as much as 5 percent relative to other areas—a giant step in the world of energy conservation" (Kaufman, 2010, p. A4).

Finally, a reminder: While constructing the campaign's message is important, a message, even if powerful, cannot alone succeed in achieving a group's ultimate objectives. Messages must always be aligned with other aspects of the campaign—its objective, key audiences, and so on. In other words, messages help to implement a campaign's overall strategy—its mode of leverage or influence—that is designed to persuade the primary decision makers who are able to act on the campaign's objective.

Another Viewpoint: Framing or Organizing?

Sociologists Robert Brulle and Craig Jenkins disagree with George Lakoff and others' views of framing and its importance for environmental advocacy. They argue that simply reframing an issue linguistically without addressing the basic causes of political and economic change won't alter entrenched power. Satirizing Lakoff's views, they write,

Social reality is defined simply in terms of how we *perceive* reality. If we just get the right frames out there, it will create political consensus, and the progressive alliance can then take power. However comforting this idea might sound, it is a form of linguistic mysticism that assumes that social institutions can be transformed by cultural redefinition alone. . . . The structure of power has to be changed as part of the process, and any rhetorical strategy that promises to be effective must link its rhetoric to a broader political strategy that includes grassroots organizing at its base. . . . Although better framing would be useful, alone it can do little. We need to move beyond simplistic analyses and clever spin tactics. What is needed is a new organizational strategy that engages citizens and fosters the development of enlightened self-interest and an awareness of long-term community interests. (pp. 84, 86).

SOURCE: Robert J. Brulle & J. Craig Jenkins. (2006, March). Spinning Our Way to Sustainability? *Organization & Environment, 19*(1), 82–87.

SUMMARY

In this chapter, we focused on the environmental advocacy campaign, its key characteristics, and the importance of constructing campaign messages that resonate with the values of a campaign's audience.

- In the first section, we defined the *advocacy campaign* as a strategic course of action, involving communication, that is undertaken for a specific purpose, and we distinguished campaigns from *critical rhetorics,* the general questioning or criticism of the status quo.
- In the second section, we outlined the basic elements of advocacy campaigns:
  - Goals and objectives
  - Identifying key decision makers and audiences
  - Developing a strategy

- And the communication tasks of an advocacy campaign; among these are
  - ○ mobilizing constituencies, and
  - ○ including a campaign "message" in all communications.
- In the third section, we described a successful environmental justice advocacy campaign—the Zuni Salt Lake Coalition's campaign against plans to strip-mine for coal near sacred Native American lands.
- Finally, we discussed the role of message construction in campaigns, including the importance of key values of an audience; we also identified two challenges:
  - ○ Overcoming the attitude–behavior gap
  - ○ Identifying values in the construction of messages that motivate key audiences

It is our hope that, when you have finished reading this chapter, you will appreciate some of the elements important in designing an advocacy campaign, as well as the role of critical rhetorics in questioning existing practices and ideologies. As a result, we hope you will feel inspired to work with others on your campus or in your community to do extraordinary things.

SUGGESTED RESOURCES

- Roger Kaye, *Last Great Wilderness: The Campaign to Establish the Arctic National Wildlife Refuge*. Fairbanks: University of Alaska Press, 2006.
- *A Fierce Green Fire: The Battle for a Living Planet* (Synopsis, trailer, and DVD at: http://afiercegreenfire.com)
- 350.org: "We're building a global movement to solve the climate crisis." See this site for current campaigns, projects, and resources for grassroots actions in as many as 188 countries.
- Simon Cottle and Libby Lester (Eds.), *Transnational Protests and the Media*. New York: Peter Lang, 2011.
- For images, film, and history of the Zuni Salt Lake and the campaign to protect it, see www.sacredland.org/zuni-salt-lake.
- Ronald E. Rice and Charles K. Atkin (Eds.), *Public Communication Campaigns* (4th ed.).Thousand Oaks, CA: Sage, 2012.

KEY TERMS

Advocacy 177

Advocacy campaign 180

Attitude–behavior gap 198

Base 189

Communication tasks (of a campaign) 183

Critical rhetoric 179

Goal (of a campaign) 184

DISCUSSION QUESTIONS

1. A common perception of strategy is that, with the worsening of environmental problems, people will wake up and begin to take action. Is this an accurate view? What would it take to wake people up to be really effective?

2. Do you, as a consumer, have power to affect environmental change? Journalist William Greider (2003) says that consumers are in a weak position and have very little actual leverage over the actions of large corporations. Do you agree?

3. Can advocates for wilderness or endangered species appeal to biospheric values and still gain acceptance from the audiences they must persuade? Or must they appeal to the egoistic concerns of individuals or to their social–altruistic values to gain a hearing or mobilize support?

4. How effective is the framing of a campaign's message? Do you agree with Brulle and Jenkins ("Another Viewpoint") that simply reframing an issue without addressing political and economic change won't alter entrenched power?

NOTES

1. For more information on the success of this campaign, see http://ecocycle.org/bestrawfree.

2. For more information about the campaign, see http://content.sierraclub.org/coal/campuses.

3. Environmental, public health, and consumer groups have continued their pressure on meat producers' uses of antibiotics. In 2014, one of the largest U.S. poultry producers announced it would "no longer use antibiotics in its hatcheries, one of the last places it was using such drugs routinely" (Strom, 2014, p. B3).

4. In describing the Zuni Salt Lake Coalition's campaign, we are indebted to meeting notes of the coalition and its campaign materials and to Andy Bessler, a coalition member and environmental justice organizer, who worked for the Sierra Club and generously shared his recollections of the campaign in a personal interview with Cox, September 24, 2003.

In 2014, Greenpeace tweeted 20 time-lapse photos, including these images of deforestation in Brazil. Tropical deforestation is a major contribution to global climate change (https://twitter.com/Greenpeace/status/502066581417885697).

CHAPTER 9

Digital Media and Environmental Activism

From Romania to Peru to Canada, [environmental] protest movements have disrupted projects in recent years, in part because activists have harnessed the power of social media and mobile technology. . . . Civil unrest can spell disaster for mining projects. . . . That is not new. What has changed is activists' ability to mobilize, a trend that echoes political upheavals that social media have helped fuel across the Middle East and North Africa.

—Blog post about activists' opposition to
Europe's biggest open-pit gold mine
(Martell & Patran, 2014, para. 1, 2)

When Greenpeace launched its "Unfriend Coal" campaign against Facebook—on Facebook itself—it persuaded thousands of users worldwide to "unfriend" the social networking giant for using coal to fuel its operations. Soon after, Facebook announced "a 2015 goal of powering 25% of the platform with renewable energy" (Pomerantz, 2012, para. 2; see also Carus, 2011). Then, Greenpeace turned attention towards Apple's iCloud and then Amazon for a similar reliance on coal.

From activists' posts on Facebook, Twitter, Instagram, and other social media to use of smartphones as an essential tool for organizing, digital media are opening new ways to circulate arguments, record events, issue calls to action, and network. In this chapter, we expand our earlier description of environmental advocacy campaigns (Chapter 8) by surveying—and giving examples of— environmental and climate justice activists' uses of digital communication technologies.

Chapter Preview

- The first section previews a range of digital communication technology uses by environmental advocates to consider how they allow us to alert, amplify, and engage. We introduce concepts of hypermediacy, remediation, and digitally mediated social networks to help us understand the dynamics of digital environmental activism.
- Part two surveys the extensive uses of digital media by environmental nongovernmental organizations (NGOs), as well as several obstacles to such uses, including NGOs' self-representation online, their mobilizing of public support, and their building "public will" for environmental protection.
- In the third section, we describe efforts by environmental NGOs to go beyond a linear or "broadcast" model of digital communication to a "network" model or the digital infrastructure and dynamics of larger-scale interactions that are enabling diverse individuals to engage with others in coordinated actions offline as well as online.
- In the final section of this chapter, we illustrate the impact of digital networking in organizing the "Peoples Climate March" on the eve of the United Nations' meeting on climate change in late 2014, when online activism moved "into the streets."

Grassroots Activism and Digital Media

From social media alerts of oil spills to community-generated hashtags to coordinate protests at UN climate summits, digital media are transforming the role and impacts of environmental and climate activism. One result has been that, "growth in networked digital communications technology . . . and use since the 1990s has helped to change the conditions for visibility in environmental politics" in many societies around the world (Lester & Hutchins, 2012, p. 848). In this section, we review some of these uses and the factors that help to explain their impact.

Alert, Amplify, and Engage

As we noted in Chapter 8, environmental advocacy campaigns may have a range of objectives. Some aim to raise awareness to new issues; still others seek to engage their supporters actively in persuading public official to act on their demands. Let's look at different uses of digital media as activists pursue these ends.

Alert

The simple acts of pointing and naming something out there in the world, we said in Chapter 3, are "the basic entry to socially discerning and categorizing parts of nature" (Milstein, 2011, p. 4). And, in doing so, our naming something as a "problem" also indicates an orientation to the world and therefore "influences our interaction with it" (Oravec, 2004, p. 3). Perhaps in no other way has digital media had such a powerful impact as this simple act of pointing to and showing some event or condition "out there" as an environmental problem.

For example, in Beijing, some Chinese citizens have taken to social media to share images and document the severity of the worsening air pollution in the city, showing people walking in face masks and grey skies. "Online posts from celebrities and ordinary people turned China's poisonous smog from a little discussed issue to a top-level concern, and helped push officials to acknowledge the problem" (Gardiner, 2014, para. 3). And in other cities around the world, social media have "given committed air-quality campaigners a powerful tool for drawing attention to an issue whose profile remains relatively low despite its big impact on urban dwellers' health" (Gardiner, 2014, para. 4).

Drawing attention to a problem, obviously, is only a beginning. Going beyond alerts that hope to raise awareness, many activists, community groups, and environmental organizations also are incorporating mobile and web-based technologies into their local activism and advocacy campaigns.

Amplify

As we noted in Chapter 4, when the *Deepwater Horizon* oil rig caught fire in 2010, killing eleven workers and spewing oil into the Gulf of Mexico, visual testimony of the disaster—including photographs, videos, and live-feeds—immediately shaped how the event was witnessed and publicized. In response, "social networking sites like Facebook and Twitter [were] alight with posts from government, the

©iStockphoto.com/Jen Grantham

| Photo 9.1 | Activists using their mobile phones to record images of a police car that had been lit on fire during a protest of the G20/G8 summits in Toronto, Canada, in 2010. |

oil industry and citizens around the globe" (Rudolf, 2010, para. 2). The disaster kept many riveted to dramatic images, streamed on countless blogs, of oil gushing from the ruptured wellhead (Fermino, 2010). As a result, environmentalists, individuals, and others hammered the main culprit, the BP oil corporation, "all over the Web, with thousands of posts on Twitter and Facebook slamming the company for its failure to prevent the disaster and its inability to stanch the flow of oil" from the well (Rudolf, 2010, para. 4).

The Sierra Club also responded by initiating an advocacy campaign to go "Beyond Oil."

In mobilizing and coordinating the activities of its members and the general public, the campaign made use of social media and its own online resources in a number of ways:

- Tweeting and live blogging during the weeks that the ruptured wellhead spewed oil, to alert and inform members
- Launching a new website (beyondoil.org) that generated thousands of public comments to President Obama
- Using Sierra Club's Convio system (a constituent services application) to organize house parties around the United States, recruiting new supporters and generating comments to public officials
- E-mailing news feeds urging reform of offshore drilling to media outlets, quoted in news blogs, editorials, and business sections of newspapers (online and print)
- Using the location-based platform Gowalla to create a trip, "BP Oil Disaster," and then cataloging sites impacted by the spill and coordinating these locations with activists visiting the Gulf Coast
- Building a separate website housing photos and stories from events protesting the spill around the United States to sustain interest and enthusiasm

Many causal factors, of course, were involved in bringing about reform of offshore oil policies in the aftermath of the Gulf oil spill. Nevertheless, the public criticisms and coordinated advocacy campaigns by environmental, public heath, and consumer groups illustrate the expanding capacity of digital media and Web 2.0 applications to generate political pressure on public officials in a way that amplifies what is happening offline.

Globally, for example, environmental activists have relied on social media to gain support and pressure public officials to curb pollution in urban areas. In Krakow, Poland, for example, "lawmakers recently banned the burning of coal to heat homes after activists pressured them with a Facebook campaign that drew 20,000 followers" (Gardiner, 2014, para. 1). The small group of activists said Facebook, Twitter, and other social media had "given them an invaluable tool for amplifying their voices, engaging the public and getting the attention of elected officials" (para. 5). One of the activists, Andrzej Gula of the Krakow Smog Alert, explained "When you get many supporters on Facebook and in social media, you become a real power, and politicians want to talk to you because they see they can reach voters" (Gardiner, para. 5).

Engage

Finally, we should note that Web 2.0 technology also has enabled another type of grassroots activism. **Self-initiating** refers to the ability of individuals, through what are often called *bottom-up* online sites, to instigate social change via social media that engage others. **Bottom-up sites** provide easy-to-use online tools that enable users to start petitions on platforms like Facebook or Twitter, for example. Dedicated sites at Moveon.org (http://petitions.moveon.org), Care2 (care2.com), and Change .org also have similar online tools. (As we're writing this chapter, Care2, for example, a site dedicated to animal welfare, human rights, environment, LGBT, health, food, and other causes, claims 7,443,410 people are initiating or engaging with its online petitions monthly.)

Although hosting a variety of issues, traditionally these websites have enjoyed a steady rise of individuals concerned about the environment. One Change.org petition, for example, succeeded in bringing pressure to bear on the Food Network television program to stop featuring shark as food:

> Sadly, according to the International Union for the Conservation of Nature, as many as one-third of all shark species are threatened with extinction or are close to becoming threatened. Because shark populations are in such dire straits, there's no reason for people to be eating these threatened swimmers. . . . Tell Food Network that you want a policy stating that shark will not be featured as an ingredient on air, in the magazine, or in recipes posted on Food Network-owned Web sites. (Belsky, 2011a, para. 2)

Fewer than 10 days after Jessica Belsky initiated her "Stop Featuring Shark" petition on Change.org, over 30 thousand people had signed, and "the media company removed all shark recipes from its Web site and committed to not use shark as an ingredient in the future" (Belsky, 2011b, para. 1).

Affordances of Digital Communication Technologies

At the same time that activists are expanding their uses of digital technologies, communication scholars are exploring the complex processes by which mobile and web-based platforms and pathways for circulation, exchange, and co-creation of content affect users/receivers' perceptions and behaviors. As environmental communication scholars Kevin DeLuca and Jennifer Peeples (2002) remind us, "new technologies introduce new forms of social organization and new modes of perception" (p. 131). In the following, we introduce three concepts—hypermediacy, remediation, and digitally mediated social networks—that clarify the impacts of activists' uses of digital technologies.

Hypermediacy

In their study of the protests against the World Trade Organization (WTO) by environmental, labor, and human rights activists in Seattle in 1999, DeLuca and Peeples used the idea of the "public screen" to describe activists' ability to adapt so effectively to "a wired society" (2002, p. 125). The public screen, they argued, is a scene of **hypermediacy**, that is, a heterogeneous space, in which representations of events "open on to other representations or other media" in which mediations themselves are multiplied (Bolter & Grusin, 1999, quoted in DeLuca & Peeples, 2002, p. 132). Such heterogeneous spaces or media include multiple and interrelated images, text, graphics, sound, and hyperlinks. Such hypermediacy allows activists what information theorist Manuel Castells (2012) observes is "constant reference to a global hypertext of information whose components can be remixed" (p. 7). (See, for example, Greenpeace's tweeting of 20 climate photos showing #ClimateChange, in the chapter opening photo).

Remediation

A second helpful concept in considering the affordances (Chapter 4) of digital communication technologies is the idea of **remediation**, which is "a defining characteristic of the new digital media" (Bolter & Grusin, 2000, p. 45). Remediation is a process where older media (such as a 35 mm film) are re-presented or refashioned in a new medium (for example, in a web video). Paired with new, digital technologies, remediation also accounts for the multiplication or going "viral" of messages, images, posts, or videos, as these are disseminated via different media and spread across digital networks.

One of the better-known, recent examples of remediation is the animated video, *The Meatrix*, a biting critique of what is called **factory farming**—the practice of large-scale, industrial agriculture known for severely restricting the movements of, and space in which chickens, pigs, calves, and other animals are kept and/or reproduce. A riff on the popular movie *The Matrix*, the video was launched online in 2003, and immediately spread. Translated into more than 30 languages, *The Meatrix* is "one of the most successful online advocacy campaigns ever" (Grace Communication Foundation, 2015, para. 1). Its website features *Meatrix II* and other *Meatrix* videos, and a 360-degree interactive tour of a factory farm; it also enables viewers to "put banners, links and graphics on [their] blog, website or social networking site" ("Spread the Word," 2014, para. 2).

In his study of *The Meatrix*, environmental communication scholar Dylan Wolfe (2009) attributed the video's ability to attract and engage more than 20 million viewers online to a "rhizomatic view of *The Meatrix* assemblage," that is, its hypermediacy and specific mode of dissemination (p. 329). In using the metaphor of a "rhizome"[2] to describe this mode, Wolfe argues that the circulation of the animated video online resembled "a distributed system, an open and shifting constellation of intertextual, disseminating, and user producing relationships" (p. 329). (For more, see themeatrix.com.)

> ### ☞ FYI Satirizing Coal via Hypermediacy and Remediation
>
> Activists throughout history have used media all of types—from pamphlets to television—to criticize corrupt politicians and other powerful interests. With Facebook, Twitter, Instagram, and other social media, the reach of public scrutiny and criticism has accelerated dramatically. Some of the most biting environmental criticism online is characterized both by hypermediacy and multiple remediations.
>
> Recently, several activists, calling their group "Coal Is Killing Kids," have been embarrassing Peabody Energy, the giant coal corporation, with a satiric website "Coal Cares" on Facebook and Twitter. The mock Peabody Energy website (coalcares.org) declares,
>
> > Coal Cares is a brand-new initiative from Peabody Energy, the world's largest private-sector coal company, to reach out to American youngsters with asthma and to help them keep their heads high in the face of those who would treat them with less than full dignity. For kids who have no choice but to use an inhaler, Coal Cares™ lets them inhale with pride. (Hecht, 2011, para. 2)
>
> The website for Coal Is Killing Kids showcases images of a young boy, stretching his arms wide at the beach, and the words, "Breathe *easier* with a Free INHALER FOR YOUR CHILD!" and satirically announces "Puff-Puff™ inhalers are available free to any family living within 200 miles of a coal plant." A link then takes you to Coal Cares' list of inhalers with cartoon images, including Dora the Explorer (ages 0–3), Batman (ages 3–8), and Justin Bieber ("The Bieber") for ages 9–13. After its launch, the website quickly went viral on other sites and blogs.
>
> A spokesperson for the activist group explained that the site was created "to highlight . . . that pollution that comes from coal plants hurts kids. Something that wasn't doom and gloom" (Palevsky, 2011, para. 7). The spoof site was created in collaboration with the Yes Men's online lab at yeslab.org.

Digitally Mediated Social Networks

Activists such as those protesting air pollution in Beijing are increasingly connecting with like-minded others via **digitally mediated social networks**. From online petitions against animal cruelty at Care2 and organizing against hydraulic "fracking" (Chapter 6) on Facebook and Tumblr to the hundreds of thousands of climate activists using 350.org or Avaaz.org's global networking sites, individuals and groups are sharing their "outrage and hope" (Castells, 2012), initiating projects, and coordinating with others to create change. Indeed, Manuel Castells has called such "interactive and self-configurable" digital social networks "a new species of social movements" (p. 15).

Such networks are local as well as global. Community activists in the Coachella Valley of Southern California, for example, are using Web 2.0 applications and social networking portals to document abandoned waste dumps and shoddy migrant housing conditions in this region with a large population of poor, farmworker families. Residents are able to log online and report suspected hazards, including unexplained fumes or pollution and environmental officials (Flaccus, 2011). As a result, residents' complaints have begun to pay off. In 2011, the EPA and state regulators "cracked down on a soil recycling plant that was blamed by air quality officials for a putrid stench that sickened dozens of children and teachers at a nearby school" (para. 2). (We have more to say about digitally mediated networks and environmental activism in the final sections of this chapter.)

Hypermediacy, remediation, and digital networking are only a few of the dynamics facilitating environmental activists' pointing to and naming something as an environmental "problem," organizing public support, and more. In the next sections, we look at some of the ways environmental and climate justice groups have integrated digital technologies into their mobilizing efforts and campaigns, and, further, the ways in which some groups are building the digital infrastructure to enlarge their efforts to attract and engage a more powerful movement.

Environmental NGOs and Digital Campaigns

Beyond individual activists and loosely coordinated digital networks, the more established environmental nongovernmental organizations (NGOs) have also quickly embraced new, digital media. Both large and small organizations—from World Wildlife Fund and Greenpeace International to groups like Asociación Pola Defensa da Ría (www.apdr.info)—a small Galician NGO concerned about the degradation of Pontevedra, Spain's *ria* or coastal inlet—have embraced a full suite of new media, from Facebook and other social networking sites to blogs, Twitter feeds, YouTube, and more.

Interestingly, both local and more global environmental NGOs share many similar purposes in using digital media, from the crafting of a group's identity, recruitment of members, and education of the public about environmental issues to the integration of digital tools as a critical part of longer-term advocacy campaigns. In this section, we look at these different uses and several case studies of such digital activism.

"Sustainable Self-Representation"

In creating awareness and urging action, environmental NGOs are also invariably creating an identity in their self-presentation online and via social media. Whether it's a local conservation group protecting a forest or estuary, or a global NGO such as Greenpeace, each organization, through its website and other digital communication, has the ability *to define itself* independent of the "gatekeeping" role of traditional news media (Chapter 5). The ability to do this stems from what Lester and Hutchins (2009) explain is the capacity of digital media for "**sustainable self-representation**," or ability to create and sustain an identity that avoids "both the fickleness of changing news agendas, the vicissitudes of reporting and editorial practices, and the contending corporate interests of large-scale news conglomerates" (p. 591).

One of the better-known examples of an environmental groups' sustainable self-representation—reinforced online and through other digital portals—is the National Audubon Society, the premiere U.S. birding organization (www.audubon.org). Indeed, the remediations of James Audubon's lush color paintings in digital formats are only one element in the rich, hypermediated space of the society's online presence. From its homepage, which features high-definition photos and links to numerous bird species' profiles, videos, interactive maps, blogs, articles, and educator resources, to its geographical search portal for identifying effects of climate change on birds in locales near you, Audubon has nurtured a reputation for being an authoritative source for all things concerned with birds.[3]

A word of caution: As online space becomes more crowded, environmental NGOs are experiencing what one digital media advisor called a potential "**content flood**" for users of social media and the Internet, that is the sheer volume of text, images, data, and other information produced via social media and the Internet (T. Matzzie, personal communication, July 14, 2011). With hundreds of thousands of content producers and online sources, across multiple platforms, and faster gigabyte pipelines for delivering content, the resulting volume renders any one environmental NGO's identity harder to sustain.

While many environmental or conservation groups use digital media primarily for educational purposes, others (as we've seen earlier) actively draw upon digital technologies to mobilize the public to take action. In doing so, digital media have greatly enabled these environmental NGOs' abilities to advance their advocacy campaigns, and it is to this function that we now turn.

Action Alerts: Environmental NGOs' Digital Mobilizing

Beyond their uses for self-representation and raising awareness, a range of digital technologies have emerged most powerfully as organizing resources for environmental NGOs, amplifying their voices in local and global public spheres. As early as the 1990s, North American environmental activists, for example, were pioneering the use of computer networks to organize their opposition to the North American Free Trade Agreement (NAFTA) (Frederick, 1992); and following Union Carbide's toxic chemical disaster at Bhopal, India, in 1984, the International Campaign for Justice launched a website for mobilizing online and offline resources in its fight for justice for the survivors of the plant explosion (Pal & Dutta, 2012).

Greenpeace's "Barbie" Viral Campaign

More recently, Greenpeace initiated an innovative social media campaign against the large toy companies Lego, Disney, Hasbro, and Mattel (maker of Barbie) for their use of product packaging from endangered rainforests. On Facebook, YouTube, and Twitter, it sent out action alerts, mobilizing public support to pressure the companies to halt their destructive practices. Greenpeace's main target was Mattel. It used Barbie's boyfriend Ken as its virtual spokesperson, by releasing a spoof video on YouTube:

> The spoof plays on Mattel's [then] current advertising campaign which involved Ken winning Barbie back after seven years apart. In the Greenpeace YouTube video, Ken discovers Barbie's deforestation habits in Indonesia and dramatically ends their recently renewed relationship! Ten days after it was first uploaded, the YouTube clip had been viewed over a million times in multiple languages. (Stine, 2011, paras. 5–6)

After its launch of the spoof video, and after hanging a huge banner at Mattel's corporate offices, Greenpeace then turned to Facebook and Twitter to enlist the public to protest the toy companies, and "directed users on the social media sites to confront Mattel via Barbie's pages and also send e-mails directly to Bob Eckert, Mattel's CEO." (Stine, 2011, para. 8)

Within a matter of months, Greenpeace's campaign was showing signs of success. Responding to public criticism, Mattel announced that it was directing its suppliers to stop buying products from a Singapore company that had clear-cut vast areas of Indonesian rainforests and, further, was investigating the deforestation allegations (Roosevelt, 2011). Overall, Stine (2011) noted, "The relative success of Greenpeace's campaign during its first two months reveals the potential of social media in ethical campaigns" (para. 4).

Again, a note of caution: As the space online for organizing has gotten crowded, environmental advocacy groups' calls to action are increasingly competing with thousands of other communications daily from social media feeds, texts, corporate ads, and more. As a consequence, some claim that we are entering an **app-centric** world, where mobile applications (on smartphones, Twitter, iPads, notebooks, and so on) are becoming our main portals for information online. As a result, NGOs mobilizing efforts are being challenged to adapt to such selective searches. These are not insurmountable obstacles, but staying in the game of making environmental content easily accessible to audiences requires more media savvy than before. One avenue that environmental NGOs have chosen is targeted e-mail; let's look at this strategic use and an example.

E-Mailing for "Clean Power"

At key moments in a campaign, environmental NGOs use e-mail, as well as social media, to mobilize public comment quickly to a pending governmental action or decision. For example, in June 2014, the U.S. Environmental Protection Agency proposed a new, clean air regulation—the "Clean Power Plan"—that limited the amount of carbon dioxide from coal-burning power plants. (For information on the proposed rule, see http://www2.epa.gov/carbon-pollution-standards.) Although there had been restrictions against pollutants like arsenic and mercury at coal-burning power plants, there were currently no national limits on these plants' carbon dioxide emissions, which are a major source of U.S. greenhouse gas emissions.

As we discuss in Chapter 12, whenever the EPA or other government agency proposes an action that potentially affects the environment, it is subject to strict requirements for public comment. So, in announcing its "Clean Power Plan," the EPA prepared to hear from the public. In response, the League of Conservation Voters (LCV), a major U.S. environmental NGO, sent an e-mail blast with the subject, "One click = lasting environmental protection" (LCV email, August 27, 2014). By clicking its hyperlink "Send a message to the EPA supporting the Clean Power Plan," users were directed to a page where they could add their name and address and send a prepared message, or use a space to "personalize your message" in support of EPA's proposal. LCV and other environmental groups also initiated online efforts, as well as TV ads, to solicit public comments and rally grassroots support at public hearings in Atlanta, Washington, DC, Pittsburgh, and other U.S. cities.

We should also note that advocacy campaigns also rely on e-mail *to follow up* the responses that are generated through Twitter and other means, such as clicking an online e-petition on a "bottom-up" site such as www.change.org. These actions have been called **clicktivism**, or taking action simply by clicking on a link online. In itself, such clicktivism, sometimes called "slacktivism," is often inconsequential. E-petitions can give the (misleading) impression that one is having an effect simply by clicking on an online link. Nevertheless, as Rutgers University Media Professor David Karpf explains, a follow-up e-mail to a user's click enables an advocacy group to encourage a "ladder-of-engagement" that leads to more volunteer involvement and actions as a campaign goes forward (Karpf, 2010).

Another Viewpoint: "The Revolution Will Not Be Tweeted"

Malcom Gladwell (2010) has argued that "the revolution will not be tweeted." His criticism of social media grows out of his comparison of the U.S. civil rights movement with some current activism. In describing the commitment of the students who practiced nonviolent civil disobedience in the lunch counter sit-ins in Greensboro, North Carolina, and in Mississippi Freedom Summer in the 1960s, Gladwell argued their success depended upon strong, personal relationships, trust, and endurance in the face of overwhelming obstacles. He then said,

> The kind of activism associated with social media isn't like this at all. The platforms of social media are built around weak ties. Twitter is a way of following (or being followed by) people you may never have met. Facebook is a tool for efficiently managing your acquaintances, for keeping up with the people you would not otherwise be able to stay in touch with. That's why you can have a thousand "friends" on Facebook. ... In other words, Facebook activism succeeds not by motivating people to make a real sacrifice but by motivating them to do the things that people do when they are not motivated enough to make a real sacrifice. We are a long way from the lunch counters of Greensboro.

SOURCE: Malcolm Gladwell, "Small Change: Why the Revolution Will Not Be Tweeted." *The New Yorker,* October 4, 2010, available at: http://www.newyorker.com/magazine/2010/10/04/small-change-3?currentPage=all

---

Gamification Apps

Mobile apps, of course, are not just for information seeking, particularly in engaging younger audiences. Some environmental groups, universities, and other interests have begun to create apps that make a "game" out of environmental challenges. **Gamification** is the encouragement of something through play, often involving point scoring, competition, and rules of play. These apps can associate environmental awareness with fun (instead of sacrifice) and motivate people who enjoy competition. For example, at Indiana University, the Office of Sustainability hosts an IU Energy Challenge with a mobile app developed on campus. Every year, for four weeks, residence halls, Greek Houses (fraternities and sororities), Academic Buildings, and Administrative Buildings engage thousands of participants to reduce their water and electricity consumption. The app allows each competitor to track its usage and awards one winner in each category at the end of the month. In the short term, it is a popular and fun campaign to raise awareness on campus. What is less studied is whether or not such a short-term competition has long-term impacts on consumption behavior once the competition is over.

Overall, digital technologies for mobilizing are typically embedded in much larger, ongoing strategic campaigns (such as clean air reform or tackling climate change). In such campaigns, digital media may be used for other purposes than simply for calls for action, and it is to these purposes that we now turn.

Online/Offline and "Public Will" Campaigns

As we've just seen, environmental NGOs' action alerts usually occur in the context of wider, strategic campaigns where such actions may be only one step in the process of achieving a larger goal. (See the discussion of tactics and strategy in Chapter 8.) As a consequence, NGOs' uses of digital media also *aim to affect attitudes and behaviors offline* as well as online—from speaking at a public hearing, organizing video screenings, volunteering to staff an information booth, influencing a public decision, and more. In this section, we look at two important concepts underlying these strategic campaigns—integrating online and offline actions in the public sphere, and the goal of marshaling a "public will" in support of campaign goals.

Online/Offline and "the Places of Social Life"

Some of the larger environmental NGOs have accumulated hundreds of thousands—in some cases, millions—of e-mail addresses, as well as visitors to their websites or Facebook page. However, many of these visitors too often *disperse*, that is, they may check out a website and then move on. The challenge for these NGOs, then, is how to retain and actively engage this potential audience in the groups' advocacy campaigns.

One obstacle to engaging visitors to remain has been one-off or singular actions that many environmental NGOs request, for example, "click to send a message"; once the visitor takes that action, the relationship ends. For this reason, it's helpful to distinguish what new media theorist David Karpf (2011) calls *political activity* in the social media world and the *political organizing* that occurs in more integrated and ongoing advocacy campaigns. The latter, while using digital technologies, also makes possible a more sustained and strategic engagement by supporters *offline* as well as online.

Key to political organizing is the ability to "carve out a new public space that is not limited to the Internet, but *makes itself visible in the places of social life* [emphasis added]" (Castells, 2012, p. 11). This new public space, Castells says, is "a hybrid space between the Internet social networks and the occupied urban space," or the places where economic or political issues are often mediated (p. 11). Let's look at an example of such online/offline organizing and a "ladder-of-engagement" one group built in an environmental advocacy campaign.

In Texas, a coalition of environmental organizations, including the Sierra Club and Environment Texas, has been waging a long-running, successful campaign against coal-burning power plants. Flavia de la Fuente, an organizer for the Sierra Club, helped to lead the coalition campaign's social media efforts in an effort to turn out supporters to testify at public hearings.

Working out of Environment Texas' office in Austin, Flavia was given access to the group's Twitter account. An official told her to "go crazy on Twitter" in building support for the coal campaign. (Flavia said that she had been on Twitter for years as a DREAM Act activist.[4]) Initially, her objective was to generate more followers on the group's Twitter feed and hold their interest. This was facilitated by breaking news, at the time, about the BP oil spill in the Gulf and subsequent incidents of power outages that affected many Texans. Flavia, therefore, started tweeting updates about these events and information for getting help with these outages.

Gradually, Flavia began to inform her Twitter followers about the underlying cause of many of these problems—Texans' heavy reliance on fossil fuels, particularly coal-burning power plants. At this point, she began to test how many potentially active followers might be out there, asking, for example, "Would you be willing to take action?" She followed up each reply with a personal e-mail, directing respondents to a Facebook campaign page for more information for taking action in local areas. As the campaign grew, Environment Texas and Sierra Club mobilized these supporters via e-mail to turn out at rallies, for speaking at public hearings, and so forth, moving them up the "ladder-of-engagement" (Karpf, 2010).

Building a Twitter following, nurturing the relationships with personal e-mails, and providing followers with information and resources in "the places of social life" all contributed to a greater solidarity of support and to their later willingness to engage, attending rallies, and the like.

Social media, of course, were not the only tools used to mobilize supporters, but were integrated with the coalition's wider campaign. The digital platforms were instrumental in encouraging supporters to act in the public spaces in which decisions to shut down or deny construction of new coal-burning plants were occurring in Texas. (For more, see beyondcoal.org.)

Digital Media and "Public Will" Campaigns

Underlying most advocacy campaigns that aim to alter public policy or adopt new programs are efforts to summon the public's "will" or the desire of many people for a change. **Public will**, once created and marshaled, can be a powerful, rhetorical force in society for infusing an action with symbolic legitimacy (Chapter 6). For example, when the American public was made aware, in the 1970s, that the bald eagle was endangered, the desire to prevent its extinction became a force for passage of the Endangered Species Act. (The American bald eagle appears on the official seal of the United States.)

It is not surprising, then, that the strategic role of public audiences is the central characteristic of what Salmon, Post, and Christensen (2003) have called **public will campaigns**. They define these campaigns as "strategic initiatives [that are] designed to legitimize and garner public support . . . as a mechanism of achieving . . . change" (p. 4). As one type of an advocacy campaign (Chapter 8), these are strategic precisely because they mobilize the newly informed public will in ways that are pivotal in achieving desired outcomes.

In recent years, public will campaigns have benefited greatly from uses of digital technologies. The rapid, viral spread of images (for example, the lone polar bear on a melting ice flow), retweets and popular hashtags (#sustainability, #fairtrade, #climate change, etc.), help establish public interest and sustain these topics in the public's mind. As a consequence, the ability of social media to disseminate and sustain powerful **memes**—an idea, phrase, or image that is widely shared in a culture—is a pivotal ingredient in nurturing the public's demand for various environmental protections. The visual meme of residents wearing face masks in Beijing and other urban areas' poisonous smog-filled air—shared on Facebook, Instagram, Twitter, and Pinterest, for example—almost certainly helped to raise awareness of dangerous air pollution and build demand for clean(er) air in these cities.

© Earth Hour Brunei/Azme

Photo 9.2 "Earth Hour" is a global environmental movement of WWF. It has grown from a symbolic lights-out event in one city to the world's largest grassroots environmental movement, crowdsourcing actions from individuals, businesses, organizations and governments to generate environmental outcomes. (See www.earthhour.org)

Memes, of course, are difficult for an advocacy campaign to initiate, since, by definition, they occur by *being shared*. Nevertheless, one way that some environmental NGOs go about building public will is by harnessing such memes, already in circulation, in powerful narratives that function to frame—and legitimize—societal goals such as "cleaning up our air." One group, the Climate Meme Project, for example, identifies popular climate-related memes that might be used to further "the shift in consciousness necessary to propel global action" (Jones, 2013, para. 6). The project is gathering memes in active use already, on Twitter, Facebook, and other social media, and "analyzing them to figure out the feelings and beliefs that people already have about global warming," in order to know to what ideas or messages might be "successful with less environmentally-focused groups" in building public will (para. 7). The intention is to "supplant memes that keep the conversation stuck like 'drill, baby drill' or 'the science isn't settled yet'" (para. 8).

Thus far, we've been describing a specific kind of digital communication—raising awareness, "action alerts," and, for the most part, communication that involves a kind of "sending to others." We want now to turn to a different dynamic for understanding the uses of digital media by environmental activists.

Multimodality and Networked Campaigns

The activities just described, that is, "sending to others," illustrate what some have called a broadcast communication model. In contrast, many environmental activists have been shifting to a "networked model" (Clark & Slyke, 2011, p. 245). While there are many green social networking sites, such as TreeHugger's Facebook links (at www .treehugger.com), that facilitate discussion of common interests, we are using the word *networked* in a different way here. Recall our earlier definition of *digitally mediated social networks* as those mobile and Internet-based sites that enable activists not only to share their "outrage and hope" (Castells, 2012), but, just as importantly, *to organize, initiate projects, and coordinate their activism with others*.

In the following sections, we explore the "multimodal" form of environmental activists' networked communication that helps to explain the ability of many groups to "scale up" or dramatically increase the size and impact of their activism. We also describe some of these multimodal forms, including the "network of networks" nature of recent global environmental and climate activism.

Environmental Activism and Multimodal Networks

With the growth of large-scale, popular social movements in recent years, such as the Arab Spring, Occupy Wall Street, and the *Indignadas* movement in Spain, many commentators have noted the distinctive role that social media play in such movements. Castells (2012) focuses on the "multimodality" of populist, networked movements. By **multimodality**, Castells refers to the multiple and intersecting forms of networked relationships that are developed in sustaining the movement. Both the Internet and mobile communication networks, he argues, are essential, but,

> the networking *form* [emphasis added] is multimodal. It includes social networks online, as well as pre-existing social networks, and networks formed during the actions of the movement. Networks are within the movement, with other movements, . . . with the Internet blogosphere, with the media and with society at large. (p. 221)

While Castells is referring to the form of networking that characterizes large-scale social movements like the Arab Spring, we believe multimodality is also relevant to a wide range of digital networking that is increasingly used by environmental activists and NGOs around the world.

As we saw in Greenpeace's "Barbie" (rainforest deforestation) campaign and Environment Texas's uses of Twitter and e-mails to turn out citizens to speak against coal power plants at public hearings, the multiple forms of social and digital networks characterizing environmental activism have enabled NGOs to connect with "the social places of life" of members and the public, marshaling their concerns, and coordinating actions in the public forums where decisions are being made. For example, activists in the nation of Jordan are relying on social media as one of their principal tools "to raise awareness about the protection of nature and rally public support for environmental causes" (Namrouqa, 2012, para. 1).

One of the major environmental campaigns in which Jordanian activists succeeded through social networks was the Halt Ajloun Deforestation Campaign. Started on Facebook, the campaign "attracted more than 5,000 supporters and succeeded in stopping the construction of a military academy in Ajloun's Bergesh Forest" (Namrouqa, 2012, para. 13). One of the supporters, Sahar Raheb, said

> she learned about the plans to cut down thousands of old trees to construct an academy in Bergesh from her friends on Facebook. . . . "One of my friends shared a photo of Bergesh showing a very beautiful green forest, and commented that those centennial trees will be uprooted, . . ." Raheb told the *Jordan Times*. "I followed the link and joined every sit-in and clean-up initiative the campaign organised, all of which I heard about from the campaign's page on Facebook." (Namrouqa, 2012, para. 14–16)

Given the growing importance of networked activism, it may be useful to map more explicitly some of the principal configurations of environmental groups' networked communication. Let's look at two: NGOs' sponsored networks and the coordinated actions of activists and environmental NGOs in networking with other networks themselves, a "network of networks."

NGOs' Sponsored Networks

As digital platforms proliferated, many of the larger environmental NGOs initially began establishing online networks that invited greater engagement of members with the groups' mission and campaigns. While activists in Jordan relied on existing social networks such as Facebook, some larger NGOs have built sophisticated websites enabling multiple forms of interaction, including joining with ongoing projects or campaigns, creating one's own local events, or coordinating with others regionally, nationally, or globally.

One of the earliest NGO-sponsored online networks, launched by World Wildlife Fund (WWF) in 2007, was Earth Hour (earthhour.org). The idea was to raise awareness globally about climate change. To do this, the Earth Hour site invited "people around the world to switch off lights for one hour, thus saving electricity" (O. R., 2013, para. 3.) Earth Hour's original website led to formation of a wider, multimodal network that connected millions of individuals around the world, online and offline, in their social networks—friends, families, peers:

> As more and more people started using social media tools [the Earth Hour] event became extremely popular. Earth Hour activists created various posters, Facebook pages, videos and twitter accounts for this event, while different people reposted this information. Earth Hour theme parties became popular among youth and students. They get together, light candles and play guitar. Even Youtube took part in this campaign by changing the background colors from white to black. (O. R., 2013, para. 3–4)

By 2014, the Earth Hour site itself had become a richly hypermediated space, with updates, images, photos, and videos of past Earth Hour local events and a portal for "Joining the Hour," with links to local start times, Earth Hour event "Starter Kits," and an "Instagram Challenge." That year, Earth Hour set a new record, with events in over 7,000 town and cities in 162 countries and territories across the globe. WWF tracked over 70 million global digital interactions and another 1.2 billion people on Twitter timelines for March 29th using the hashtag #EarthHour (www.earthhour.org/2014-outcomes).

Act Locally!

Turn Out the Lights During "Earth Hour"!

Earth Hour, sponsored by WWF, encourages people to turn out the lights locally, as a way to raise awareness about climate change. The event occurs at the same, local time, in countries across the world. For example, Earth Hour occurred on March 25th, from 8:30–9:30 (local time) in 2015.

Organize your friends or an organization you're involved with, to turn out the lights where you live, work, or attend classes. For information about dates, local times, and ideas about activities, check www.earthhour.org.

How can you use this event—and the ability to join with millions globally—to educate, share information, and plan actions to address your impact on climate change?

Other environmental, peace, and social justice activist groups such as Avaaz.org and climate justice 350.org have been among the innovative developers of social networking sites. A small group of students from Middlebury College in Vermont, along with Environmental Studies Professor Bill McKibben, for example, launched

350.org's site as its main platform for organizing students, indigenous groups, and other civil society groups globally to call attention to the risks of climate change and the urgent need for solutions. Indeed, social networking is central to 350.org's organizing activities, with some observers describing the site as "more of a social media network than an activist group" (DeLuca, Sun, & Peeples, 2011, p. 153), although this is perhaps an unfair assessment. (We describe its role below, as a principal organizer of the People's Climate March.)

Network of Networks: Global Environmental Activism

Integrating Local Activists Into Global Networks

A key characteristic of large-scale coordinated actions such as 350.org's climate organizing is the complex, multimodality of these groups' communication networks. In fact, it may be more accurate to speak of a "network of networks" in some cases. As an example, consider the non-profit Bank Information Center (BIC), whose mission is "Applying Local Voices to Democratizing Development" (www.bicusa.org). BIC itself is a global network of civil society groups in developing countries that aims "to influence the World Bank and other international financial institutions (IFIs) to promote social and economic justice and ecological sustainability" (www.bicusa.org).

BIC illustrates the way that globally networked NGOs sometimes interact with other, loosely organized environmental and other groups' networks to coordinate local actions. A very visible example of this occurred in 2011, when BIC and its network of local partners launched an "International Day of Action" demanding that the World Bank end its lending for fossil fuels. The campaign was announced online and via social media, as well as with press releases and NGOs' own websites and e-mail blasts. The International Day of Action, March 11th, brought together a global network of local civil society organizations, climate justice groups, and environmental NGOs for a one-day series of protests outside World Bank offices in London, Berlin, Paris, Johannesburg, Rome, Washington, DC, and other cities.

In addition to its networked organizing and local onsite rallies and protests, the International Day of Action made extensive use of social media, asking supporters, for example, to tweet @WorldBank, use the hashtag #WBDayOfAction, e-mail their friends, and post comments on the World Bank's Facebook page and its blog demanding that the Bank phase out its lending for coal plants and other fossil fuel projects around the world. Sample tweets calling on the World Bank included "'Free Us From #fossilfuels' #WBDayOfAction," and "How long until @WorldBank updates its energy strategy to live up to its pro-poor, pro-climate rhetoric? #WBDayOfAction." (For more about this campaign, see www.bicusa.org/world-bank-fossil-fuel-lending-action.)

Digital Infrastructure for Local/Global Campaigns

Some environmental NGOs do more than network with other campaigns. These NGOs enable activists to jump-start and carry out their own campaigns via the digital infrastructure and networks that these larger groups maintain. For example, WWF's

Earth Hour website (above) provides a link for crowdsourcing both newly launched campaigns and ongoing campaigns hosted by Earth Hour (for example, stopping a coal plant in Greece). (Check out www.earthhour.org/crowdsourcing.) For example, the group Shark Savers is networking with WWF to campaign for an end to the abusive practice of shark fins on menus across Asia. These options are possible because WWF provides the digital infrastructure that enables users to start, organize others, and solicit participation in actions via the WWF website.

One of the largest global NGOs providing digital, web-based infrastructure for organizing is Avaaz.org. The name "Avaaz" means "voice" or "song" in many languages, including Farsi, Hindi, and Bosnian. Launched in 2007, Avaaz.org's mission is simple and powerful:

> Organize citizens of all nations to close the gap between the world we have and the world most people everywhere want. . . . Avaaz empowers millions of people from all walks of life to take action on pressing global, regional and national issues, from corruption and poverty to conflict and climate change. Our model of internet organising allows thousands of individual efforts, however small, to be rapidly combined into a powerful collective force. (www.Avaaz.org/en/about.php)

The wireframes for Avaaz.org, and other NGOs' similar platforms, typically provide pathways for self-initiating activists and groups to organize, launch initiatives, and enable people globally to engage with their projects or campaigns, online and in their local communities. A **wireframe** is simply a way to visualize the structure, function, and navigational steps of a website, allowing a user, for example, to navigate from a homepage via links to specific issues or campaigns and from there to a sign-up menu or a portal for networking with others, etc. Such platforms spotlight issues, expose wrong doings, and give visibility to local or regional groups' campaign websites.

The local/global reach of Avaaz.org can best be seen in a recent victory—the 20-year Maasai struggle in East Africa to keep their community's traditional lands. (Tanzanian government officials had wanted to evict Maasai from some of their traditional lands to allow a big game hunting company to bring in tourists for sports-hunting of wildlife.) The Avaaz.org online community launched a petition (and more) calling on the Tanzanian president to stop the evictions:

> Avaaz members funded hard-hitting adverts in local papers calling out the government, got CNN and Al Jazeera on the ground and reporting the story and supported traditional leaders to camp outside the Prime Minister's office for three weeks, forcing him to listen to their case.
>
> When in October reports surfaced that the government was set to renege on its pledge, we sprung back into action with over 2.3 million people signing the petition and taking to social media forcing the Tanzanian President to promise on Twitter that the government would never "evict the Maasai people from their ancestral land." (Avaaz.org, 2014, para. 2–3)

In response to mounting pressure, the Tanzanian Prime Minister visited the community and promised not to evict the Maasai people from their lands.

While the Maasai campaign was local, the ability of environmental and climate networking sites such as Avaaz.org, 350.org, and others to greatly enlarge the numbers of people who are willing to participate may be seen in recent actions to "sound an alarm" on climate change.

Scaling Up: The "People's Climate March"

They came—climate activists, environmentalists, scientists, labor unions, indigenous groups, public officials, students, the UN Secretary General, and more—to the People's Climate March, the "largest climate change protest in history" (Friedman, 2014, para. 2). On a clear and sunny Sunday, September 21, 2014, an estimated 400,000 individuals marched in the heart of New York City, with thousands of others marching in London, Berlin, Paris, Delhi, Rio de Janeiro, Melbourne, in war-torn Aleppo, Syria, in Papua New Guinea, and in 162 countries worldwide (Alter, 2014, para. 3; Foderaro, 2014, p. A1).

Prior to the People's Climate March, the organizers launched a website (http://peoplesclimate.org) and a visually stunning video, *Disruption* (http://watchdisruption.com), widely disseminated online. *Disruption* was a visual narrative, not only of the devastating impacts of climate change, but the multiplicity of voices—those experiencing the disruptions of typhoons, flooding, droughts, and more; scientists warning of further effects to the planet; and organizers planning an event to demand that global leaders take action. The video included a countdown to the march, its planning, and an urgent call for activists to move, in massive numbers, offline and "into the streets."

South Bend Voice/Flickr. Used under Creative Commons license https://creativecommons.org/licenses/by-sa/2.0/

Photo 9.3 Over 400,000 individuals walking in the heart of New York City during the People's Climate March to demand action on climate change, on September 21, 2014.

On the day of the march, the events were streamed live: people walking peacefully; music playing; people holding signs and banners ("There Is No Planet B," "Jobs, Justice, Clean Energy," and others); balloons and white paper doves floating above the marchers; environmentalists, Capuchin monks, student groups, artists, environmental justice activists, and more, marching together. "We need to act now. . . . We only have one atmosphere and we of the Marshall Islands only have one land to call 'home,'" said Kathy Jetnil-Kijiner, from the Marshall Islands, before the march (quoted in Visser, 2014, para. 8). The climax of the march came at 1:00 in the afternoon:

> All along the route, crowds had been quieted for a moment of silence. . . . Then at exactly 1 p.m., a whistle pierced the silence, setting off a minute-long cacophony intended as a collective alarm on climate change. There were the beats of the drums and the blaring of horns, but mostly it was whoops and cries of the marchers. (Foderaro, 2014, p. A17)

The People's Climate March came the day before world leaders convened a climate summit at United Nations headquarters to begin laying the basis for a new international climate agreement to be adopted at the end of 2015. The march, however, was not meant as a protest against the United Nations, but a sounding of the alarm and demonstration of the urgent need for action. As one of the marchers said, she wanted "the day's energy to translate into local and global policies and to keep up the pressure" on the United States, China, the European Union, India, and other negotiators (Friedman, para. 5).

The idea for a large-scale climate event came on May 21, 2014, in "A Call to Arms," an essay by 350.org's founder Bill McKibben on *Rolling Stone* magazine's website: "This is an invitation . . . to come to New York City. An invitation to anyone who'd like to prove to themselves, and to their children, that they give a damn about the biggest crisis our civilization has ever faced" (McKibben, para. 1). Initial planning for the march came from over a dozen environmental, labor, social justice, LGBTQ, faith, women, student, and other groups, including, most prominently, 350.org.

Following a meeting in early 2014 in New York City, organizers launched an online, networking platform (http://peoplesclimate.org), "to let everyone everywhere join groups like these to work on outreach, projects and actions related to the march" (People's Climate, 2014, para. 3). Then, in July of 2014, the People's Climate group kicked off "a one week long Facebook Invite Blitz to get over 200,000 people invited" to the event (People's Climate, 2014, July 30 e-mail, para. 3). Meanwhile, groups such as 350.org, Greenpeace, Avaaz.org, Sierra Club, WWF, and the New York City Environmental Justice Alliance began reaching out through their own networks to other groups. (For example, 400,000 people pledged on Avaaz's site to participate in one of the global events timed with the People's Climate March.)

News of the march spread across other media as well—from major newspapers such as *The Guardian* (with its own hyperlinks) to ads on the New York subway and posters in the London Tube. Ultimately, over 1,500 groups took part in organizing for the march, according to People's Climate, the principal group doing much of the planning. One reason the People's Climate March succeeded may be the group's early decision to organize through "a participatory, open-source model [http://peoples climate.org/partners], because this is a 'movement of movements' moment."

Two days after the People's Climate March, U.S. president Barack Obama spoke to delegates at the Climate Change Summit at UN Headquarters. Acknowledging the earlier march, he said,

> The alarm bells keep ringing. Our citizens keep marching. We cannot pretend we do not hear them. We have to answer the call. We know what we have to do to avoid irreparable harm. . . . We have to work together as a global community to tackle this global threat before it is too late. We cannot condemn our children, and their children, to a future that is beyond their capacity to repair. Not when we have the means—the technological innovation and the scientific imagination—to begin the work of repairing it right now. (The White House, 2014, para. 5–6)

As we write, world leaders have begun the daunting task of negotiating a new international agreement to address the causes and impacts of global climate change. Their goal is an agreement to be signed in December 2015, at the next UN climate summit in Paris. Meanwhile, environmental, climate justice, and other activists are continuing to share information and coordinate across a multimodal "network of networks" to encourage a more sustainable world.

SUMMARY

This chapter expanded upon our earlier (Chapter 8) account of environmental advocacy campaigns by describing the role of activists' uses of digital media and the communication dynamics underlying these uses. The first section considered how activists employed digital technologies to alert, amplify, and engage and introduced concepts of hypermediacy, remediation, and digitally mediated social networks to help us understand how these uses of digital media succeed in campaigns.

Part two surveyed the uses of digital media by environmental nongovernmental organizations (NGOs), in NGOs' self-representation online, their mobilizing of public support, and their building "public will" for environmental protection. We also cautioned that the increasing content flood online can sometimes make it difficult for environmental NGOs to reach viewers or mobilize supporters with calls for action.

In the third section, we described efforts by NGOs to go beyond a linear or "broadcast" model of digital communication to a "network" model—the digital infrastructure and dynamics of larger-scale interactions that are enabling diverse individuals to engage with others in coordinated actions offline as well as online.

We illustrated the effectiveness of such digitally mediated social networks, in the final section of the chapter, in organizing for the "People's Climate March" at the United Nations in 2014, as hundreds of thousands of people moved offline, "into the streets," in New York City and other urban centers around the world.

SUGGESTED RESOURCES

- In 2014, Greenpeace declared a campaign victory after the toy company Lego announced it would not renew its co-promotion contract with Shell Oil. Greenpeace's main campaign tool was a viral video that was viewed almost six million times, named "Everything Is Not Awesome," and an online petition of approximately one million names. Find out more about the campaign and how Greenpeace is continuing to draw attention to drilling in the Arctic through digital media here: http://www.greenpeace.org/international/en/press/releases/LEGO-ends-50-year-link-with-Shell-after-one-million-people-respond-to-Save-the-Arctic-campaign.
- Alexandra Segerberg and W. Lance Bennett, "Social Media and the Organization of Collective Action: Using Twitter to Explore the Ecologies of Two Climate Change Protests," *The Communication Review*, 14(3) [2011], pp. 197–215.
- The Center for Story-based Strategy (formerly smartMeme) is a consulting firm on how organizations and institutions can reframe stories and controversies through changing their message.
- Check out People's Climate March (http://peoplesclimate.org) for updates on global planning around climate change as well as an example of a "network of networks" model for digital organizing. References to Peoples Climate are to links at this site, sometimes updated or deleted.

KEY TERMS

App-centric   218

Bottom-up sites   213

Clicktivism   219

Content flood   217

Digitally mediated social
 networks   215

Factory farming   214

Gamification   219

Hypermediacy   214

Meme   222

Multimodality   223

Public will   221

Public will campaigns   222

Remediation   214

Self-initiating   213

Sustainable
 self-representation   216

Wireframe   227

DISCUSSION QUESTIONS

1. Have you ever attended an environmental rally, protest, public hearing, or other event as a result of an e-mail, social media alert, or an online announcement? What impact, if any, do you believe your and others' participation in such events have?

2. Do you think there should be a hierarchy between activism online and offline? Are they in concert or conflict? Is one possible without the other today? What do you think are the limitations or challenges digital media pose to environmental communication?

3. There is so much information disseminated today through more media than ever before. How do you manage the content flood? Do you use an online aggregate news app? Do you follow Instagram more than your campus newspaper? How do you decide which sources to engage?

4. Did you join the 2014 People's Climate March in New York or in another city? Do such large protest events in the streets matter? How can they shift debates or enable action to occur?

5. Participants in the People's Climate March in 2014 wanted countries to sign a new international climate agreement. Are we getting there? How would you evaluate "success"?

NOTES

1. A note of caution: In a study of Sea Shepherd Conservation Society's anti-whaling campaign, Lester (2011) found that, while the campaign's visibility involved "a complex flow of information and meanings across various 'old' and 'new' media form," it nevertheless still remained reliant on traditional media to showcase its images and news feeds via distributed news and cable television (p. 124; see also, Lester & Hutchins, 2009, p. 580 ff.).

2. A rhizome is an underground stem of certain kinds of plants that spreads or creeps, sending out roots or shoots from nodes; rhizomes can grow new roots or, if split, are capable of growing new plants. The most common example is ginger root.

3. Although Audubon has been criticized for accepting funds from corporate sponsors (www .sourcewatch.org), it manages to sustain a credible identity for birding for many reasons, in part due to its capacity to sustain its impressive self-definition online.

4. The DREAM Act grants conditional residency to illegal immigrant students, in good standing, who arrived in the United States as children and who graduated from a U.S. high school.

"Put your fist in the air for those babies in the South Bronx with asthma. Put your fist in the air for those grandmas in Oakland dealing with cancer. Put your fist in the air for all those . . . in the Native American tribes . . . who are dealing with cancer, with flooded bad water or bad air. Put your fist in the air for humanity. And let them know that we say: Can't stop. Won't stop." (Rev. Yearwood, February 17, 2013, Forward on Climate Rally, Washington, DC, USA)

CHAPTER 10

Environmental Justice and Climate Justice Movements

To anyone who continues to deny the reality that is climate change, . . . I dare you to go to . . . the islands of the Caribbean and the islands of the Indian Ocean and see the impacts of rising sea levels. Not to forget the massive hurricanes in the Gulf of Mexico and the eastern seaboard of North America. And if that is not enough, you may want to pay a visit to the Philippines right now.

—Philippines delegate Naderev (Yeb) Saño,
speaking to the COP19 Climate Change
Summit in Poland, November 11, 2013

In 2013, Filipino climate commissioner, Naderev (Yeb) Saño, caught global attention with his moving speech. One of the strongest storms ever recorded, Typhoon Haiyan, had killed at least 10,000 people and devastated his country that week. Making connections among climate disaster victims and survivors globally, Saño pled with global governments to take heed: "What my country is going through as a result of this extreme climate event is madness. The climate crisis is madness. We can stop this madness" (Saño, 2013, para. 10).

Within the United States, grassroots voices increasingly have been demanding environmental justice as well. As with climate justice, these communities point out the disproportionate burden of costs specific communities have carried while a smaller number of elites have profited.

Chapter Preview

- In the first section of this chapter, we describe the emergence of the environmental justice movement in response to patterns of *environmental racism*, as well as how that movement has developed a discourse, become institutionalized, and achieved some success.
- The second section identifies barriers faced by people who attempt to speak out against environmental injustices. This section considers two public forums through which environmental justice advocates have attempted to gain recognition in the public sphere:
  - The stigmatization of certain voices as inappropriate in their communication styles when speaking in technical forums
  - The marginalization of frontline communities who use *toxic tours* in response, calling attention to the sights, sounds, and smells of environmental inequalities
- In the third section, we describe the climate justice movement, whose discourse seeks to connect the dots between environmental, social, and economic global struggles and whose actions range from online networks to a rise in civil disobedience.

As discussed in Chapter 2, efforts to redefine the meaning of *environment* by grassroots, multiracial struggles for social, economic, and ecological justice in the United States and globally have changed how many of us understand environmental matters today. Our hope is that when you have finished reading the chapter, you'll understand how the movements for environmental justice and climate justice are also movements for a more democratically inclusive world.

Environmental Justice: Challenging *a Place Apart*

As used by community activists and scholars studying the movement, the term environmental justice refers to (a) calls to recognize and halt the disproportionate burdens imposed on working-class and people of color communities by environmentally harmful conditions, (b) more inclusive opportunities for those who are most affected to be heard in the decisions affecting their communities, and (c) a vision of environmentally healthy, economically sustainable, and culturally thriving communities. This grassroots movement redefined *environment* to encompass where we live, work, play, and learn. Although this movement now includes urban gardens and green spaces, it began out of a broader, emerging focus on toxic pollution.

The Beginnings of a "New" Movement

While credit is due to anti-toxic advocates in the 20th century who raised awareness about public health, a less frequently acknowledged source of critical rhetoric (Chapter 8) about increased toxic pollution production are the polyvocal voices of people of color that also initiated grassroots struggles across North America. The history of Native American genocide and colonization, for example, cannot be delinked from a history of land removal, resource exploitation, and toxic chemical exposure (particularly through uranium mining and atomic bomb tests). Likewise, struggles of immigrant and migrant farmworkers always have involved critical rhetorics of working conditions in fields, including health concerns about pesticide exposure. Perhaps one of the most famous successful advocacy campaigns of the 20th century was the United Farm Workers' 1965–1970 strike and call to boycott table grapes; Cesar Chavez insisted on organizing Latino/a and Filipino/a workers together to establish the first union contracts for farmworkers, including better pay, benefits, and protections.

In the late 1960s and 1970s, African American civil rights groups, churches, and environmental leaders also tried to call attention to the particular problems of urban communities and the workplace. Dr. Martin Luther King Jr. went to Memphis, Tennessee, in 1968 to join with African American sanitation workers who were striking for wages and better work conditions—an event that sociologist and environmental justice scholar/advocate Robert D. Bullard (1993, 1994) called one of the earliest efforts to link civil rights and environmental health concerns. (*Note:* Today, many call Bullard the "father of environmental justice"—http://drrobertbullard.com.) And the 1971 Urban Environment Conference (UEC) was a coalition of labor, environmental, and civil rights groups that tried "to help broaden the way the public defined environmental issues and to focus on the particular environmental problems of urban minorities" (Kazis & Grossman, 1991, p. 247).

Despite these early attempts to bring environmental, labor, and civil rights leaders together to explore common interests, national environmental groups in the 20th century largely failed to support communities of color and working-class communities. In her account of efforts to stop the construction of a 1,600-ton-per-day solid waste incinerator in a south central Los Angeles neighborhood in the mid-1980s, Giovanna Di Chiro (1996) reported,

> These issues were not deemed adequately 'environmental' by local environmental groups such as the Sierra Club or the Environmental Defense Fund. . . . [T]hey were informed that the poisoning of an urban community by an incineration facility was a "community health issue," not an environmental one. (p. 299; see also Alston, 1990)[1]

The racism, economic elitism, and sexism of the broader U.S. culture often marginalized these grassroots efforts (Pezzullo & Sandler, 2007). Finding these individual community efforts had a great deal in common, in the early 1980s a multicultural movement of coalitions across these struggles was born.

Toxic Waste and the Birth of a Movement

When the state government of North Carolina discovered that PCB (polychlorinated biphenyl) chemicals had been illegally dumped along miles of highways, officials decided to bury the toxic-laced soil in a landfill in the predominantly African American and poor Warren County, despite needing to waive two U.S. EPA restrictions to do so. Rather than quietly accept this decision, local residents and supporters from national civil rights groups tried to halt the state's plan by filing two lawsuits, which failed. In 1982, when trucks carrying the PCB-contaminated soil began driving into Warren County, a multi-racial coalition began placing their bodies in the middle of the roads leading to the landfill and more than 500 arrests for acts of nonviolent civil disobedience occurred. The events of that summer have been narrated countless times as a "symbolic center," a "milestone," and a story of origin for the environmental justice movement (Pezzullo, 2001).

Prompted by protests in Warren County and elsewhere, in the 1980s and 1990s federal agencies and scholars began to confirm patterns of disproportionate exposure to environmental hazards (U.S. GAO, 1983). The United Church of Christ's Commission for Racial Justice (Chavis & Lee, 1987), for example, found the following:

- Race proved to be the most significant among variables tested in association with the location of commercial waste facilities. . . . Although socioeconomic status appeared to play an important role in the location of [these] facilities, race still proved to be more significant (p. xiii).
- Three out of every five Black and Hispanic Americans lived in communities with uncontrolled toxic waste sites (p. iv).
- Approximately half of all Asian/Pacific Islanders and American Indians lived in communities with uncontrolled toxic waste sites (p. xiv).

A follow-up report, *Toxic Wastes and Race at Twenty, 1987–2007,* revealed that "racial disparities in the distribution of hazardous wastes are greater than previously reported" in the original 1987 study (Bullard, Mohai, Saha, & Wright, 2007, p. x).

We Speak for Ourselves: Naming Environmental Racism

As communities began to feel the impacts of this disproportionate burden, they began to name their experiences. One powerful phrase seized upon by activists to describe their communities' plights was *environmental racism*. By linking civil rights discourse with environmental discourse, a new vocabulary emerged. Dr. Benjamin Chavis of the United Church of Christ's Commission for Racial Justice defined environmental racism as

> racial discrimination in environmental policy-making and the enforcement of regulations and laws, the deliberate targeting of people of color communities for toxic waste facilities, the official sanctioning of the life-threatening presence of poisons and pollutants in our communities, and the history of excluding people of color from leadership in the environmental movement. (quoted in Grossman, 1994, p. 278)

While Chavis and other civil rights leaders highlighted the "deliberate" targeting of people of color communities, others pointed out that discrimination also resulted from the *disparate impact* of environmental hazards on people of color communities. The 1964 Civil Rights Act used the term **disparate impact** to recognize discrimination in the form of the disproportionate burdens that some groups experience, regardless of the conscious intention of others in their decisions or behaviors. In other words, racial discrimination results from the accumulated impacts of unfair treatment, which may include more than intentional discrimination or deliberate targeting. Asian American communities also raised attention toward procedural justice concerns, such as translation (Sze, 2011).

In other cases, activists themselves began to call the conditions imposed on low-income communities a form of *economic blackmail*. For example, Bullard (1993) explained, "You can get a job, but only if you are willing to do work that will harm you, your families, and your neighbors" (p. 23). This **economic blackmail** is a false choice presented between financial worth and environmental protection that deflects attention away from the fact that jobs can be provided while meeting basic health and environmental standards.

Naming the problems associated with environmental injustices was important. Rose Marie Augustine (1993), who we will discuss later, attended a workshop for community activists in the Southwest. She said that, for the first time, "I heard words like 'economic blackmail,' 'environmental racism.' Somebody put words, names, on what our community was experiencing."

Courtesy of Asian Communities for Reproductive Justice

| Photo 10.1 | Founded in Oakland, California, in 1999, Asian Communities for Reproductive Justice (ACRJ) connects environmental justice goals with rights explicitly related to gender and sexuality. Developing an intersectional approach to the precautionary principle, this organization uses the metaphor of "looking both ways" to bridge environmental and reproductive justice concerns (de Onís, 2012). |

As protests mounted against such patterns and the failure of the mainstream environmental movement to address the problems, activists began to insist that people in affected communities be able to "speak for ourselves" (Alston, 1990). Social justice activist Dana Alston (1990) argued, in her book *We Speak for Ourselves,* that environmental justice "calls for *a total redefinition of terms and language* [emphasis added] to describe the conditions that people are facing" (quoted in Di Chiro, 1998, p. 105). Indeed, what some found distinctive about the critical rhetoric of the new movement was the way in which it transformed "the possibilities for fundamental social and environmental change through processes of redefinition, reinvention, and construction of innovative political and cultural discourses" (Di Chiro, 1996, p. 303). Environmental attorney Deehon Ferris put it more bluntly when she said, "We're shifting the terms of the debate" (Ferris, 1993).

One important shift in the terms of debate occurred in 1990, when the SouthWest Organizing Project (SWOP) publicly criticized the nation's largest environmental groups, specifically those that belonged to the "Group of Ten."[2] Called "the single most stirring challenge to traditional environmentalism" (Schwab, 1994, p. 388), the letter ultimately was signed by more than one hundred civil rights and community leaders. The letter accused the mainstream groups of racism in their hiring and policies, as well as a lack of accountability toward Third World communities within the United States and abroad. Coverage of the letter in the *New York Times* and other newspapers "initiated a media firestorm" and generated calls for "an emergency summit of environmental, civil rights, and community groups" (Cole & Foster, 2001, p. 31).

Building the Movement for Environmental Justice

Such a gathering came when delegates from local communities and national leaders from social justice, religious, environmental, and civil rights groups met in Washington, DC, for the **First National People of Color Environmental Leadership Summit**, in October 1991.[3] The summit was a "watershed moment" in the history of the nascent environmental justice movement (Di Chiro, 1998, p. 113). For three days, activists shared stories of grievances and attempted to compose a collective critique of the narrow vision of the environment and the exclusion of people of color from decisions that affected their communities. One participant declared, "I don't care to join the environmental movement, I belong to a movement already" (quoted in Cole & Foster, 2001, p. 31).

By placing concerns about toxic wastes and other environmental dangers into a civil rights frame, they were able to "characterize the distribution of environmental hazards as part of a broader pattern of social injustice, one that contradicts the fundamental beliefs of fairness and equity" (Sandweiss, 1998, p. 51). For example, Janice Dickerson, an African American activist, provided this testimony in a video shown during the summit:

> From the perspective of the African American, it's a civil rights matter; it's interwoven. Civil rights and the environment movement are both interwoven. Because, again, we are the most victimized.... There's no difference in a petro-chemical industry locating two, three hundred feet from my house and killing me off than there is when the Klan was on the rampage, just running into black neighborhoods, hanging black people at will. (Greenpeace, 1990)

By drawing on the "morally charged terrain" of the civil rights movement, the summit participants dramatically shifted the terms of public debate about the environment (Harvey, 1996, p. 387). Many of the speakers at the summit also urged participants to demand political representation and to speak forcefully to public officials, corporations, and the traditional environmental movement. At the summit, Chavis explained,

> *This is our opportunity to define and redefine for ourselves. . . .* What is at issue here is our ability, our capacity to speak clearly to ourselves, to our peoples, and forthrightly to all those forces out there that have caused us to be in this situation. (*Proceedings,* 1991, p. 59)

On the last day, participants did so in a dramatic way by adopting 17 Principles of Environmental Justice, an expansive vision for their communities and the right to participate directly in decisions about their environment.

The principles began: "Environmental justice affirms the sacredness of Mother Earth, ecological unity and the interdependence of all species, and the right to be free from ecological destruction" (*Proceedings,* 1991, p. viii). The principles developed an enlarged sense of the environment to include places where people lived, worked, and played and enumerated a series of rights, including "the fundamental right to political, economic, cultural, and environmental self-determination of all peoples" (p. viii).

The inclusion of the right of self-determination was especially important to the emerging movement. Many of the summit's participants had criticized the officially sanctioned decision making in their communities for failing to provide meaningful participation "for those most burdened by environmental decisions" (Cole & Foster, 2001, p. 16). In adopting the principles, they insisted that *environmental justice* not only referred to the right of all people to be free of environmental poisons but that at its core is the inclusion of all in the decisions that affect their health and the well-being of their communities.

Afterward, the Southern Organizing Conference for Social and Economic Justice applauded the "new definition of the term 'environment'"; the group invited community activists "to build a new movement" using the Principles of Environmental Justice adopted at the summit (personal communication, June 2, 1992). Deehon Ferris (1993) of the Lawyers' Committee for Civil Rights in Washington, DC, observed that "as a result of on-the-ground struggles and hell-raising, 'environmental justice' [emerged as] a hot issue. . . . Floodgates [opened] in the media." Urban planning scholar Jim Schwab (1994) observed "the new movement had won a place at the table. The Deep South, the nation, would never discuss environmental issues in the same way again" (p. 393). The *National Law Journal* reported that the movement had gained "critical mass" (Lavelle & Coyle, 1992, p. 5). As the movement grew, it achieved more success.

Institutionalization of Environmental Justice

In 1993, the movement persuaded the Environmental Protection Agency (EPA) to set up a **National Environmental Justice Advisory Committee (NEJAC)** to ensure a voice in the EPA's policy making for environmental justice groups. The committee was chartered to provide advice from the environmental justice community and recommendations to the EPA administrator. For example, NEJAC produced advisory reports on the cleanup of brown fields (polluted urban areas), mercury contamination of fish, and new guidelines for ensuring participation of working-class and people of color residents in decisions about permits for industries wanting to locate in their communities.

The movement also achieved an important political goal when U.S. president Clinton issued Executive Order 12898, Federal Actions to Address Environmental Justice in Minority Populations and Low-Income Populations, in 1994. **Executive Order 12898 on Environmental Justice** instructed each federal agency "to make achieving environmental justice part of its mission by identifying and addressing . . . disproportionately high and adverse human health or environmental effects of its programs, policies, and activities on minority populations and low-income populations in the United States" (Clinton, 1994, p. 7629). Although the Clinton administration began to implement the Executive Order, the succeeding administration of George W. Bush allowed it to lay dormant for eight years (Office of the Inspector General, 2004).

Under the Obama administration, however, the Executive Order on Environmental Justice received new life. The EPA's first administrator under Obama, Lisa Jackson, hosted a White House forum on environmental justice and launched community meetings across the country. "Now, it's time to take it to the next level," Jackson said, adding that the Obama administration would focus on the creation of "green jobs" in disadvantaged communities (quoted in "Obama Revives Panel on Environmental Justice," 2010, p. 2A). On the twentieth anniversary of Executive Order 12898, the EPA launched **Plan EJ 2014** as a set of strategies to recommit to and reinvigorate environmental justice efforts through legal, scientific, information, and resource development and communication. (See http://www.epa.gov/environmentaljustice/plan-ej/index.html.)

In addition, the mainstream U.S. environmental movement itself underwent changes as a result of the critique of an environmentalism that stood apart from the places where people lived. Pezzullo and Sandler (2007), for example, observed that "much has changed within . . . and happened around" the mainstream and environmental justice movements (p. 12). Dialogue between leaders of the mainstream green groups and the environmental justice community led in some cases to collaborations. Greenpeace, Sierra Club, Earth Island Institute, and Earth Justice have been particularly active in their support of environmental justice concerns.

Since these initial moments, environmental justice has served as a broad umbrella term for a wide range of topics beyond toxic pollution, including urban gardens, city planning decisions, logging debates, and more (Pezzullo & Sandler, 2007). As Bullard (2014) observes, the environmental justice movement has grown significantly over the past two decades: "The number of people of color environmental groups that support the Environmental Justice Movement has grown from 400 groups in 1994 to more than 3,000 groups and a dozen networks in 2013." A subsequent gathering—the **Second National People of Color Environmental Leadership Summit**—was held in Washington, DC, from October 23 to 26, 2002. Highlighting women's roles as leaders in the movement, the second event was even larger, attracting more than 1,400 participants. Out of this Summit, Principles of Collaboration also were written to help guide coalition work between the environmental justice movement and environmental organizations (reprinted in Sandler & Pezzullo, 2007).

Nevertheless, the movement for environmental justice also would confront new obstacles and a need to identify new ways to communicate to pursue their vision.

Challenging Indecorous Voices and Sacrifice Zones

An important theme in the discourse of environmental justice is the right of individuals in at-risk communities to participate in decisions affecting their lives. In this chapter, we want to focus on two barriers North American environmental justice advocates have challenged in the hope of gaining recognition: *indecorous voices* and *sacrifice zones*.

Dismissing the "Indecorous" Voice

When someone is judged as having an **indecorous voice**, she or he is deemed as inappropriate or unqualified for speaking in official forums, which is based in the assumption that ordinary people may be too emotional or ignorant to testify about chemical pollution or other environmental issues in settings that privilege "rational" or logical rhetorical appeals. This perceived need to split environmental science or technical decision making from the experiences and feelings of everyday people can further perpetuate environmental racism and injustices. So, what are the expectations in technical public forums?

Decorum and the Norms of Public Forums

In some ways, the unstated rules that operate in many technical forums addressing environmental problems reflect something similar to the ancient principle of *decorum*. **Decorum** was a virtue of style in the classical Greek and Latin handbooks on rhetoric and is usually translated as *propriety* or *that which is fitting* for the particular audience and occasion. The Roman rhetorician Cicero, for example, wrote that a wise speaker is one who is "able to speak in any way which the case requires" or in ways that are most "appropriate"; he proposed, "let us call [this quality] decorum or 'propriety'" (Cicero, 1962, XX.69).

With exposure to chemical contamination and official denial or resistance, affected residents often become frustrated, disillusioned with authority, and angry. Under such norms of decorum, residents sometimes confront a painful dilemma. On the one hand, to enter discussions about toxicology, epidemiology, or the technical aspects of environmental science in a technical forum is tacitly to accept the discursive boundaries about the type of knowledge valued in that space; many residents do learn a great deal about the science impacting their lives and can provide feedback in this manner, using everything from surveys to mapping public health conditions in order to provide evidence of costs. Yet, although often reasonable evidence, data are not always persuasive. On the other hand, to share stories about one's personal expertise about one's family's health and to ask questions can transgress powerful boundaries about what is reasonable or acceptable decorum and may leave one more open to being dismissed as unreasonable and not thinking about the big picture. To illustrate this dilemma, the following are two stories typical of environmental justice leaders attempting to be heard these public forums.

Rose Marie Augustine's Story: "Hysterical, Hispanic Housewives"

On the south side of Tucson, Arizona, where Latin Americans and Native Americans are the main residents, chemicals from several industrial plants had seeped into the groundwater table. This contaminated the wells from which some 47,000 residents drew their drinking water. One of the residents, Rose Marie Augustine, described her own and her neighbors' fears: "We didn't know anything about what had happened to us. . . . We were never informed about what happens to people who become contaminated by drinking contaminated water. . . . We were suffering lots of cancers, and we thought, you know, my God, what's happening?" (Augustine, 1991). EPA officials later confirmed the severity of the toxic chemicals that had been leaching from nearby Tucson industrial plants into their well water and listed this site as one the nation's priority Superfund sites for cleanup (Augustine, 1993).

Prior to the area's listing as a Superfund site, residents from the south side had tried to get local officials to listen to their concerns. Augustine (1993) reported that, when residents met with officials, the officials refused to respond to questions about the health effects of drinking their well water. She said that, when residents persisted, one county supervisor told them "the people in the south side were obese, lazy, and had poor eating habits, that it was our lifestyle and not the TC [toxic chemicals] in the water that caused our health problems." Augustine said that one official "called us 'hysterical Hispanic housewives' when we appealed to him for help."

Augustine did not let local officials discourage her. Despite the barriers she faced, she became a national environmental justice leader and eventually helped communities beyond her own. Yet, the biases of technical forums for public participation continue to persist. Our next story shares a similar tale.

Charlotte Keys's Story: "The Evidence Is in My Body!"

When she began to speak up for her community, Charlotte Keys was a young, African American woman in the small town of Columbia in southern Mississippi whom we both know through our advocacy work with the Sierra Club (Robbie as president in the mid-1990s; Phaedra interviewed her for her book, *Toxic Tourism*, 2007). Keys and her neighbors lived next to a chemical plant owned by Reichhold Chemical, which had exploded and burned years earlier. The explosion and fire spewed toxic fumes throughout the neighborhood. Residents also suspected some of the barrels of chemicals that had been abandoned by the company had leached into the yards of nearby homes and into tributaries of Columbia's drinking water sources.

Many of Keys's neighbors complained of unusual skin rashes, headaches, and illnesses. Officials from EPA and Columbia's mayor initially dismissed the residents' complaints as unsubstantiated; no health assessment was ever conducted. Reichhold spokesperson Alec Van Ryan later acknowledged to local news media, "I think everyone from the EPA on down will admit the initial communications with the community were nonexistent" (quoted in Pender, 1993, p. 1).

Ultimately, Keys organized her neighbors to speak with local officials and at public meetings. One such meeting occurred with officials from the federal Agency for Toxic Substances and Disease Registry (ATSDR), who traveled to Columbia to propose a health study of the residents. However, the ATSDR officials proposed only to sample the neighbors' urine and hair and test these for recent, acute exposure to toxins. Keys and other residents objected. They explained their exposure had initially occurred much earlier, when the plant exploded, and had been ongoing since then. Having done their homework, they insisted that the appropriate test was one that sampled blood and fatty tissues for evidence of long-term, chronic exposure. Keys urged the ATSDR officials to adopt this approach because, she said, "The evidence is in my body!" (C. Keys, personal correspondence, September 12, 1995).

The ATSDR officials refused this request, citing budgetary constraints. In turn, the residents felt stymied in their efforts to introduce the important personal evidence of their long-term exposure to chemicals that they believed was evident in their bodies. The meeting degenerated into angry exchanges and ended with an indefinite deferral of the plans to conduct a health study.[5] Like Augustine, Keys did not give up and her persistence helped not only her community but also to build a stronger movement for environmental justice.

Unfortunately, the tension between the ATSDR and the residents of Columbia, Mississippi, is not unusual. Too often, agency officials dismiss the complaints and recommendations of those facing the greatest risk of toxic chemical exposure, believing that such people are emotional, unreliable, and irrational. For example, in an early study of public comments on the EPA's environmental impact studies, political scientist Lynton Caldwell (1988) found that

> public input into the . . . document was not regarded by government officials as particularly useful. . . . The public was generally perceived to be poorly informed on the issues and unsophisticated in considering risks and trade-offs. . . . Public participation was accepted as inevitable, but sometimes with great reluctance. (p. 80)

We also have overheard agency officials complain, after hearing reports of family illness or community members' fears, "This is very emotional, but where's the evidence?" "I've already heard this story," or simply, "This is not helpful."

To be clear, we are not suggesting that the indecorous voice results in a person's rhetorical incompetence or a failure to find the "right words" to articulate a grievance. Instead, we're suggesting that the arrangements and procedures of power may

Courtesy of the Sierra Club

Photo 10.2	The Connect the 9 Community Bikeride took place in April 2014 as a biking tour, organized by the Sierra Club, Global Green USA, the Green Project, Bike Easy, and the Lower Ninth Ward Center for Sustainable Engagement and Development (CSED). Sierra Club Environmental Justice organizer Darryl Malek-Wiley noted "After Katrina, community members decided they wanted to make their neighborhoods sustainable, and that doesn't just mean greener. It also means restoring natural areas like the cypress-tupelo wooded swamp in Bayou Bienvenue, which is within the city limits in the Lower Ninth Ward."

undermine the respect accorded to such individuals by narrowly defining the acceptable rhetorical norms of environmental decision making.

Sacrifice Zones: Out of Sight, Out of Mind?

Environmental justice communities often describe themselves as **sacrifice zones** (Bullard, 1993); that is to say that the neighborhoods that bear the disproportionate burden for unsustainable practices, such as toxic pollution, do not share in meaningful benefits and are part of the "cost" of not changing course. Such neighborhoods often exist outside of dominant viewpoints, making their sacrifices easier to ignore for those who don't live in those communities.

One particularly striking form of communication used more and more by environmental justice groups to connect local communities and wider publics to resist this barrier to recognition is what grassroots activists call **toxic tours.** In *Toxic Tourism: Rhetorics of Pollution, Travel, and Environmental Justice,* Phaedra C. Pezzullo (2007) defines these as "non-commercial expeditions into areas that are polluted by toxins. . . . More and more of these communities have begun to invite outsiders in, providing tours as a means of educating people about and, it is hoped, transforming their situation" (p. 5). (For a vivid example of one such toxic tour in the "Cancer Alley" area of Louisiana, see Pezzullo, 2003.)

Often, these outsiders are journalists, government officials, corporate representatives, environmental allies, religious groups, and other supporters who—in personally experiencing the conditions of a community under environmental stress—are more likely to be persuaded to identify and, therefore, to care and publicize their experiences more widely. On a toxic tour of southern California's Coachella Valley, for example, a director of the state Department of Toxic Substances Control, commented of an arsenic-laced well: "There's the human reaction of imagining myself raising my family in a mobile home where the water looks like it came out of a ditch" (Flaccus, 2011). Unlike EPA or other agency inspections of toxic sites, toxic tours highlight "discourses of . . . contamination, of social justice and the need for cultural change" (Pezzullo, 2007, pp. 5–6). Although environmental advocates have taken reporters and others into natural areas such as Yosemite Valley and the Grand Canyon for the past century to build support for their protection, this use of toxic tours of places of sacrifice is more recent.

Witnessing Environmental Injustices in the Maquiladoras

Some years ago, the two of us and other environmental leaders participated in a toxic tour together outside of Matamoros, Mexico, south of the U.S. border near Brownsville, Texas. This area, which is part of the *maquiladora* zone or manufacturing area, has large numbers of (largely unregulated) industrial plants. These plants relocated to this area from the United States as a result of the North American Free Trade Agreement (NAFTA). Living in nearby *colonias* (crowded, makeshift housing on unoccupied land), the *maquiladora* workers and their families are subject to severely contaminated air and water, abysmal sanitation, and unsafe drinking water; many suffer from a number of illnesses. As Peterson, Peterson, and Peterson argue,

> Both unsustainable development and environmental injustice are chronically acute on borders between comparatively affluent and poor nations (for example, United States/Mexico, Costa Rica/Nicaragua, South Korea/North Korea), where long-time residents and mushrooming immigrant populations are prone to differential treatment, differential access to political systems, and differential conceptions of justice. (2007, pp. 189–190)

The tour through the crowded *colonias* was organized by the Sierra Club and its Mexican allies to introduce leaders from environmental groups to the threats to human health from pollution and unhealthy living conditions for workers and their families. As we walked through the unpaved streets by the workers' homes, we felt overpowered by the sights, smells, and feel of an environment under assault. Strong chemical odors filled the air, small children played in visibly polluted streams by their homes, while others, barely older, scavenged in burning heaps of garbage for scraps of material they could sell for a few pesos.

Being in a community harmed by such hazards opens visitors' senses of sight, sound, and smell and this awareness builds support for the community's struggle: "Odorous fumes cause residents' and their visitors' eyes to water and throats to tighten . . . a reminder of the physical risk toxics pose." One toxic tour guide observed that toxic tours give visitors "firsthand" evidence of "the environmental insult to residents [of having polluters so close to their homes], as well as the noxious odors that permeate the neighborhood" (Pezzullo, 2004, p. 248).

As toxic tours show, the environmental justice movement continues to confront real-world, on-the-ground challenges to publicize communities' grievances that are often excluded otherwise. Beyond this, environmental justice activists insist the movement embodies "a new vision borne of a community-driven process whose essential core is a transformative public discourse over what are truly healthy, sustainable and vital communities" (National Environmental Justice Advisory Council Subcommittee on Waste and Facility Siting, 1996, p. 17).

The Global Movement for Climate Justice

The environmental justice movement has expanded globally. Here, we want to turn our attention to how the movement for environment justice has been embraced by a vibrant and growing global movement for climate justice, which overlaps and exceeds the previous movement's goals. The largely dispersed, grassroots movement for climate justice views the environmental and human impacts of climate change from the frame of social justice, human rights, and concern for indigenous peoples. In this section, we describe the construction of a new frame of climate justice, as well as the efforts of communities affected by climate change to gain a voice internationally.

Climate Justice: A Frame to Connect the World

Climate scientists and advocates for climate justice generally agree "the greatest brunt of climate change's effects will be felt (and are being felt) by the world's poorest people" (Roberts, 2007, p. 295). Without immediate climate action, the Intergovernmental Panel on Climate Change (IPCC) forecasts that "hundreds of millions of people in developing nations will face natural disasters, water shortages and hunger due to the effects of climate change" (Adam, Walker, & Benjamin, 2007, para. 5). "In 2012, one person every second was displaced by a climate or weather-related natural disaster," according to the Environmental Justice Foundation's 2014 report, *The Gathering Storm: Climate Change, Security and Conflict.* As they note, "with millions of people displaced each year by rapid-onset climate-related hazards and an unknown number fleeing slow-onset environmental degradation, a changing climate presents pressing operational and geopolitical challenges to a number of states" (p. 5).

A Cruel Irony: Impacts of Climate Change

Human rights groups and environmental scholars also charge that the voices of those most affected by climate change are often not part of the conversation about solutions. Dale Jamieson (2007) notes that seventy million farmers and their families in Bangladesh will lose their livelihoods when their rice paddies are inundated by seawater. Yet despite the vast number of people around the world who will suffer from climate change, most of them are not included when decisions are made. (p. 92)

He adds, for this reason, "participatory justice is also important at the global level" (p. 92).

There is also a *cruel irony* in this exclusion. As former *New Yorker Times* reporter Andrew Revkin (2007) observed, "In almost every instance, the people most at risk from climate change live in countries that have contributed the least to the buildup of carbon dioxide and other greenhouse gases linked to the recent warming of the planet" (para. 2). That is, the **cruel irony** of the climate crisis is that those who have contributed the least to the energy policies and demands that cause climate change and have profited the least from these industries have been and will continue to be impacted the most by climate change. Further, as Jacqueline Patterson, executive director of the NAACP's Climate Justice Initiative, notes, our collective response to extreme weather events remains inequitable:

> What's significant about disasters like Hurricane Katrina or even the flooding and tornadoes in Alabama and Mississippi a couple years ago is you start to see the differential impact and the differential response on communities of color. With Hurricane Katrina, it was more a story about the differential response, not as much of a story about climate change. (as quoted in Hsieh, 2014)

As the effects of climate change—particularly in vulnerable areas of Asia and Africa—began to be experienced by local communities and regions, nongovernmental organizations (NGOs) from these nations began to build alliances and coordinate with activists in Europe and the United States. One of the most consequential developments of this alliance would be the elaboration of climate justice as a new frame in their construction of messages about global warming and in the mobilizing of others.

Framing Climate Crises as Unethical and as Human Rights Exigencies

The phrase *climate justice* was apparently used first in academic literature by Edith Brown Weiss (1989) in her study *In Fairness to Future Generations: International Law, Common Patrimony, and Intergenerational Equity.* A more movement-oriented demand for climate justice, however, may have been voiced first in the mid-1990s by Tom Goldtooth, the founder of the Indigenous Environmental Network; it was further developed in a 1999 CorpWatch report and was the basis for a resolution at the Second People of Color Environmental Leadership Summit in 2002 (Tokar, 2010, pp. 45–46).

In Bali, Indonesia, in August 2002, a coalition of international NGOs, including the National Alliance of People's Movements (India), CorpWatch (United States), Greenpeace International, Third World Network (Malaysia), Indigenous Environmental Network (North America), and groundWork (South Africa) crafted one of the first declarations redefining climate change from the perspective of environmental justice and human rights. Meeting alongside the UN delegates preparing for the Bali session earlier, the coalition developed the **Bali Principles of Climate Justice.** The principles pledged to "build an international movement of all peoples for Climate Justice" (Bali Principles of Climate Justice, 2002, para. 19). The effort would be based on certain principles, echoing values of the 1991 Principles of Environmental Justice while introducing a new focus on climate change:

1. Affirming the sacredness of Mother Earth, ecological unity and the interdependence of all species.

2. Climate Justice insists that communities have the right to be free from climate change, its related impacts and other forms of ecological destruction.

3. Climate Justice affirms the need to reduce with an aim to eliminate the production of greenhouse gases and associated local pollutants.

4. Climate Justice affirms the rights of indigenous peoples and affected communities to represent and speak for themselves. (para. 20)

One consequence of the Bali Principles, as well as other declarations, was to shift "the discursive framework of climate change from a scientific-technical debate to one about ethics focused on human rights and justice" (Agyeman, Doppelt, & Lynn, 2007, p. 121).

Another important moment in this shift came on October 28, 2002, when more than 1,500 individuals—farmers, indigenous peoples, the poor, and youth—from more than 20 countries marched for climate justice in New Delhi, India (Roberts, 2007). The culmination of the summit was the **Delhi Climate Justice Declaration.** The declaration concluded,

> We, representatives of the poor and the marginalized of the world, representing fishworkers, farmers, Indigenous Peoples, Dalits, the poor, and the youth, resolve to actively build a movement . . . that will address the issue of climate change from a human-rights, social justice, and labour perspective. (Delhi Climate Justice Declaration, 2002, para. 12)

The declaration expressed the attendees' resolve to "build alliances across states and borders to oppose climate change inducing patterns and advocate for and practice sustainable development" (para. 12).

Other international gatherings and declarations have followed, including the Durban Declaration on Carbon Trading (2004), a critique of proposed market schemes for trading greenhouse gas emissions; the People's Declaration for Climate Justice (Sumberklampok Declaration) (2007) in Bali, Indonesia, an effort to influence the start of negotiations among the world's nations of a new post-Kyoto treaty; and the large, international mobilizations at the UN climate conferences from Berlin, Germany, in 1995 to annual gatherings, with the 2015 one hosted in Paris, France.

Overall, as opposed to a more general climate change framework primarily focused on scientific-technical information, biocentric reports, and anthropocentric causes, a climate justice frame communicates a value in ethical appeals, human rights, and impacts that evoke exigency to act.

Another Viewpoint: From Fossil Fuels to a "Just Transition"

Black Mesa Water Coalition has mobilized a national network in the United States of environmental justice communities, which are challenging the fossil fuel industry for what they call a "just transition" away from unsustainable industries and toward renewable energy. They emphasize that they are not just calling for a divestment in fossil fuels, but also a reinvestment in more sustainable energy sources and the communities most impacted by these choices. Watch their eight-minute video to listen for how they are making connections between patterns of environmental injustice and building support for climate justice jobs, policies, and ways of living: The Climate Justice Alliance (2014), *Our Power Film/Black Mesa Water Coalition*. Available at http://vimeo.com/84751170.

Mobilizing for Climate Justice

As a result of their exclusion from the official forums, climate justice activists have sought to create alternative structures for communication. In 2014, Anjal Appadurai, a climate activist who helped organize a protest in Vancouver of Canadian Prime Minister Stephen Harper with signs for "CLIMATE JUSTICE NOW," emphasized the significance of marginalized voices getting heard in global arenas: "Activism takes many forms, but the primary motive is to have the voice heard" ("Climate Change Protestors," 2014). Let's look at some of the ways the climate justice movement is attempting to communicate these ideas and perspectives.

Transnational Organizing

At global summits and elsewhere, climate activists emphasize the need to create lines of communication across borders and within regions to build a movement for climate justice. Given the geographic distance of the many involved, it perhaps is not surprising that the climate justice movement itself is sustained largely online through social networking sites and LISTSERVs that help to mobilize activists for actions at sites such as the Copenhagen conference. For example, the India Climate Justice Forum is hosted by the India Resource Center, a project of Global Resistance, whose goal is "to strengthen the movement against corporate globalization by supporting and linking local, grassroots struggles against globalization around the world" (indiaresource.org).

The climate justice movement also has initiated new social networking sites, blogs, and information sites of its own. For example, the London-based Rising Tide Coalition for Climate Justice (risingtide.org.uk) consists of environmental and social justice groups from around the world, especially in Europe (Roberts, 2007). There is also a Rising Tide North American network (2008, risingtidenorthamerica.org). Rising Tide grew out of the efforts of groups who came together to organize events alongside a UN climate conference in The Hague in 2000. Other prominent online networks include Environmental Justice Climate Change Initiative (ejcc.org); It's Getting Hot in Here (itsgettinghotinhere.org,), and 350.org.

The number 350 in climate justice discourse refers to 350 parts per million of carbon dioxide as the level in the atmosphere that is safe for the global climate. Like many other online sites, 350.org provides daily updates, video uploads from local activists, and analyses of official proposals. The site also has played a prominent role in the past several years, mobilizing climate justice activists ahead of the UN climate conferences across the globe, and annually since then, in same-day, international displays of concern. Their website claims to involve grassroots activists from 188 countries and "Email is how 350 connects" (350.org). Co-leader Naomi Klein describes how activists from a community in a country like El Salvador have found the link between their history of struggles for social justice to resonate with climate activism, because both are about a desire for "deepening democracy":

> It's a community that was born out of the civil war, a community of refugees . . . and they bring their revolutionary history—and their history of fighting for economic justice—to the climate fight. They're finding ways to respond to climate change that really transform their community in every way, from housing to health care. . . . Climate change is the human-rights struggle of our time. And it's too important to be left to the environmentalists alone. (Stephenson, 2012)

As a result of its online and offline organizing, the climate justice movement brings together coalitions of students, antinuclear and social justice activists, indigenous peoples, academics, opponents of carbon trading, religious groups, and others.

This transnational organizing perhaps was best witnessed so far in the global events on September 21, 2014, when hundreds of thousands of people around the world rallied to "sound the alarm" for the need for climate action. In New York City, the march itself was organized to represent the importance of organizing across communities and concerns, with the following messages represented in six parts of the march (http://peoplesclimate.org/lineup/):

1. Frontlines of Crisis, Forefront of Change (indigenous, environmental justice and other sacrifice zone communities)

2. We Can Build the Future (labor unions, families, students, elders, and more)

3. We Have Solutions (just transition possibilities, including renewable energy and water justice)

4. We Know Who Is Responsible (calling out those holding back progress)

5. The Debate Is Over (scientists and interfaith communities, and more)

6. To Change Everything, We Need Everyone (LGBTQ, people representing states and countries around the world)

Strategic Campaigns

Yet, others have questioned whether the reliance on mass action alone is equal to the scale and complexity of the global economic system that is responsible for climate change: "We really do need a longer term plan," acknowledged Patrick Bond, the director of the Center for Civil Society in South Africa, a plan that "will make the gains we've taken, on the streets and in the communities . . . actually real. How can they be turned into good public policy?" ("Only Political Activism," 2010, p. 186). Radical environmentalist Eirik Eiglad (2010) similarly asked, "How can we see beyond the current . . . focus on the major climate summits [like Copenhagen or Cancun], important as they may be?" (p. 10).

While efforts so far have failed to persuade the U.S. Congress and many other governments that contribute to greenhouse gases to act in significant ways on climate change, some environmental groups are pursuing more strategic campaigns outside of the political process to reduce greenhouse gases directly. Perhaps the most successful of these efforts has been the Sierra Club's Beyond Coal campaign against coal-burning power plants (Chapter 8). As a result of the targeting of state utility commissions, banks, and energy companies themselves, the campaign and its supporters have blocked or cancelled more than 150 proposed plants in the past five years. Yet, even as plants are cancelled in the United States, coal-fired power plants continue to be built as other nations pursue their own economic paths to development. Continuing to study such acts of mobilization, Cox (2010) argues, is essential not just to improve the movement's ability to achieve a particular end, but also to build alliances that can adapt to changing political and climatic situations.

Others, therefore, acknowledge that slowing then reversing climate change will require a basic restructuring of the global economy. What is needed, they argue, are "broad and visionary alliances with people and movements around the world to begin the fundamental transformation of society" (Petermann & Langelle, 2010, p. 187). How to achieve this—and what strategic leverage is required to affect global economies—are challenging questions that are still being debated. (See "Act Locally! How Can You Make a Difference on Global Climate Change?")

Act Locally!

How Can You Make a Difference on Global Climate Change?

"Think globally, act locally" has been a saying in the environmental movement since Earth Day. But how can an individual act locally on a problem like global climate change? We talk about sustainability efforts more in the next chapter, but it is important to remember that you can become more involved in addressing climate change:

1. Investigate what your college or university is doing to reduce its energy consumption or operate in a more sustainable way and find out if you can become involved in these initiatives.

(Continued)

(Continued)

2. Use an app or online carbon footprint tracker to calculate (and reduce) your own output of carbon dioxide. Check out the numerous trackers that are available, like Footprint Tracker at http://carbontracker.com, and iPhone, iPad, and Android all have carbon footprint apps for doing this. Better yet, initiate a crowd sourcing contest to find the top three personal carbon footprint trackers for your campus or group.

3. Call or go online to your local electric utility company for its energy savings guide or request an energy audit, identifying ways to save energy in your apartment or house.

4. Find opportunities to speak at public hearings on renewable energy, such as a bill in your state legislature requiring a percentage of electricity in your state be produced from renewable energy or a town council's public comments on a proposal for a light rail or new bike lanes.

Every year, the exigency of the climate crisis increases. Repeatedly, the IPCC and those who cover it claim that communication is one of the major areas requiring attention. As we discussed in Chapter 6, while global scientists have consensus on the phenomenon that is climate change, its main causes, and its potential consequences, they feel their inability to communicate this information in compelling ways in an age of uncertainty and risk is their main challenge. In particular, "the IPCC has come in for significant criticism for the way in which it has communicated the complex uncertainties inherent in climate science" (Climate Outreach Network, 2014, p. 1). The National Climate Assessment website that allows one to view how climate change may impact any place in the world is one attempt to improve climate communication (National Climate Assessment, 2014).

A Revival of Civil Disobedience

An act of civil disobedience is a peaceful form of protest that violates laws and accepts legal consequences (such as arrest) in order to point out ongoing injustices (Chapter 2). Many global activists who choose civil disobedience as a tactic draw inspiration from Mahatma Gandhi, a leader at the turn of the 20th century in the struggle for Indian civil rights in South Africa and later in the overthrowing of British colonial rule over India. In contrast to more negative perspectives, Gandhi's philosophy optimistically underscores the potential to redeem an institution or opponent through such nonviolent protest.

Although climate justice activists engage the range of public advocacy tactics we have discussed, it is notable that an increasing number of people support climate justice advocates using civil disobedience to participate in the public sphere. According to the Yale Project on Climate Change Communication's and George Mason University Center for Climate Change Communication's 2013 survey report, almost one in four U.S. citizens would "support an organization that engaged in non-violent civil disobedience against corporate or government activities that make

Photo 10.3	A strong advocate for climate action, United Nations Secretary General Ban Ki-Moon from Korea is presiding over the UN summits about potential international climate change agreements from 2007–2015, perhaps the most crucial years of climate change negotiations in history.

global warming worse," and one in every eight people say "they would be willing to personally engage in non-violent civil disobedience against corporate or government activities that make global warming worse" (Yale, 2013, p. 4).

In order to dramatize the exigence of climate injustices and the moral obligation to respond, a revival of civil disobedience has begun. In 2008, Nobel Peace Prize winner and former U.S. vice president Al Gore had urged the next generation of climate activists to use civil disobedience:

> If you're a young person looking at the future of this planet and looking at what is being done right now, and not done, I believe we have reached the stage where it is time for civil disobedience to prevent the construction of new coal plants that do not have carbon capture and sequestration. (Vitello, 2008)

That year, as an undergraduate at the University of Utah, Tim DeChristopher heard a scientist of the IPCC on climate change explain the exigence of the climate crisis and decided to join local activists who were protesting the auctioning off of public lands to fossil fuel corporations. He decided to register for the auction and outbid the corporations. Once he had won several bids and was identified as someone incapable of paying for these public lands, he was arrested. DeChristopher spent two years in jail, was released in 2013, and became a national icon for a new generation of climate justice activists. Upon going to jail, he declared,

> At this point of unimaginable threats on the horizon, this is what hope looks like. In these times of a morally bankrupt government that has sold out its principles, this is what patriotism looks like. With countless lives on the line, this is what love looks like, and it will only grow. (Peaceful Uprising, n.d.).

Since his release from jail, DeChristopher joined Harvard Divinity School and became part of Divest Harvard, a group of students attempting to convince their administration to divest from fossil fuel investments and reinvest in renewable energy. On May 1, 2014, one undergraduate, Brett A. Roche, was arrested in an act of civil disobedience, though all charges were dropped. Frederick E. Small, a Harvard Divinity School graduate and a Unitarian minister based in the area, said, "I don't think Harvard is scared of Brett [Roche] physically. I think Harvard is scared about Brett morally" (Clarida, 2014). The next day, seven Washington University students attempted to enter the quarterly meeting of their Board of Trustees in protest of Peabody Energy CEO Greg Boyce being a member and consequently were arrested. Caroline Burney, a member of the group, Students Against Peabody, clarified that the civil disobedience was connected to broader struggles for climate justice:

> Today's arrests are part of a larger fight against Peabody Coal in St. Louis, across the country, and around the world. We're here for ourselves and for all of the other communities that Greg Boyce and Peabody Coal have destroyed, including Rocky Branch, Illinois and Black Mesa, Arizona. Our fight will continue here in St. Louis at next week's Peabody shareholder's meeting and in Black Mesa at the end of May. (Biggers, 2014)

Many express a growing sense of not just frustration, but urgency, and a moral calling to respond. As Mithika Mwenda, secretary general of the Pan African Climate Justice Alliance, said, "If rich industrialised countries continue to block these talks we . . . will build our forces to hold them to account. We will not accept delay and we will demand our governments withdraw from an unsatisfactory outcome" (Vidal, 2013).

Not just youth are choosing civil disobedience. Forty-eight leaders were arrested on February 13, 2013, in front of the White House as part of their ongoing campaigns to call on U.S. president Barack Obama to reject the Keystone XL tar sands pipeline. Among those arrested for blocking the sidewalk were civil rights leader Julian Bond, climate scientist James Hansen, 350.org founder Bill McKibben, and the executive director of the Sierra Club, Michael Brune, whose participation marked the first use of civil disobedience in the organization's 127-year history. (See Brune, 2013, for the Sierra Club's rationale for engaging in CD.) The action came the day before the Forward on Climate Rally (featured in the opening photograph of this chapter).

On September 22, 2014, a day after the largest climate march in history, at least 2,000 global activists from young to old risked arrest organized under the title of "#FloodWallStreet" declaring,

> Wearing blue to represent the sea that surrounds us, we rise to the steps of the NY Stock Exchange at 12:00 pm, flooding the area with our bodies in a massive sit-in—a collective act of nonviolent civil disobedience—to confront the system that both causes and profits from the crisis that is threatening humanity. (http://floodwallstreet.net/)

Communicating the sense of urgency only seems to build.

SUMMARY

In this chapter, we identified some of the communication practices of the grassroots movements for environmental justice and climate justice, as well as their challenges.

- In section one, we described the emergence of a critical rhetoric of environmental justice as people of color and working-class communities challenged mainstream environmental groups' organizational biases and discourse about the environment as a place apart from humans. The environmental justice movement has

  o demanded a halt to the inequitable burdens often imposed on poor and people of color communities,
  o called for more opportunities for those who are most affected by environmental injustices to be heard in the decision making of corporations and public agencies, and
  o articulated a vision of environmentally healthy and economically sustainable communities.

- Section two introduced the barriers and responses to public participation:

  o *Indecorous voices*: communication practices that dismiss the voices of some as "inappropriate" or "emotional," as well as persistent environmental justice leaders who continue to speak up.
  o *Sacrifice zones*: communities that bear the disproportionate burden of unsustainable practices and are hidden from mainstream sight, as well as the use of *toxic tours* to call attention to the sights, sounds, and smells of environmental oppression.

- In section three, we described the global movement for climate justice and the ways in which it has reframed the threat of climate change as a matter of ethics, human rights, and exigence to act. We also compared advocacy choices to mobilize online, imagine new alternatives, and commit acts of civil disobedience.

Although the ecological, cultural, and political challenges we face have no easy answers, the environmental justice and climate justice movements have mobilized grassroots communities globally in ways that enable us to imagine the topic of our next chapter: "sustainability."

SUGGESTED RESOURCES

- The U.S. EPA (2014) has launched a 20th Anniversary Video Series. Watch a few videos and attempt to find overlap and differences in the struggles and language choices used. See http://www.epa.gov/environmentaljustice/events/20th-anniversary.html.
- Plan EJ 2014 notes several resources communities may use to gather information and to make choices.
- Many of these tools are digital mapping resources that continue to be developed. Compare and contrast some of the following to see how accessible and persuasive these tools are for addressing environmental injustices: Community-Focused Exposure and Risk Screening Tool (http://www.epa.gov/heasd/c-ferst/); Community Cumulative Assessment Tool (http://www.epa.gov/research/healthscience/health-ccat.htm); Tribal-Focused Environmental Risk and Assessment Tool (http://www.epa.gov/heasd/research/tferst.html).
- Toxic tours—or a version of them—increasingly are offered online. Communities for a Better Environment is a nonprofit organization that shares videos, stories, and images on their website. See http://www.cbecal.org/get-involved/toxic-tours/.
- *Disruption*, 2014, is an online video created to mobilize communities for the People's Climate March. Watch it to consider the ways the climate justice movement is creating a new frame that responds to barriers faced in relation to climate science communication, environmental injustices, and more. See http://watchdisruption.com/.
- There is a great deal of music inspired by environmental justice (see, for example: Mos Def's song, "New World Water" or Indie Rock band Dutch Holland's "Toxic Tour") and climate justice (see, for example: http://music.peacefuluprising.org/album/folk-songs-for-climate-justice) efforts. Choose a song to analyze for its rhetorical appeals or make a list of ten songs that move you the most.

KEY TERMS

Bali Principles of Climate Justice   250

Cruel irony   249

Decorum   243

Delhi Climate Justice Declaration   250

Disparate impact   238

Economic blackmail   239

Executive Order 12898 on Environmental Justice   242

First National People of Color Environmental Leadership Summit   240

Indecorous voice   243

DISCUSSION QUESTIONS

1. Environmental historian Robert Gottlieb poses a challenging question: "Can mainstream and alternative groups find a common language, a shared history, a common conceptual and organizational home?" (Gottlieb, 2003, p. 254). What do you think? Are a common language and shared history needed for environmentalists and activists from diverse neighborhoods to work together?

2. What effect has the climate justice movement had to date in influencing action on climate change? How effective is mass action in the street or global, same-day protests like the Forward on Climate Rally? Do you think civil disobedience makes a persuasive rhetorical appeal about the exigence of the climate crisis we face?

3. 350.org activist and journalist Naomi Klein (2014) argues a few key cultural shifts should be the focus of future efforts:

   a. "Climate change demands that we consume less, but being consumers is all we know."

   b. "Climate change is slow, and we are fast."

   c. "Climate change is place-based, and we are everywhere at once."

   d. "Climate pollutants are invisible, and we have stopped believing in what we cannot see."

   Discuss these four challenges. How are our identities, sense of timing, sense of place, and ability to believe based in our relationship with communication? Can you find media examples about climate justice that reflect these four challenges or responses to them?

4. Philippines delegate Naderev (Yeb) Saño's speech on the first day of the COP19 Climate Change Summit in Poland, November 11, 2013, is a rich rhetorical text that can be found readily via YouTube. What types of rhetorical appeals does he make to move his audience? How does he draw on ethos, pathos, and logos? (For more context, a link to the speech, and an interview with Saño, see Amy Goodman and Juan González, "Stop This Madness": Filipino Climate Chief Yeb Saño Begins Hunger Fast to Protest Global Inaction, Democracy Now! Available at http://www.democracynow.org/2013/11/12/stop_this_madness_filipino_climate_chief).

NOTES

1. Di Chiro (1996) noted, "Eventually, environmental and social justice organizations such as Greenpeace, the National Health Law Program, the Center for Law in the Public Interest, and Citizens for a Better Environment would join [the] Concerned Citizens' campaign to stop [the proposed facility]" (p. 527, note 2).

2. The Group of Ten were the Environmental Defense Fund, Friends of the Earth, the Izaak Walton League, the National Audubon Society, the National Parks and Conservation Association, the National Wildlife Federation, the Natural Resources Defense Council, the Sierra Club, the Sierra Club Legal Defense Fund (now Earth Justice), and the Wilderness Society. They remain predominantly led by white men (Goldenberg, 2014).

3. Cox was fortunate to have the opportunity to attend and participate in the sessions that included leaders of traditional environmental organizations. Pezzullo attended and participated in the Second National People of Color Environmental Leadership Summit.

4. Portions of this section are derived from a paper Cox presented at the Fifth Biennial Conference on Communication and the Environment (Cox, 2001b).

5. Reichhold Chemical ultimately offered to assist community members by helping to fund a health study and a community advisory panel to assist in decisions about the polluted site.

"We are caught in the paradox that we cannot save the world without saving particular places. But neither can we save our places without national and global policies that limit predatory capital and that allow people to build resilient economies, to conserve cultural and biological diversity, and to preserve ecological integrities" (Orr, 1994, p. 170). When U.S. president Obama used his executive authority in 2014 to create the largest marine sanctuary in the world, he cited his experience growing up in Hawaii as part of what motivated him to do so.

CHAPTER 11

Sustainability and the "Greening" of Corporations and Campuses

Sustainability . . . is not an end point, not a resting place, but a process.

—Peggy F. Bartlett and Geoffrey W. Chase, 2004, p. 7

We have defined *sustainability* as the capacity to negotiate environmental, social, and economic needs and desires for current and future generations. But how does one learn what environmental educator Orr calls "the arts of inhabitation"? When should we focus on our individual impacts on the environment and when should we turn our attention to structural change? And who is this "we"?

To attempt to respond to these questions, in this chapter, we introduce you to the fundamental facets of sustainability, critical ways to interpret and to assess corporate "green" products and services, and some of the ways college campuses are responding to the challenges and opportunities posed by unsustainable practices.

Chapter Preview

- In the first section of this chapter, we clarify how sustainability encompasses the 3 Es (environment, economics, and social equity) or the 3 Ps (people, prosperity, and the planet).
- In the next two sections, we describe a "free market" discourse that underlies much of corporate environmental communication. Then, we examine three communication practices of corporations that are used in "green marketing": (a) product advertising; (b) image

(Continued)

(Continued)

enhancement; and (c) corporate image repair. Through this discussion, we make distinctions between "greenwashing" (the corporate use of deceptive advertising) and "green consumerism" (consumer efforts to buy products to try to live more sustainably through eco-products, fair trade, and sustainable tourism).

- In the final section, we offer a brief overview of sustainability efforts on college campuses, which promote a wide variety of exciting ways to become involved in sustainability efforts.

Although we can identify core values of sustainability, there is no "one size fits all" model. Each ecosystem and community is unique. It is important, therefore, as we go through this chapter, to note overall values and goals expressed in broader discourses of sustainability and observe how they vary in particular contexts with the specific means available. Further, as Bartlett and Chase remind us (above), sustainability is never over, it is an ongoing process that requires continual reflection, dialogue, education, networking, publicity, and reassessment.

Sustainability: An Interdisciplinary Approach

Environmental challenges are rarely met by a single discipline or perspective. Fish-kills in a river may require knowledge of hydrology, soil erosion, and ichthyology (the study of fish). All of those environmental sciences, however, only cover physical relations. But they cannot account for the attitudes of farmers upstream or the ways indigenous people use the river for subsistence fishing or the value tourists in the region place on the aesthetics of the site or the location of the river near an e-waste recycling facility. Interest in sustainability, as a result, has emerged to begin to address these differing perspectives.

Perhaps the most famous definition of "sustainability" was offered in 1987 through The **Brundtland Commission Report** (1987) of the United Nations, which defined "sustainable development" as that which "aims to promote harmony among human beings and between humanity and nature," and, importantly, that which "meets the needs of the present without compromising the ability of future generations to meet their own needs" (sections 81, 1). Since then, sustainability has become a very popular discourse, used by governments, corporations, and communities across the globe. As such, we have begun to fine-tune sustainability as more than an intergenerational, environmental perspective, but one that must recognize that environmental challenges cannot be met by scientific expertise or any one perspective alone.

As noted in Chapter 2, the rhetorical appeal of sustainability "lies in its philosophical ambiguity and range" (Peterson, 1997, p. 36). Nevertheless, the movement for sustainability generally encompasses three goals or aspirations that are called the **Three Es**: Environmental protection, Economic health, and Equity (social justice) or the **Three Ps**: People, Prosperity, and the Planet. That is, "sustainability" explicitly notes a commitment not just to the environment, but also to social justice and economic well-being.

Consider, for example, how U.S. president Barack Obama's speeches addressing the environment switched from "greening the economy" and "green jobs" to moral language about "future generations" and "our children." Environmental reporter Chris Mooney (2013) insightfully notes that it is interesting to consider how the initial statements were responding, in part, to hard economic times and the latter to extreme weather events that have caught the nation's attention. Yet, we want to raise this example to show how either approach depends on a sustainable perspective. That is, the president never speaks just about the environment or planetary goals; he always links those goals to the other two Es and Ps: economics/profits (jobs and new green industries needed) and equity/people (the fate of our children and how ecological disasters already are ruining homes of everyday, hard-working people). The solution requires that we recognize that "saving the planet" is interconnected with saving people and our livelihoods.

As popular as sustainability discourses are in political speeches and throughout the public sphere today, sustainable policies often still struggle to be enacted, nationally and internationally. As environmental justice scholar Julian Agyeman (2005) points out, there is a **sustainability gap**: "In almost all areas of sustainability, we know scientifically what we need to do and how to do it; but we are just not doing it" (p. 40). So, how do we motivate more people to live more sustainable lives? Corporations and campuses are two significant sets of institutions attempting to define and to promote visions of sustainability. Let's look at their choices.

Free Market Discourse and the Environment

Before looking more closely at the diverse forms of corporate communication, it's important to appreciate the ideological premises and sources of persuasion that underlie much of these appeals. Corporate advocacy regarding the environment does not occur in a vacuum. Instead, it often draws upon an ideological *discourse* (Chapter 3) that circulates a set of meanings about business and the proper role of government. This is a discourse about the nature of economic markets and the role of government and is particularly evident in the opposition of corporations to environmental standards.

Behind the discourse of much corporate environmental communication is the belief in a "free market," a phrase that is usually meant to refer to the absence of governmental restriction of business and commercial activity. A **discourse of the free market**, therefore, sustains the idea that the private marketplace is self-regulating and ultimately promotes the social good. As a result, a discourse of free markets constructs a powerful *antagonism* (Chapter 2), that is, a questioning of environmental rules, taking such rhetorical forms as, "We need to get 'big government' off our backs," and "Companies will find the best solutions when left to themselves."

At the core of this rhetoric is the belief held by many business leaders that environmental protection can be secured by the operation of the marketplace through the unrestricted or unregulated buying and selling of products and services. Such faith in the market assumes that "the public interest is discovered in the ability of private markets to transform the individual pursuit of self-interest into an efficient social allocation of resources" (Williams & Matheny, 1995, p. 21). In the sections that follow, we see the use of *free market discourse* as a rationale for some corporations' opposition to government-imposed standards for environmental performance or public health.

Corporate Sustainability Communication: Reflection or Deflection?

As popular support for the environment has increased, many industries have worked to improve their environmental performance and to link their goods, services, or brand with "green" initiatives. In this section, we highlight three common communication strategies: *green marketing, greenwashing,* and *green consumerism.*

U.S. corporations spend several billion dollars a year on public relations related to sustainability, or what is sometimes called *green marketing.* **Green marketing** is a term used to refer to a corporation's attempt to associate its products, services, or identity with environmental values and images. We use the term *green marketing* more specifically in this chapter to refer to corporate communication that is used for three purposes: (1) product advertising, (2) image enhancement, and (3) image repair.

We look at charges that these marketing efforts are a form of **greenwashing**, an attempt to promote the appearances of products and commodity consumption as environmental or "green," while deliberately disavowing environmental impacts. We'll also consider the growing trend of green consumerism, efforts to buy products to try to live more sustainably. Recalling Kenneth Burke's *terministic screens* in Chapter 1, we are continuing to hone our ability to identify what communication about "sustainability" is *reflecting* and what it might be *deflecting.*

Green Product Advertising

The most familiar form of green marketing is the association of a company's products with popular images and slogans that suggest a concern for the environment. Such **green product advertising** is the attempt to sell commodities "that are presumed to be environmentally safe" for retail, "designed to minimize negative effects on the physical environment or to improve its quality," and an effort "to produce, promote, package, and reclaim products in a manner that is sensitive or responsive to ecological concerns" (American Marketing Association, 2014).

Selling Green

The list of products that are marketed as "green" or "eco-friendly" can be lengthy: coffee, cars, water filters, clothing, cars, computers, allergy pills, breakfast cereals, lipstick, and children's toys are only a few examples. In one year alone, the U.S. Patent Office received more than 300,000 applications for environmentally related "brand names, logos, and tag lines" (Ottman, 2011). Not surprisingly, the brands for these products are often accompanied by visual images of mountain peaks, forests, clear water, or blue skies.

What is being sold at the same time as a product, of course, is often a relationship with the natural environment:

> An advertisement for an SUV shows the vehicle outdoors and . . . ads for allergy medications feature flowers and "weeds." In them, the environment per se is not for sale, but advertisers are depending on qualities and features of the nonhuman world to help in . . . selling [their] message. (Corbett, 2006, p. 150)

In green advertising, the environment offers a seemingly limitless range of possibilities for such images or identifications. From Jeep ads encouraging urbanites to escape to less traveled forest roads to "all-natural" or "organic" breakfast foods, green ads rely on evocative appeals to nature as a powerful rhetorical frame.

A majority (84%) of U.S. consumers are buying green products—environmentally friendly clothing, foods, cleansers, personal-care products, and more (Ottman, 2011, p. 9). And, many are looking for eco-labels in stores. Not surprisingly, Roper's Green Gauge poll has shown a related trend, "a growing tendency towards 'pro-cotting'— buying products from companies perceived as having good environmental track records" (quoted in Ottman, 2003, para. 4). As a play off of *boycott* (noted in Chapter 8 as an environmental advocacy mode to foster corporate accountability), the term **buycott (girlcott or pro-cott),** is "is a concerted effort to make a point of spending money—as well as to convince others to make a point of spending money—on a product or service in the hopes of affirming specific condition(s) or practice(s) of an institution" (Pezzullo, 2011, p. 125). Roper's latest Green Gauge survey finds consumers from at least 25 countries "are still thirsty for greener options" in their stores ("Green Marketing Insight," 2011, para. 2).

Examples of green product ads may be unlimited, but the underlying communication frames for such advertising draw on some common themes. What environmental communication scholar Steve Depoe (1991) first identified over two decades ago still applies. There are three basic frames for green advertising: (1) nature as backdrop (e.g., SUV ads showing mountain terrain), (2) nature as product (e.g., everything from "all-natural" raisins to "natural flavors"), and (3) nature as outcome (e.g., bamboo bowls or a compost bins), through which products do not harm and may even claim to improve the environment.

Another Viewpoint: When Customers Claim Sustainable Is Not Salable

In 2010, the snack company Frito-Lay decided to sell their multi-grain Sun Chips in a 100% compostable package. They launched their new packaging before Earth Day with a green marketing campaign and boasted the bag could disintegrate in 14 weeks. The ad ended with the tagline "That's our small step. What's yours?" They then ended with a website URL for people to respond (greenimpact.com, now defunct).

(Continued)

(Continued)

This product would seem like a sustainable victory to reduce waste and to encourage everyone to become more sustainable. Next time you are at a grocery store, look at how many plastic bags are sold. It's a large number.

Yet, customers complained the Sun Chip packaging was too loud and rejected the company's attempt to become more sustainable. As a result of the strong customer reaction, Frito-Lay took the bags off the market and did not bring them back until they could design a quieter compostable bag one year later in 2011.

Since most of us probably would not consider chips the quietest foods we eat, it is interesting to consider this initial marketing failure, as well as the subsequent solution that pleased customers and still sold chips for the company. Are there any barriers for you to buying more sustainable products?

For a video with a sound test and review of the old bags engaging this viewpoint, see Kyle VanHemert, 2010, "SunChips' New 100% Compostable Bag Is Hilariously, Ear-Damagingly Loud," Gizmodo at http://gizmodo.com/5616427/sunchips-new-100-compostable-bag-is-hilariously-ear-damagingly-loud.

Regulating Green Labels and Guidelines

Environmentally friendly labels on products include words such as *organic, nontoxic, GMO free, biodegradable, all natural, free range, fair trade, recycled, ozone-friendly,* and so forth. They seem increasingly common and on a wider range of products daily.

As such, a word of caution about green advertising may be in order. This practice is largely unregulated. With the exception of *organic* (which is regulated by the U.S. Department of Agriculture), most environmentally friendly labels and green product claims in the United States are governed only by voluntary guidelines. The FTC does have regulatory oversight over green marketing claims and publishes its "Guidelines for Environmental Marketing Claims," or "Green Guides." In past years, however, these guidelines have been quite vague, and compliance has been voluntary for corporations. In her analysis of 247 print ads in mainstream British and American magazines, Lauren Baum (2012) found "Firms advertising in US magazines were significantly more likely to employ misleading/deceptive environmental claims than UK firms and magazines. These findings suggest that without increased environmental advertisement regulation, greenwashing will persist as a disingenuous means for corporate reputation enhancement" (p. 423).

The U.S. Department of Agriculture does set standards for organic products under its National Organic Program. (See www.ams.usda.gov/nop.) However, there is no uniform standard for other labels such as *free range* or *cage free*. In fact, "one company's free range label might mean that the animal went outside for 15 minutes a day, while another's might mean that the animal roamed a 10-acre field all its life" ("Consumers Beware," 2006, p. 23A; see also Foer, 2010).

As a result of a loose regulatory landscape, green product advertising may signal a wide range of meanings—from unsubstantiated claims to accurate information about the qualities of the product or the behavior of the corporation. In fact, the business blog *GreenBiz.com* ("Marketing and Communications," 2009) once trumpeted the opportunities being opened by "the lack of standards for determining what it means to be a green product—or a green company" (para. 1). It also noted that, with the number of consumers who want to buy green as well as the popularity of eco-labeling, an opportunity existed "for just about anything to be marketed as green, from simple packaging changes to products and services that radically reduce materials, energy, and waste" (para. 1).

Green product advertising is making use of the practice of environmental seals or **eco-label certification programs.** The presence of an eco-label on a product ostensibly signals an independent group's assurance to consumers that the product is environmentally friendly or produced in a manner that did not harm the environment. For example, the Environmental Protection Agency's (EPA's) *Energy Star* label on light bulbs and appliances is meant to signal their energy savings. And, the Forest Stewardship Council's *FSC-certified* label on wood products indicates that the wood is from an ecologically managed forest.

The use of eco-labels or other certification seals also has grown in recent years. Currently, there are now more than 400 green certification labeling systems. One problem has been that such labels may make broad, unqualified claims about a product. As a result, the Federal Trade Commission (FTC) has written guidelines to help consumers assess these claims. (See "Act Locally! Check for Deceptive Green Marketing Claims.")

Act Locally!

Check for Deceptive Green Marketing Claims

Have you ever wondered if the label on certain foods, personal care products, or clothing means these products are truly green or environmentally friendly? What does it mean to buy a product that is labeled as recycled, biodegradable, or nontoxic?

The Federal Trade Commission's (n.d.) "Guides for the Use of Environmental Marketing Claims" (Section 260.7) states, "It is deceptive to misrepresent, directly or by implication, that a product, package or service offers a general environmental benefit."

The FTC has the power to bring law enforcement actions against false or misleading environmental marketing claims under Section 5 of the FTC Act, which prohibits unfair or deceptive acts or practices.

How is the FTC using its power? Have the FTC's guides given clearer definitions for green advertising claims? Has it taken actions against deceptive green ads? Check out http://www .consumer.ftc.gov/articles/0226-shopping-green. Then, answer the following questions:

- How accurate or helpful are the FTC guidelines for *recycled, biodegradable, compostable, renewable energy,* or *carbon offsets?*
- What enforcement actions against deceptive environmental ads or labels has the FTC taken recently? For specific corporations or products, you can click on the FTC's "Scam Alerts" button, at their homepage: http://www.consumer.ftc.gov/. For example, you will find recent fraudulent activity they discovered after natural disasters with solicitations for donations.
- Overall, how effective do you think the FTC's "Green Guides" are in discouraging false or misleading environmental labels or advertising? Had you ever heard of them before? Can you find them at local stores where you shop?

The only enforcement power the FTC has is Section 5 of the Federal Trade Commission Act. This provision prohibits deceptive practices in advertising. Although its "Guidelines for Environmental Marketing Claims" are voluntary, the FTC can prosecute a corporation if it can be proved the corporation acted deceptively in its claims about a product. And the FTC has brought several enforcement cases alleging false or unsubstantiated environmental claims. For example, in 2009, the commission brought charges against three companies for "making false and unsubstantiated claims that their products were biodegradable" and charged four clothing and textile businesses with "deceptively labeling and advertising these items as made of bamboo fiber, manufactured using an environmentally friendly process, and/or biodegradable" (Federal Trade Commission, 2010, pp. 19–20).

As a result of what some believe are unclear or weak enforcement of guidelines for green marketing claims, a number of independent groups have arisen to monitor or verify environmental advertising claims. The most prominent group is SourceWatch (sourcewatch.org), which monitors corporate funding and behavior, generally. Other groups monitor specific claims, for example, that eggs are from *cage-free* chickens and that vegetables and meat are *natural, free range,* or *humanely raised.*

A few animal welfare groups issue their own labels that certify when products meet certain criteria. For example, the American Humane Society has the American Humane® Certified program (formerly known as the Free Farmed program); its certified label assures consumers of third-party, independent verification that producers' care and handling of farm animals meet science-based animal welfare standards. And Whole Foods has been developing an "animal-compassionate" program that will require that animals be raised "in a humane manner" (cage free, and so on) until they are slaughtered (Martin, 2006).

Finally, Canada has banned the use of eco-labels with "vague claims implying general environmental improvement" (Sustainable Life Media, 2008, para. 1). Instead, Canada's Competition Bureau released a set of guidelines that require companies to stick to "clear, specific, and accurate" claims that have been substantiated. Sheridan Scott, Commissioner of Competition, stated, "Businesses should not make environmental claims unless they can back them up" (Sustainable Life Media, 2008, para. 2).

Green Image Enhancement

Corporations navigate a constantly changing business environment. Doing so often requires an investment of resources to ensure that the public maintains a positive image of a corporation's identity and performance. Promoting a positive, environmental image for a corporation relies on the practice of **image enhancement.** This is the use of PR (public relations) to improve the brand or ethos of the corporation itself by associating it with positive environmental messages, practices, and products.

As environmental values became increasingly popular in the United States and other countries, many corporations refined their communication with the public, bolstering their identities as environmentally responsible corporate citizens. This communication is particularly important when media, the government, or environmental groups question a corporation's behavior or intentions.

In this section, we look at two image enhancement campaigns: (1) the coal industry's touting of "clean coal" as a form of energy and (2) Walmart's "sustainability" initiatives to eliminate waste and reduce its carbon footprint.

Redefining Coal as "Clean Coal"

One of the more visible image enhancements recently has been the coal industry's multimillion-dollar Clean Coal advertising campaign. High-quality-production ads appeared thousands of times on TV, radio, billboards, and online. Starting in 2002, the ads declared that "advancements in clean coal technologies are effectively making our environment cleaner," and assured listeners that "new coal-based power plants built beginning in about 2020 may well use technologies that are so advanced that they'll be virtually *pollution-free* [emphasis added]" (quoted in SourceWatch, 2009, para. 4).

These ads are not trying to sell a product—a ton of coal or a new power plant. Instead, they are intended to reassure lawmakers and opinion leaders that the coal industry is vital to America's energy future and that, because coal is "clean"—that is, coal-burning power plants can produce electricity without causing pollution—the industry should not be regulated. Why? What is changing in the business environment that the U.S. coal industry feels it must navigate in order to survive?

Fundamentally, the business environment for coal is changing. The EPA has proposed tougher standards for pollutants like sulfur dioxide and mercury from coal-burning power plants and is currently proposing to regulate for the first time carbon dioxide (CO_2), a greenhouse gas and a major source of climate change. And, as we saw in Chapter 8, environmental groups have been blocking new coal plants and major financial institutions have begun imposing stricter requirements on loans for the construction of such plants.

As a result, the coal industry invested heavily in a multimillion-dollar "clean coal" ad campaign in an attempt to forestall new regulations on coal-burning power plants. In 2008 alone, industry groups like the American Coalition for Clean Coal Electricity (ACCCE) spent $35 million to $45 million on image advertising, "most of it on television ads aired during the 2008 campaigns—pitching 'clean coal' as a new environmentally friendly fuel" (LoBianco, 2008, para. 2; see also Mufson, 2008). ACCCE, formerly called Americans for Balanced Energy Choices, is a public relations group for coal mining companies, coal transport (railroads), and coal-electricity producers (SourceWatch, 2009). Many of the TV and billboard ads directed viewers to more detailed information on ACCCE's sophisticated website, AmericasPower.org.

One of the ACCCE (2008) image enhancement TV ads was titled "Adios." It showed an elderly couple sitting on their porch, a young woman driving her convertible, two workers going into a factory, kids waving, and a family at the beach. A (male) voice declares,

> We wish we could say farewell to our dependence on foreign energy. And we'd like to say "adios" to rising energy costs. But first, we have to say "so long" to our outdated perceptions about coal. And we have to continue to advance new clean coal technologies to further reduce emissions, including the eventual capture and storage of CO_2. If we don't, we may have to say "goodbye" to the American way of life we all know and love. Clean coal. America's power. (America's Power, 2009)

The ad draws on the strategy of invoking social norms to suggest ordinary Americans—like "us"—are saying "so long" to "outdated perceptions about coal" and to suggest that coal can continue to power "the American way of life."

The year the ACCE's image enhancement campaign was launched, a public opinion poll found that "72 percent of opinion leaders nationwide support the use of coal to generate electricity, a significant increase over the past year and the highest level of support since the group began polling nearly 10 years ago" (*Business Wire*, 2008, para. 3). In the short term, the campaign may have worked to some degree. Nevertheless, as we discussed in Chapter 8, environmental advocates have pushed back and, as this book goes to press, "clean coal" seems to be a phrase waning in popularity.

Walmart's "Sustainability" Campaign

On October 24, 2005, the CEO of Walmart, Lee Scott, stood before a packed audience at Walmart's home office in Bentonville, Arkansas. Speaking to employees and via a live video feed to its 6,000 stores worldwide and 62,000 suppliers, Scott announced that Walmart was "going green in the biggest way imaginable, embracing sustainability" (Humes, 2011a, p. 100). Walmart was an unlikely corporation to appear to adopt environmental values. Known as the "Bully of Bentonville," the company had been a target for years, attracting "endless bad press, protests, political opposition, investigations, labor problems, health insurance problems, zoning problems, and more than ever, environmental problems" (p. 13). The initiative that Scott would announce on that day, however, would grow over the subsequent years to be one of the most ambitious, but often doubted, green corporate initiatives ever.

In his Bentonville speech, Scott asked rhetorically, "What if we used our size and resources to make this country and this earth an even better place for all of us?" (quoted in Humes, 2011a, p. 102). He then went on to announce what would become Walmart's long-term environmental goals:

1. To be supplied 100% by renewable energy

2. To create zero waste

3. To sell products that sustain people and the environment (Walmart, 2011, p. 1)

Scott's announcement created news headlines and puzzled the company's competitors. While a number of environmental groups expressed interest in Walmart's new, environmental goals, most of them reserved judgment, "waiting to see if the company's actions matched its lofty rhetoric" (Humes, 2011a, p. 104). Other reactions were harsher, ranging from to skepticism that such a large retailer could ever be sustainable to charges of *greenwashing* (Mitchell, 2007).

In the years following, however, Walmart would go on to announce and begin implementing a series of sustainability initiatives—lowering the carbon footprint of its stores, increasing the fuel efficiency of its transportation fleet, and seeking to curb wasteful packaging. In 2006, the company introduced "a green rating system designed to push their 60,000 worldwide manufacturing vendors to reduce the amount of packaging they use by 5%, to use more renewable materials, and to slash energy use" (Ottman, 2011, p. 172).

Then, in 2009, Walmart went further in announcing it would create a Sustainability Index. It would be a scorecard that measured the sustainability of Walmart's entire chain of suppliers. The index would measure a supplier's carbon footprint, use of natural resources, energy efficiency, water use, and so forth. Ultimately, Walmart hoped to translate the data on a supplier's performance into a score for each of its products, "a simple rating for consumers about the sustainability of products—the ultimate dream of green-minded shoppers" (Makower, 2010, para. 6). As a consequence, Walmart intended that the Sustainability Index would influence a wider circle of behaviors by its suppliers by determining which of their products qualified to get onto the store's shelves (Ottman, 2011, p. 168). (For more information about the Sustainability Index, see http://walmartstores.com.)

Public reactions were mainly positive at the time. A *New York Times* editorial praised the index as "a sound idea" and added, "Given Wal-Mart's huge purchasing power, if it is done right it could promote both much-needed transparency and more environmentally sensitive practices" ("Can Wal-Mart Be Sustainable?" 2009, para. 2). Brian Merchant (2009), blogging at Treehugger.com, admitted, "It's getting harder to hate Wal-Mart." While the giant corporation's Sustainability Index "raises plenty of questions," if it succeeds, it "could literally change the face of retail forever" (para. 1). Two years later, Edward Humes (2011b), in a *Los Angeles Times* op-ed, was even more positive: "I'm no apologist for Wal-Mart. The giant retailer is still ripe for criticism on a number of fronts, from hiring and labor issues to its impact on local businesses and communities. But in this area of sustainability, Wal-Mart got it right" (para. 5).

There have also been doubts, and not a little criticism, from others. Some critics reminded Walmart that it cannot separate its sustainability goals from related concerns for the company's impacts on people and the environment. Writing in the *Penn Political Review,* Melissa Roberts (2011) argued that, while environmentalists had pushed for years for corporations to include sustainability in their bottom lines,

> true environmentalism cannot separate environmental and social responsibility. If a company buys organic cotton but underpays farmers so that they have to use unsustainable farming practices [or] burn through the Amazon rainforest, there is no net gain for the environment. (para. 3)

Likewise, criticism is made of Walmart's anti-union activities with their own employees (Mieszkowski, 2008).

Other criticism came from some who believed that Walmart and other big box retailers were fundamentally fueling an unsustainable economy. As they saw it, "no matter how genuine the effort and positive the result, sustainability only distracts the press and public from the fact that the real problem is the big-box economy itself" (Humes, 2011a, p. 227; see also Humes 2011b).

Whether Walmart's sustainability initiatives ultimately succeed at image enhancement or helping foster a more sustainable world remains to be seen. In 2013, environmental organizations (including Friends of the Earth, Greenpeace, Rainforest Action Network, and the Sierra Club) wrote a letter to Walmart claiming their efforts have fallen short (Sheppard, 2013). Since, Walmart appears to have renewed its public commitment to sustainability. President and CEO Doug McMillon acknowledged the road to sustainability remains a process: "A great deal of innovative work is happening every day, but there are still too many gaps and missed opportunities" (Green Retail Decisions, 2014). Given the impact of transnational corporations on energy, food, and transportation systems, what is clear is that environmental advocates cannot afford to ignore what Walmart does or does not do next.

Green Corporate Image Repairs

One form of corporate communication that is pivotal to a company's success in difficult times is the repair or recovery of its credibility. This practice of **image repair** (also called *crisis management*) is the use of PR to restore a company's credibility after an environmental harm or accident. Corporations that engage in wrongdoing often face a crisis of public trust and possible legal or economic repercussions. Corporate image repair, therefore, attempts to minimize the harm and any public perceptions that might "cause the organization irreparable damage" (Williams & Olaniran, 1994, p. 6). Image repair is vital to a company's continued operations, but the practice can be controversial, especially when a corporation's communication is viewed as insincere.

As we write in 2014, BP is still struggling to recover from the blow to its credibility from the 2010 rupture of its oil wellhead in the Gulf of Mexico. Called "the worst man-made environmental disaster in U.S. history," the explosion "killed 11 workers, causing more than 200 million gallons of oil to spew into the Gulf of Mexico" (Democracy Now!, 2014).

Immediately after the event, the Obama administration launched a criminal and civil investigation into the corporation, and civil lawsuits piled up. Still, BP initiated a number of efforts to try to restore its credibility. After the accident, the company began running TV ads and full-page ads in major newspapers like the *Wall Street Journal, USA Today,* and the *Washington Post,* as well as local ads in the Gulf States.

One ad in the *New York Times* carried a photo of workers placing a boom to protect the shoreline from the oil slick. The tagline read, "BP has taken full responsibility for cleaning up the spill in the Gulf of Mexico. . . . We will make this right" ("We Will Make This Right," 2010, p. A11).

BP's crisis management during the oil spill, however, quickly became a case study of ineffective image repair. The corporation's credibility certainly was not helped by its CEO, Tony Hayward, in the immediate aftermath of the disaster. Hayward complained publicly about how much time he was spending on the disaster, saying, "I would like my life back," and he sought to play down the spill's ecological impacts. The Gulf is "a big ocean," he said; and "the environmental impact of this disaster is likely to be very, very modest" (MSNBC.com, 2010, para. 4).

In 2014, petitions continue to be delivered to the EPA to ask for BP to be held accountable. According to Jaclyn Lopez, staff attorney with the Center for Biological Diversity, the petition was "signed by over 50 organizations representing millions of Americans. It was independently signed by another 66,000 people who all recognize that BP needs to be held accountable by being prohibited from future contracts with the American government" (Democracy Now!, 2014). Questioning BP's credibility to run operations years after the accident suggests that its image repair remains in crisis.

AP Photo/Patrick Semansky

| Photo 11.1 | Former BP CEO Tony Hayward, center, speaks at a news conference in Port Fourchon, Louisiana, during the worst offshore oil spill in U.S. history. |

Greenwashing and the Discourse of Green Consumerism

Corporate practices of "green" advertising, image enhancement, and image repair have not been without their critics. In this section, we describe two criticisms in particular: (1) the charge that corporate green marketing is a form of *greenwashing* and (2) a discourse of *green consumerism*, the belief that purchasing environmentally friendly products can help save the Earth.

Corporate Greenwashing

In an earlier study of corporate opposition to environmental regulations, Jacqueline Switzer (1997) noted that often corporate "public relations campaigns—called 'greenwashing' by environmental groups—[are] used by industry to soften the public's perceptions of its activities" (p. xv; see also Corbett, 2006). *Greenwashing* is an attempt to promote the appearances of products and commodity consumption as environmental or "green," while deliberately disavowing environmental impacts. Since the number of products with "green" claims seems to increase annually, it may be useful to explore the idea of greenwashing further.

Environmental groups routinely use the term *greenwash* to call attention to what they believe is deception by a corporation—an effort to mislead or divert attention from a corporation's poor environmental behavior or products. For example, Greenpeace (2008) gave the oil company BP its Emerald Paintbrush award in

> recognition of the company's attempts to greenwash its brand over the course of 2008, in particular its multimillion dollar advertising campaign announcing its commitment to alternative energy sources . . . [and its use of] slogans such as "from the earth to the sun, and everything in between." (para. 4)

Using internal BP documents, Greenpeace claimed that, in 2008,

> the company allocated 93 per cent ($20bn) of its total investment fund for the development and extraction of oil, gas and other fossil fuels. In contrast, solar power (a technology which analysts say is on the brink of important technological breakthroughs) was allocated just 1.39 per cent, and wind a paltry 2.79 per cent. (para. 5)

BP, on the other hand, insists that the company is committed to developing new, renewable energy sources (BP, 2009). (For rhetorical analyses of BP and ExxonMobile greenwashing, see also Plec & Pettenger, 2012; Smerecnik & Renegar, 2010.)

How, then, can someone tell if a corporate advertisement is greenwashing or the report of a legitimate environmental achievement? Most critics point to a basic standard of deception. Has the ad conveyed information or an impression that is countered by factual evidence? Many times, the truthfulness of a claim may be difficult for the ordinary consumer to determine. In other cases, there are groups such as SourceWatch (www.sourcewatch.org) that monitor the statements and behavior of corporations, provide information about their compliance with environmental regulations, and even evaluate specific marketing campaigns. As mentioned in Chapter 4, a life-cycle-assessment approach requires we consider a wide range of interrelated activities.

In a more general effort, the environmental marketing firm TerraChoice has identified the main patterns of greenwashing and calls them "Sins." (See "FYI: The Seven Sins of Greenwashing.")

☞ FYI   The Seven Sins of Greenwashing

The environmental marketing firm TerraChoice has identified patterns used by companies that greenwash that it called the Seven Sins of Greenwashing. Here are the sins:

1. *Sin of the Hidden Trade-Off*: Suggesting a product is "green" based on a single environmental attribute (the recycled content of paper, for example, which may not account for chlorine use in bleaching).

2. *Sin of No Proof*: An environmental claim that cannot be verified by easily accessible supporting information or by a reliable third-party certification (facial or toilet paper, for example, that claims recycled content).

3. *Sin of Vagueness*: A claim "so poorly defined or broad that its real meaning is likely to be misunderstood by the consumer" (such as "all-natural," which isn't necessarily "green" if you consider natural elements like arsenic and mercury).

4. *Sin of Worshipping False Labels*: "A product that, through either words or images, gives the impression of third-party endorsement where no such endorsement exists; fake labels, in other words."

5. *Sin of Irrelevance*: "An environmental claim that may be truthful but is unimportant or unhelpful for consumers seeking environmentally preferable products" (CFC-free is an example, since CFCs are illegal).

6. *Sin of the Lesser of Two Evils*: "A claim that may be true within the product category, but that risks distracting the consumer from the greater environmental impacts of the category as a whole" (e.g., organic cigarettes).

7. *Sin of Fibbing*: "Environmental claims that are simply false," such as those claiming to be Energy Certified but are not.

SOURCE: TerraChoice Environmental Marketing Inc. (2010).

The practice of greenwashing is actually quite high in the United States and Canada. TerraChoice (2010) claims, "more than 95% of consumer products claiming to be green were found to commit AT LEAST ONE [emphasis in original] of the 'Sins of Greenwashing.'" What labels do you see the most? Which ones do you trust?

Discourse of Green Consumerism

Green marketing and discourses based on the free market raise another important question for students of environmental communication: Can consumers minimize damage to, or even improve, the environment by their purchase of certain products? That is, can we reduce air pollution, end the clear-cutting of national forests, or protect the ozone layer by buying recycled, biodegradable, nontoxic, and ozone-free products? Many people appear to think so. As we saw earlier, Roper's Green Gauge poll reported consumers' tendency toward *pro-cotting,* or buying products from companies that are perceived as having good environmental track records. Irvine (1989) was the first to refer to this "use of individual consumer preference to promote less environmentally damaging products and services" as **green consumerism** (p. 2). This is the belief that, by buying allegedly environmentally friendly products, consumers can do their part to help make the world a little more sustainable.

Why is the idea of green consumerism popular? Most of us do not wish to harm the environment, and we suspect that most of us believe we can consciously choose to lessen our impact on the Earth through our actions. While we both have bought reusable bags for our groceries, we both feel more skeptical about shampoos that offer a "truly organic experience" despite including ingredients with artificial colors and fragrances. Both of us believe at some level, that what we buy or do not buy reflects our values to some degree.

Helping to sustain this commonly held belief is a wider set of beliefs, buttressed by green advertising, that invite a specific identity. In her provocative book, *The Myth of Green Marketing: Tending Our Goats at the Edge of Apocalypse,* Smith (1998) argues that green consumerism is not simply an act—the purchase of a certain product—but a discourse about the identity of individual consumers. (In Chapter 2, we described *discourse* as a pattern of knowledge and power communicated through linguistic and nonlinguistic human expression.) Smith explains that our purchasing does not occur in a vacuum but is "an act of faith"; that is, "it is based on a belief about the way the world works" (p. 89). Our actions have effects, and among these is the effect of our purchasing on producers of products. In other words, the discourse of green consumerism assures us that, when we buy more sustainable products, our buying can affect the actions of large corporations, such as persuading coffee companies to use organic, fair trade practices, and, as a result, can transform our more unsustainable habits of pesticide pollution and exploitative labor practices.

Smith argues that the idea of green consumerism resonates with us because the act of purchasing is cloaked in an aura of other, authoritative discourses that buttress our identities as purchasers. She explains that our belief that we can do well for the environment by green shopping is underwritten by certain discourses that encode our buying with significance. Two discourses in particular assign meaning to our purchasing decisions: the discourses of market forces and of participatory democracy.

First, green advertising affirms the belief that the market can be an avenue for change; that is, that by doing our bit, we contribute to the *free market* theory of economics, and as "all the little bits are counted, the consequence will be a net good"

(Smith, 1998, p. 157). Second, the discourse of participatory democracy nurtures the belief that, in a liberal democracy, each of us is entitled to a voice in deciding about issues that matter to us. Thus, "Customers vote at the cash register" (quoted in Smith, p. 156). In each case, consumers are encouraged to believe that their purchases exercise a democratic influence: "Voting" through what we do or do not buy affects retailers directly in determining which products succeed and which are in disfavor, and it affirms the consumer's identity as someone who acts responsibly toward the Earth.

The discourse of green consumerism can be an attractive magnet, pulling one toward a persuasive identity as a purchaser. "Green consumerism makes sense," explains Smith (1998). "That is why people are attracted to it; they are not irrational, immoral, or uninformed. Quite the opposite: they are . . . moral in their desire to do their bit" (p. 152). It is hard to argue against the fact that buying and using a reusable water bottle, for example, usually is cleaner, cheaper, healthier, and less polluting than buying, drinking, and tossing bottled water every day.

Nevertheless, Smith believes that green consumerism also poses a danger by co-opting a more skeptical attitude toward the social and environmental impacts of excessive consumption. In a provocative charge, Smith claims that green consumerism serves to deflect serious questioning of a larger **productivist discourse** in our culture, one that supports "an expansionistic, growth-oriented ethic" (p. 10). Indeed, whether green consumerism can be a real force in the marketplace or a subtle diversion from the questioning of our consumer society is a question that invites serious debate in our classes and in research by environmental scholars.

Courtesy of the University of Oregon

Photo 11.2

The buildings we live and work in communicate values and possibilities to us, as well as shape our financial costs (through energy bills, for example) and impact on ecological systems (such as carbon footprint). The world's primary certification system of sustainable architecture is LEED (Leadership in Energy and Environmental Design). This image shows a platinum-LEED building on the University of Oregon campus, including sustainability features such as rooftop solar panels, use of local building materials, and climate-appropriate landscaping.

In summary, the practice of green marketing is now widespread. It involves subtly and skillfully associating corporations' products, images, and behaviors with environmentally friendly values. As we saw, this effort to construct a green identity can serve any of three purposes: (1) product promotion (sales), (2) corporate image enhancement, and (3) image repair in the aftermath of negative publicity about a company. As we see in the following section, college campuses are becoming living learning laboratories for thinking through what differences individual choices and structural choices can make physically and culturally.

Communicating Sustainability on and Through Campuses

Higher education in the United States and, arguably, globally is going through a radical reassessment period, brought on by the emergence of new media technologies and global unsustainable crises (economic stress, cultural divisions, and ecological challenges). Where we learn, through which media, and about what are all under radical reassessment at this point in history. As such, it perhaps is not a surprise that sustainability initiatives rapidly are developing on campuses through energy, food, materials, governance, investment, wellness, curriculum, interpretation, and aesthetics (Thomashow, 2014). Communication studies have much to offer these initiatives. We highlight two areas worthy of our attention: curriculum, or what we are teaching and learning, as well as infrastructure (or materials), the physical elements where we are teaching and learning.

Communicating Sustainability Curricula

As teachers and consultants, we (Robert and Phaedra) both have been to many meetings about sustainability curricular goals and assessment to talk about what we think should be taught and how we can judge whether or not students have learned those subjects. Inevitably, communication is listed as a key outcome for sustainability education whether or not we are a part of the discussion. A core competency desired by employers and expected by interdisciplinary faculty teaching about sustainability, according to a recent report of the American College Personnel Association (ACPA, n.d.), is "the ability to generate support for change through strong communication skills, consensus building strategies, and with openness to the ideas and struggles of others" (n.d., p. 22).

Common curricular goals for sustainability educators include the following:

- Improved oral and written communication skills for a variety of audiences and goals
- Capacity to work well independently and in small groups or organizations
- Ability to navigate ambiguity, risk, and resilience in ways that account for particular contexts and constraints
- Ability to develop argumentation and advocacy (how to research, organize, and analyze information and ideas, as well as articulate them)

- Fostering critical thinking about practices in relation to a range of media texts
- Engagement in related rhetorical concepts of democracy, citizenship, culture, and community

All of these competencies are developed in the wheelhouse of communication studies, particularly as they apply to sustainability.

The International Environmental Communication Association (IECA) provides an online resource of syllabi showing a range of courses that focus on these sustainable teaching goals, including environmental advocacy, environmental journalism, communicating sustainability, environmental conflict resolution, health literacy, media and the natural environment, writing for the environmental industry, public relations in natural resources, and much more (for the full list, see http://theieca.org/courses).

For what we are teaching and learning in sustainability, however, it must be clarified that a sustainability approach is not just focused on environmental facets. What differentiates a **sustainability curriculum** from an environmental curriculum is that the former must engage all three facets of sustainability (the Three Es or the Three Ps). So, an environmental scientist cannot teach a course on mapping climate changes through maps and claim that course is central to sustainability unless she or he also engages economic and social impacts and/or constraints. Further, while most schools have developed degree programs in an interdisciplinary approach to sustainability, more and more appear to be considering how sustainability should be taught across college curriculums, whether or not sustainability is the final focus (for more on this approach at Northern Arizona University, see Chase & Rowland, 2013).

It bears repeating that each campus has its own mission and location because sustainability must be contextually imagined; so, a sustainability curriculum at the University of Texas in El Paso should not look the exact same as one at the University of Ghent in Belgium or Songdo Global University in Korea. Clinical psychologist, author, and Spelman College president Beverly Daniel Tatum notes that in 2002 she was not sure what the mission statement of her school, a prestigious, historically black all women's school in Atlanta, Georgia, had to do with sustainability. Today, she has no hesitation. The mission is for their students to improve the world, but also to empower students: "An ethic of conservation, an ethic of self-care, and an ethic of commitment to our mission: together these values will help ensure a sustainable Spelman" (Tatum, 2013, p. 162). The College of Menominee Nation, a tribal school in Wisconsin, expands the three dimensions of sustainability usually noted to six in order to reflect their mission, which includes land and sovereignty, technology, and institutions (Van Lopik, 2013, p. 107).

Communication Through Infrastructure

In addition to sustainability shaping the content of what is being taught at universities today, it also is shaping the infrastructure. **Sustainable infrastructure** refers to the physical elements needed to allow an organization to operate (including everything from the chair you sit on in class to the light bulbs used in the stairwells to the energy

Albright Garden/Flickr. Used under Creative Commons license
https://creativecommons.org/licenses/by/2.0/

Photo 11.3	Sustainable campus initiatives include reducing energy use in dorms, reducing waste at university sports events, reselling student furniture and other secondhand items at the end of the year, and much more. This is an image of a community garden at Albright College, where they use permaculture to teach students about agriculture, food, working with a team, and leadership skills. Does your campus have a garden yet? Does it share produce with dining services or the local food banks? Is it a part of your sustainability curriculum?

source used to heat water in dorms). The Association for the Advancement of Sustainability in Higher Education (AASHE) (2005–2012) focuses on eight categories of campus operations: buildings, climate, dining services, energy, purchasing, transportation, waste, and water (see aashe.org/resources/campus-operations-resources/).

Considering how sustainable infrastructure can help an institution become more sustainable is important, of course, but as students of environmental communication, it also is important to recognize the communicative dimensions of these physical choices. Put another way, where we learn not only has physical consequences, but also rhetorical ones. Scholars who study this mode of communication call it **material rhetoric**, the interpretation of how humans and the physical world (including buildings and bodies) constitute meaning.

Communication scholar Carole Blair (1999) has been at the forefront of theorizing how rhetoric is not just symbolic or instrumental, but also material. She provides five insightful questions that rhetorical critics might ask of a text (broadly understood as an object or event that is interpreted) materiality:

1. "What is the significance of the text's material existence?" (pp. 30–37)

2. "What are the apparatuses and degrees of durability [or vulnerability] displayed by the text?" (pp. 37–38)

3. "What are the text's modes or possibilities of reproduction or preservation?" (pp. 38–39)

4. "What does the text do to (or with, or against) other texts [in its surrounding environment]?" (pp. 39–45)

5. "How does the text act on person(s)?" (pp. 45–50)

When it comes to sustainable infrastructure, a material rhetoric approach can help us take into account how we learn from where we are learning. Consider how your campus, for example, does or does not communicate sustainability as a value: Do you have accessible recycling or composting bins? Are your uses of energy and water on campus motion-activated or not to conserve money and resources? Is your campus bicycle-friendly? Are there educational signs pointing out the sustainable features of buildings and water fountains on your campus? Can you earn course credit for service-learning opportunities to improve infrastructure in low-income neighborhoods near your campus? Overall, how does your campus's infrastructure encourage you to think and to act more sustainably or not?

No matter whether your campus is known to be one of the leaders of sustainability or one that is lagging behind, there always is more to be done when one understands that sustainability remains an ongoing process.

SUMMARY

In this chapter, we focused on corporations and campuses as key institutions that are shaping our future as sustainable or not every day. We emphasized that sustainability is a process that involves making connections between specific, local places and larger social, cultural, and physical conditions globally.

- In the first section of this chapter, we clarified how sustainability involves an interdisciplinary approach that accounts for the 3 Es (environment, economics, and social equity) or the 3 Ps (planet, profit, people).
- In the next two sections, we described a free-market discourse underlying much of corporate environmental communication. Then we explored in more detail the practice of corporate "green marketing" through three ways of communicating: (1) product advertising; (2) image enhancement; and (3) corporate image repair. We also shared debates over greenwashing and green consumerism.
- Finally, we highlighted a range of sustainability efforts happening on and enabled through college campuses, which promote a wide variety of exciting ways to become involved in sustainability efforts, including through curricular approaches and infrastructure.

Given the popularity of sustainability initiatives today, it is clear that many corporations and campuses have come to appreciate the environmental values embraced by the general public, consumers, and the media, as well as recognize the ways communication is vital to bringing about a more sustainable world. Nevertheless, other institutions and individuals appear to be appropriating the language of sustainability for its persuasive appeals without grappling with the more complex, dynamic, and substantial challenges of an unsustainable world. It is important for all of us to develop our ability to clarify the grounds on which we judge competing claims of sustainability. In the next chapter, we focus on public participation in government forums to further consider the ways we can make spaces for multiple voices to be heard and negotiate the most viable sustainable choices agreed upon in particular contexts.

SUGGESTED RESOURCES

- A regular topic of *TED Talks* is sustainability (www.ted.com/topics/sustainability). One presented by Jason Clay, a World Wildlife Federation vice president, is about his work with multinational corporations to transform global food markets to become more sustainable: http://www.ted.com/talks/jason_clay_how_big_brands_can_save_biodiversity.
- In 2005, Dow Chemical launched an image enhancement campaign called "The Human Element" (youtube.com/watch?v=vsCG26886w8). A countercampaign spoofing the original was launched one year later (youtube.com/watch?v=lbpuSPL-FNU&feature=related). Watch both. Compare the soundtracks, words, and images.
- In 2001, Chevron launched an image enhancement campaign called "Human Energy." The tagline was: "WE AGREE. DO YOU?" It involves pro-environmental messages such as, "Oil companies should put their profits towards good use" and "Protecting the Planet is Everyone's Job." Then, audiences can click on a tab below these statements to concur: "I agree." According to their global counter on the website, over 500,000 clicks have indicated support for one of these statements. The same day, the Rainforest Action network and professional tricksters the Yes Men launched a spoof site with a fake press page and press statement, focusing on their grievances of environmental injustice in Ecuador. Discuss what you perceive to be the value and/or the risks of this environmental advocacy tactic.
- For more on countercampaigns challenging image enhancement campaigns, see media studies scholar Kembrew McLeod's 2014 book, *Pranksters*. For crisis communication advice on how corporations might adapt to such campaigns, see Gidez, 2010.
- The documentary film, *Food Inc.* (2008), directed by Robert Kenner and featuring Michael Pollan (*The Omnivore's Dilemma*) and Eric Schlosser (*Fast Food Nation*), shares perspectives on green marketing and sustainability initiatives from the voices of farmers, businesses, everyday parents, and more.

- Go to the Association for the Advancement of Sustainability in Higher Education (AASHE)'s website (aashe.org/) to see how your campus is doing, compares to others, or could be doing something different in relation to sustainability. They have a Sustainability Tracking, Assessment, and Rating System™ (STARS) represented through a global, interactive digital map at https://stars.aashe.org/.

KEY TERMS

The Brundtland Commission Report 264

Buycott (girlcott or pro-cott) 267

Discourse of the free market 265

Eco-label certification programs 269

Green consumerism 266

Green marketing 266

Green product advertising 266

Greenwashing 266

Image enhancement 270

Image repair 274

Material rhetoric 282

Productivist discourse 279

Sustainability curriculum 281

Sustainability gap 265

Sustainable infrastructure 281

Three Es and Three Ps 264

DISCUSSION QUESTIONS

1. Search online for examples of sustainability diagrams. What types of examples can you find? A range of sustainability efforts incorporate graphic designs with different words and colors overlapping, cartoons, and photographs. What are the rhetorical constraints (limitations or possibilities) of communicating sustainability through a diagram? Try to design a diagram of your own. What choices did you make?

2. Can green consumerism help to protect the environment? Can we have some effect—even if small—on air pollution, clear-cutting of our national forests, or global warming by buying products that are biodegradable, nontoxic, recyclable, reusable, and so forth?

3. Can Walmart or other multinational corporations ever be sustainable? How credible is Walmart's claim to be helping the environment while making a profit? What evidence do you find credible for assessing their green labels and claims? What additional evidence would you like to have?

4. Some universities have sustainability offices that focus on infrastructure of the university (energy, food sources, etc.) and separate academic sustainability programs that focus on curriculum; others combine them to coordinate efforts. Do you think it is important to keep these two missions separate? Why or why not?

PART V

Citizen Voices and Environmental Forums

Public participation requires research of relevant information and rhetorical judgments about what to say to whom and how. In a 2010 U.S. congressional hearing on whether or not to reform the 1976 Toxic Substances Control Act, the American Chemistry Council president and chief executive officer Calvin M. Dooley held up his BlackBerry to illustrate technological progress made possible by limited regulation of the chemical industry. In the image above, Environmental Working Group president Kenneth A. Cook asked to borrow the same device to contend: "This ought to be as safe as a pesticide" (Hogue, 2010).

CHAPTER 12

Public Participation in Environmental Decisions

There can be no participation without communication.

—Arvind Singhal, 2001, p. 12

Public participation is a seminal concept in the environmental policy decision-making and natural resource management arenas. As an area of research and practice, public participation ranges from relatively formal activities like public hearings and litigation procedures to informal events such as community workshops and field trips.

—Gregg Walker, 2007, p. 106

One of the most striking features of environmental communication in the public sphere has been the increase in direct, democratic participation by ordinary citizens, environmental groups, scientists, businesses, and others in government decisions that impact their lives. Environmental historian Samuel Hays (2000) noted that people have been "enticed, cajoled, educated, and encouraged to become active in learning, voting, and supporting [environmental] legislation . . . as well as to write, call, fax, or e-mail decision makers at every stage of the decision-making process." All this has been "a major contribution to a fundamental aspect of the American political system—public participation" (p. 194).

In this chapter, we focus on developments in the United States and other nations that strengthen the public's legal rights to be involved in democratic decisions about the environment. Such involvement by the public often has been the critical element in a range of environmental efforts, for example, to protect threatened wildlife habitats, to achieve cleaner air and water, and to ensure a safer workplace.

Chapter Preview

This chapter describes some of the legal guarantees and public forums for communication that enable citizens to participate actively in democratic decisions about the environment.

- The first two sections of this chapter focus on two legal rights and practices that embody the ideal of environmental public participation:
  - The right to know
  - The right to comment
- Also in section two, we describe one challenge ordinary citizens sometimes face to their right to comment—Strategic Litigation against Public Participation, or SLAPP lawsuits; these occur when some corporations attempt to silence or intimidate individuals who criticize their companies in public comments for harming the environment.
- The last section of the chapter describes the growth internationally of democratic provisions for public participation, particularly through the approach of participatory communication.

Public participation is grounded in the belief that "those who are affected by a decision have a right to be involved in the decision-making process" (*Core Values*, 2008). It is also an assumption that such involvement—although this outcome is not guaranteed—can have some influence on the decisions affecting one's life or the society one lives in. Public participation, therefore, is viewed as a core characteristic of a democracy. This has been especially true of environmental decisions.

Here, we define public participation more specifically as the ability of individual citizens and groups to influence decisions through (1) the **right to know** or ability to have access to relevant information, (2) the **right to comment** or the opportunity to publicly address the agency or entity that is responsible for a decision about actual or potential harms and benefits one perceives as a result of that decision, and (3) the right of standing or the legal status accorded a citizen who has a sufficient interest in a matter and who may speak in court to protect that interest.

These rights reflect more basic, democratic principles: (1) the *right to know* reflects the principle of **transparency**, or openness of governmental actions to public scrutiny, (2) the *right to comment* publicly reflects the principle of *direct participation* in decision making, and (3) the *right of standing* assumes the principle of **accountability**, that is, the requirement that political authority meet agreed-upon norms and standards. (These principles are summarized in Table 12.1.) We focus on the right of standing in Chapter 14; for now, we focus on the rights to know and to comment.

Right to Know: Access to Information

One of the strongest norms of democratic societies is the principle of transparency. Simply put, this is a belief in openness in government and the right of citizens to know information important to their lives. Internationally, the principle of transparency was applied specifically to environmental concerns at the United Nations Conference on Environment and Development, informally known as the Earth Summit, in 1992. Principle 10 of the Rio Declaration (1992) states that "environmental issues are best handled with participation of all concerned citizens" and that individuals "shall have appropriate access to information concerning the environment that is held by public authorities . . . and the opportunity to participate in decision-making processes." The principle of transparency gained further recognition in the *Declaration of Bizkaia* (1999), which proclaimed that transparency requires "access to information and the right to be informed. . . . Everyone has the right of access to information on the environment with no obligation to prove a particular interest."

Recognition of the principle of transparency also illustrates the importance of information—and who controls it—in influencing environmental policies. As Hays (2000) observed, political power lies increasingly in an ability to understand the complexities of environmental issues, and "the key to that power is information and the expertise and technologies required to command it" (p. 232).

By the late 20th century, moves to ensure transparency in decision making led to new **sunshine laws**. These laws required open meetings of government bodies in

Table 12.1	Rights of Public Participation in Environmental Decisions		
Legal Right	Democratic Principle	Modes of Participation	Examples of Laws
Right to Know	Transparency	• Written requests for information • Access to documents online (including databases) • Public notification of hearings Environmental Impact Statements	• Freedom of Information Act • Toxic Release Inventory • Clean Water Act • Food Quality Protection Act • Aarhus Convention • Environmental Impact Assessment
Right to Comment	Direct participation	• Testimony at public hearings Participation in citizen advisory committees • Written comment (letters, e-mail)	• National Environmental Policy Act • Aarhus Convention • Environmental Impact Assessment
Right of Standing* *See Chapter 14	Accountability	• Plaintiff in lawsuit • Amicus brief (third party) in legal case	• Clean Water Act • Court rulings, including *Sierra Club v. Morton* (1972)

order to shine the light of public scrutiny on their workings. The U.S. Congress threw open the doors to government records more generally. The Clean Water Act of 1972 for the first time required federal agencies to provide information on water pollution to the public.

Globally, two types of laws in particular have provided important guarantees of the public's right to know or their ability to have access to relevant information. In the United States these are the Freedom of Information Act and the Emergency Planning and Community Right to Know Act.

Freedom of Information Act

As you know from writing research papers to deciding which movie to watch, to be able to make a decision, gathering information usually is the first step. For environmental decisions, finding out timelines, seeing maps of proposed projects or existing impacts, gaining access to technical information about what precisely is involved, and more can be useful data to gather before forming an opinion about an environmental decision.

As a result of public pressure for greater access to information, the U.S. Congress passed the **Freedom of Information Act (FOIA)** in 1966. The act guarantees that any person has the right to see the records of the federal government. Federal agencies and departments whose records are typically requested by environmental reporters, scholars, and organizations include the U.S. Forest Service, Fish and Wildlife Service, Bureau of Land Management, Department of Energy, and the EPA, among others. Upon written request, an agency is required to disclose records relating to the requested topic unless the agency can claim an exemption from disclosure. (For a description of these exemptions, see www.usdoj.gov.) FOIA also grants requesting parties who are denied their request the right to appear in federal court to seek the enforcement of the act's provisions.

In 1996, the Congress amended FOIA by passing the **Electronic Freedom of Information Amendments**. The amendments require agencies to provide public access to data online. This is done typically by posting a guide for making a FOIA request on the agency's website. (See "FYI: How to Make a Request Under the Freedom of Information Act.") Individual states have adopted similar procedures governing public access to the records of state agencies.

☞ FYI   How to Make a Request Under the Freedom of Information Act

For information on FOIA, consult the U.S. Department of Justice guide (www.justice.gov/oip/foia_guide09.htm) or the Reporters Committee for Freedom of the Press's "Draft a FOIA Request" (www.rcfp.org/foia). Also, see the EPA's website (www.epa.gov/foia) for requesting documents under the FOIA.

To request information from another agency, see its website. For example, if you want to know what the U.S. Forest Service office in your area has done to enforce the Endangered Species Act (ESA) in a recent timber sale, go to the Forest Service's website for FOIA requests (www.fs.fed.us/im/foia/). There, you will find instructions for submitting your request for information. The Forest Service site also includes a sample FOIA request letter (www.fs.fed.us/im/foia/samplefoialetter.htm).

Under FOIA, individuals, public interest groups, journalists, and others routinely gather information from public agencies as they monitor their decisions and enforcement of permits. For example, a local River Guardians group might be interested in knowing what a mining company plans to do if its application to mine for gravel near a local river is approved by the Army Corps of Engineers. (The Corps is the federal agency responsible for permits under the Clean Water Act.) Although the application itself is public, the mining company's actual proposal may not be readily available; as a result, the River Guardians group can request this information by filing a FOIA request.

Individual citizens living in a community contaminated with toxic chemicals also may use FOIA to gather information for a *tort* or legal action against the polluter. An **environmental tort** is a legal claim for injury or a lawsuit, such as those depicted in the films *Erin Brockovich* and *A Civil Action*. Under federal law, the EPA is required

tejasbarrios.org

Photo 12.1 The Right to Know is a valued democratic principle at the foundation of the environmental justice movement. Access to information about what chemicals are in one's air or water is an essential step in participating in decisions that will impact one's health and the planet.

to maintain records for companies that handle hazardous waste, including notices of permit violations. As the group prepares its legal case, it can request these documents from the EPA under the agency's procedures for complying with FOIA. Using its right to access these documents, then, a community can file a FOIA request. If successful, they then have important (often incriminating) information that they can use in their tort action.

The Right to Know in the Post-9/11 Era

In the immediate aftermath following the attacks on the United States on September 11, 2001 ("9/11"), the U.S. Congress and the executive branch moved quickly to give new authority to federal law enforcement agencies and intelligence services. However, civil libertarians, public interest groups, and environmentalists discovered that these actions had troublesome implications for the public's right to know. Historians Gerald Markowitz and David Rosner (2002) reported "in the wake of the September 11 attacks, the Bush administration acted to restrict public access to information about polluting industries and restricted journalists' and historians' access to government documents previously available through the Freedom of Information Act" (p. 303).

Information about environmental topics available to the public before 9/11 were privatized. For example, *USA Today* reported,

> When United Nations analyst Ian Thomas contacted the National Archives . . . to get some 30-year-old maps of Africa to plan a relief mission, he was told the government no longer makes them public. When John Coequyt, an environmentalist, tried to connect to an online database where the Environmental Protection Agency lists chemical plants that violate pollution laws, he was denied access. (Parker, Johnson, & Locy, 2002, 1A)

In fact, in the eight months following the 9/11 attacks, the federal government removed hundreds of thousands of public documents from its websites; in other cases, access to material was made more difficult. For example, documents reporting accidents at chemical plants, previously available online from the EPA, were now to be viewed only in government reading rooms (Parker, Johnson, & Locy, 2002).

When the Obama administration came into office, many journalists and civil libertarians were hopeful that the president would reverse the Bush era trend toward secrecy in government. Indeed, on his first full day in office, January 21, 2009, President Obama issued a memorandum calling on all government agencies to "usher in a new era of open government." His memo directed the U.S. attorney general to issue comprehensive new guidelines for administrating the FOIA that adopted "a presumption in favor of disclosure" (White House, 2009).

The Obama administration record under the FOIA has been uneven, however. Several environmental agencies, such as the EPA and the Departments of Energy and the Interior, on the one hand, have moved to reduce the backlog of FOIA requests and increase transparency. In 2010, the EPA began launching online searchable databases for information held by the agency, reducing the need for FOIA requests. For example, it established ToxRefDB, a database allowing scientists and the public to search thousands of toxicity results and the potential health effects of chemicals (see http://epa.gov/ncct/toxrefdb). On the other hand, a study by George Washington University found that only "a minority of agencies" had responded to the president's memorandum with concrete changes in their FOIA practices; less responsive were the CIA, Department of State, and NASA ("Sunshine and Shadows," 2010). It will be interesting to see if this next presidential administration will be as influenced by the events of 9/11 or if new political contexts will frame environmental decision making.

While the FOIA has a mixed history, it has nevertheless been an invaluable tool for requesting information about the environment for journalists, environmental organizations, and residents in communities affected by toxic chemicals. A second U.S. law that provides vital information about industrial pollutants in local communities is the Emergency Planning and Community Right to Know Act.

Emergency Planning and Community Right to Know Act

Thousands of people were killed or injured when two separate Union Carbide plants released toxic chemicals: one more severe accident in Bhopal, India, in 1984 killed thousands of people in three days in one of the worst global chemical disasters in history, and another lesser known, but still notable leak from a pesticide plant in West Virginia in 1985 injured hundreds. These two incidents fueled public pressure for plans to prepare for chemical emergencies, as well as accurate information about the production, storage, and release of toxic materials in local communities. Responding to this pressure, the U.S. Congress passed the **Emergency Planning and Community Right to Know Act (EPCRA)** in 1986. The law requires industries to report to local and state emergency planners the use and location of specified chemicals at their facilities, as well as emergency notification procedures. (For the text of this law and description of its provisions, see http://www2.epa.gov/epcra/what-epcra.)

The Toxic Release Inventory

The Right to Know Act also requires the EPA to collect data annually on any releases of toxic materials into the air and water by designated industries and to make this information easily available to the public through an information-reporting tool, the **Toxic Release Inventory (TRI)**. The goal of the TRI "is to empower citizens, through information, to hold companies and local governments accountable in terms of how toxic chemicals are managed" (United States Environmental Protection Agency, 2010d).

In the years since the TRI debuted, the EPA has expanded its reporting and now collects data on approximately 581 different chemicals (Environmental Protection Agency, 2010). The EPA regularly makes these data available through online tools such as its TRI Explorer (www.epa.gov/tri/), although the data tend to lag by two years. Other public interest groups also use the TRI database to offer more user-friendly e-portals for individuals wanting information about the release of toxic materials into the air or water in their local communities. (For an example, see "Act Locally!: What Toxic Chemicals Are in Your Community?")

Act Locally!

What Toxic Chemicals Are in Your Community?

If you live in the United States, use the TRI to check for the presence of toxic chemicals in the air, soil, or water in the community where you or your family or friends live, work, or attend school. If you do not, try a zip code of a community of interest (examples to compare are Detroit, Michigan, 48204, and Fairfax Station, Virginia, 22039, a commuting location for Washington, DC).

To access the TRI database, use the EPA's TRI Explorer (www.epa.gov/tri/) or its Envirofacts website (www.epa.gov/enviro) or we recommend the more user-friendly Scorecard (www.scorecard.org), which provides environmental justice ratings in English and Spanish. Sponsored by Environmental Defense, Scorecard also makes it possible for you to contact the polluters in your area or e-mail state and federal decision makers.

Scorecard also links you to volunteer opportunities and environmental organizations in your area. Also see the EPA's Enforcement and Compliance History Online (ECHO) (www.epa .gov/echo). This site allows you to know whether the EPA or state governments have conducted inspections at a specific facility, whether violations were detected or enforcement actions were taken, and whether penalties were assessed in response to environmental law violations.

Many community activists as well as city and state governments believe that the TRI is the single most valuable information tool for ensuring community and industry safety. Sometimes, disclosure of information by itself may be enough to affect polluters' behavior. For example, political scientist Mark Stephan (2002) found that public disclosure of information about a factory's chemical releases or violations of its air or water permit may trigger a **shock and shame response**. This occurs when community members find out that a local factory is emitting high levels of pollution; their public expression of shock, including media coverage, may cause officials at the polluting facility to feel shame and, as a result, respond by improving conditions. Stephan conceded, however, that another explanation might be that, rather than feeling shame, company officials fear a backlash from citizens, interest groups, or the market, so change occurs as a result (p. 194).

In West, Texas, in 2013, a fertilizer plant exploded, killing 15 people and making national headlines. In 2014, a building in Athens, Texas, burned. It was storing ammonium nitrate, so journalists requested Community Right to Know Act information on the building. But, journalists were denied information as the attorney general ruled that such data could reveal the location of weapons facilities and, therefore, cause a security risk; as a result, only first responders and officials are allowed that information. When journalists followed up with government officials, they were stonewalled. A director with the Center for Effective Government, Sean Moulton responded: "I just think it's irresponsible and incredibly short-sighted to have officials involved in emergency planning responsible for keeping the public safe, and to say the public isn't going to be involved in that process" (Shipp, 2014). As this book goes to press, the conflicting state and federal guidelines have not been resolved.

Calls for the Right of Independent Expertise

After searching a toxic database, your reaction might be "Great, I just read that there is a high percentage of Volatile Organic Compounds in my community. What does that mean?"

An important supplement to the TRI is communities' access to independent experts to aid their understanding of the chemicals found at hazardous sites. Because the effects from exposure to toxic chemicals involves complex issues, advocates from communities with toxic waste sites long have sought access to sources of expertise to aid them in understanding the effects of these chemicals. Often, there is a disparity in the expertise that is available to government agencies or industry, on the one hand, and what is available to local citizens. That is, citizens—in affected communities— lack training in toxicology or other environmental sciences that would allow them to assess the government's findings.

In response to this gap, Congress enacted the **Technical Assistance Grant (TAG) Program** in 1986. The TAG program is intended to help communities at **Superfund sites**. Superfund sites are abandoned chemical waste sites that have qualified for federal funds for their cleanup. Decisions about the cleanup of these sites are usually based on technical information that includes the types of chemical wastes and the technology available. The purpose of the TAG program is to provide funds for citizen groups to pay for technical advisors to explain these technical reports, as well as EPA's cleanup proposals for the sites; the grants also allow these advisors to assist local citizens in participating in public meetings with the EPA.

Overall, the public's access to information about their environments has been greatly aided by laws such as the FOIA and the Emergency Planning and Community Right to Know Act. The tools available to the public under these laws are a major advance for the democratic principle of transparency, as well as particularly important resources for communities to learn about environmental hazards in their backyards.

Right to Comment

Information gathering is a significant first step in making decisions. Lack of information or unreliable sources can create barriers to sound decision making. Yet, all opinions also draw from personal experience. You might be able to find out the name of the chemicals in your local air if you access that information, but you are the expert in whether or not breathing that air is causing you difficulty or sending more of your neighbors to the doctor.

Town hall meetings and the right of citizens to speak directly to their government are long-standing traditions in democracies. When it comes to U.S. environmental law, this tradition got a significant boost in 1970, the year millions of citizens first celebrated Earth Day. Signed into law in that year, the **National Environmental Policy Act (NEPA)** was a landmark statute requiring all federal agencies take into account their environmental impact in decision making and guaranteeing that the public would have an opportunity to speak directly to U.S. government agencies before these agencies could proceed with any actions affecting the environment. At its core, the public's right to comment promised citizens a kind of "pre-decisional communication" would occur between them and the agency responsible for any decision that could potentially harm the environment; that is, before it could act, the agency must first solicit and hear citizens' views (Daniels & Walker, 2001, p. 8).

Public comment typically takes the form of in-person, spoken testimony at public hearings, exchanges of views at open meetings, written communications to agencies (e-mails, letters, and reports), and participation on citizen advisory panels.

In this section, we focus on the right to comment provided by NEPA. We also examine one of the most common forums for public participation, the *public hearing*. (Again, we describe citizens' advisory panels and more informal collaboration approaches for resolving environmental conflicts in the next chapter.)

National Environmental Policy Act (NEPA)

The core authority for the public's right to comment directly on federal environmental decisions, as we just noted, comes from the NEPA. Two NEPA requirements are intended to inform members of the public and then give them an opportunity to communicate regarding a proposed federal environmental action. These are (a) a detailed statement of any environmental impacts, that is, information on what could happen and (b) concrete procedures for *public comment*.

Environmental Impact Statements

As implemented by the Council on Environmental Quality, NEPA requires federal agencies to prepare a detailed **Environmental Impact Statement (EIS)** for any proposed legislation or major actions "significantly affecting the quality of the human environment" (Council on Environmental Quality, 1997). Such actions range from constructing a highway to setting standards for greenhouse gas emissions. Regardless of the specific action that is proposed, EISs typically describe at least four things: (1) the environmental impact of the proposed action, (2) any adverse environmental effects that could not be avoided should the proposal be implemented, (3) alternatives to the proposed action, and (4) how the public was given an opportunity to participate in the decision making.

Furthermore, NEPA requires that an EIS clearly communicate its meaning to the public:

> Environmental impact statements shall be written in plain language and may use appropriate graphics so that decision makers and the public can readily understand them. Agencies should employ writers of clear prose or editors to write, review, or edit statements, which will be based upon the analysis and supporting data from the natural and social sciences and the environmental design arts. (National Environmental Policy Act, 1969, Sec. 1502.8)

NEPA, therefore, not only requires communication between federal agencies and the public to foster democratic decision making, but also specifies that any information provided in an EIS should be expressed in a way that the general public might be expected to understand.

When federal agencies prepare a flawed EIS, they may be subject to legal action; or worse, they may allow harmful practices to go forward. For example, drilling for oil in deep water poses serious risks of environmental damage, as we discovered in the BP *Deepwater Horizon* oil spill in the Gulf of Mexico in 2010. Some scholars believe this could have been prevented if the federal agency in charge—the Minerals Management Service (MMS)—had fulfilled its NEPA responsibilities. Instead, in the BP *Deepwater Horizon* case, "the MMS approved the drilling and operating permits without undergoing full NEPA analysis," instead accepting the oil company's assurances that "there was very little risk of a blowout," and if did occur, they had the tools to prevent a disaster. They didn't. In the future, some argue, agencies should require a "realistic worst-case analysis in deepwater drilling" in preparing their EISs (Flatt, 2010, p. 7A).

Public Comment on Draft Proposals

As noted, NEPA requires that, before an agency completes a detailed statement of environmental impact, it must "make diligent efforts to involve the public" (Council on Environmental Quality, 1997, Sec. 1506.6 [a]). That is, the agency must take steps to ensure that interested groups and members of the public are informed and have opportunities for involvement prior to a decision. As a result, each federal agency must implement specific procedures for public participation in any decisions made by that agency that affect the environment. For example, citizens and groups concerned with natural resource policy ordinarily follow the rules for public comment developed in accordance with NEPA by the U.S. Forest Service, the National Park Service, the Bureau of Land Management, or the Fish and Wildlife Service. Community activists who work with human health and pollution issues are normally guided by EPA and state rules. The states are relevant because the EPA delegates to them the authority to issue air and water pollution permits for plants and construction permits and rules for managing waste programs (landfills and the like).

The requirements for public comment or communication under NEPA typically occur in three stages: (1) notification, (2) scoping, and (3) comment on draft decisions.

These steps are guided by the rules adopted by the Council on Environmental Quality (CEQ) to ensure that all agencies comply with the basic requirements for public participation that are implied in the NEPA statute itself.

The process normally starts with publication of a **Notice of Intent (NOI)** to the public. This announces an agency's intention to prepare an EIS for a proposed action. The NOI is published in the *Federal Register* and provides a brief description of "the proposed action and possible alternatives" (Council on Environmental Quality, 2007). The NOI may also be announced in the media and in special mailings to interested parties.

Typically, a notice describes the proposed regulation, management plan, or action and specifies the location and time of a public meeting or the period during which written comments will be received by the agency (Council on Environmental Quality, 2007, p. 13). An example of a NOI was the U.S. Department of Interior's proposal to list polar bears as a threatened species under the ESA. As you learned in Chapter 4, polar bears have seen their ice habitats shrink as the Arctic Ocean continues to warm.

The NOI under NEPA also describes an agency's **scoping** process. *Scoping* is a preliminary stage in an agency's development of a proposed rule or action, including any meetings and how the public can get involved. It involves canvassing interested members of the public about some interest—for example, a plan to reallocate permits for water trips down the Colorado River in the Grand Canyon—to determine what the concerns of the affected parties might be (Council on Environmental Quality, 2007). Such scoping might involve public workshops, field trips, letters, and agency personnel speaking one-on-one with members of the public.

Finally, NEPA rules require agencies to actively solicit public comment on the draft proposal or action. Public comments usually occur during public hearings and in written comments to the agency in the form of reports, letters, e-mails, postcards, or faxes. The public also may use this opportunity to comment on the adequacy of any EIS accompanying the proposal, or it may use the information in the EIS to assess the proposal itself.

In response, the agency is required to assess and consider comments received from the public. It must then respond in one of several ways: (a) by modifying the proposed alternatives, (b) by developing and evaluating new alternatives, (c) by making factual corrections, or (d) by "explain[ing] why the [public] comments do not warrant further agency response" (Council on Environmental Quality, 2007, Sec. 1503.4).

The success of NEPA's public participation process obviously depends on how well agencies comply with the law's original intent. For example, in its study of NEPA's effectiveness, the CEQ observed, the

> success of a NEPA process heavily depends on whether an agency has systematically reached out to those who will be most affected by a proposal, gathered information . . . from them, and responded to the input by modifying or adding alternatives throughout the entire course of a planning process. (Council on Environmental Quality, 1997, p. 17)

If, for example, notice only is extended through a local newspaper that costs money and a federal website that most people do not check daily, the opportunity for public comment is limited. This is one reason that face-to-face communication remains vital to debating environmental decisions in the public sphere, in addition to digital communication.

Public Hearings and Citizen Comments

At the heart of NEPA, as we've just seen, is the guarantee of the public's right to comment on proposed environmental decisions. Although soliciting written comments, particularly using digital technology, is practiced increasingly by federal agencies to foster communication in the public sphere, public meetings remain significant as well. In particular, NEPA may require **public hearings.** These are face-to-face meetings that solicit public input on a decision before an agency takes action that might significantly affect the environment. In this section, we look at the kind of communication that typically occurs in such public hearings.

Communication at Public Hearings

The public hearings required by NEPA usually occur in the communities that may be most directly impacted by an agency's action. Typically, the agency will announce

U.S. Department of Agriculture/Flickr

| Photo 12.2 | At public hearings, some people come to listen in order to become more informed, some come to show that the issue is important to them by merely being present, and some speak. This image is of citizens signing a petition outside of a public town hall meeting in Lubbock, Texas, to discuss a Food, Farm and Jobs Bill, as well as immigration. |

its proposed action, notify the public of the times and locations of the planned hearings, and then conduct the hearings, allowing interested parties to express their opinions at a microphone in a room open to the public. Both supporters and opponents of a proposal may attend, and both sides usually speak at public meetings. Journalists sometimes attend to report on what was said to an even wider audience.

Public hearings can become an important mobilizing goal for a grassroots movement. Cherise Udell (2007), the president and founder of Utah Moms for Clean Air, describes a successful process of mobilizing people for a public hearing succinctly:

> Breathing Salt Lake City's dirty air during a winter inversion is the same as smoking half a pack of cigarettes. The image of my baby with a cigarette dangling from her toothless mouth was enough to move me to action. Utah Moms for Clear Air was born that day with a simple but heartfelt email to about 100 moms inviting them to join together to make Utah's air cleaner and safer. The response has been phenomenal. In three short weeks, Utah Moms for Clean Air is almost 300 strong, and counting. Utah Moms has already made our voices heard at the Air Quality Board hearing on the health affects of pollution, and held our first public meeting in which over 100 moms (and dads) attended. (para. 2)

Persuading people to attend is one step. Convincing people to speak and figuring out how to speak in a compelling way is another. For example, we both have attended public hearings on a permit to construct a new coal-burning power plant in our respective states. At both, the hearing room was crowded with agency staff, lawyers, employees of Duke Energy (the electric utility company in both cases), health professionals, students, representatives from environmental organizations, parents with their young children, religious leaders, reporters, and many others. As usual, there were sign-in sheets for those of us wishing to speak. Before inviting comments from the public, the presiding official called on agency staff to provide technical information on the proposed permit. Then, others in the audience were given three minutes each to comment orally or to read a statement. Some individuals read from prepared statements, others spoke extemporaneously. In both hearings attended, the comments were overwhelmingly in opposition to the permit for the power plant, though the legislative body approved the plant in one of our states.

The comments themselves at public meetings may be polite or passionate, restrained or angry, or informed or highly opinionated. The range of comments and emotions reflects the diversity of opinions and interests of the community itself. Officials may urge members of the public to speak to the specific issue on the agenda, but the actual communication often departs from this, ranging from individuals' calm testimony to emotionally charged stories of their families' experiences to criticism of opponents or public officials.

Some people may denounce the actions of the agency or respond noisily, even angrily, to proposals affecting their lives or communities, but seem ignored. On the other hand, an individual's quiet testimony may be emotionally powerful. At the public hearing on the coal-burning plant in North Carolina, a young mother told of her and her husband's borrowing money to install a solar panel, so concerned were they about global warming and their children's futures. Weeping quietly, she pleaded with the officials to deny the permit for the power plant. At the one in Indiana, many of the seats were filled with workers given the option to take the day "off" with pay and a free lunch if they would get on the company bus to attend the meeting.

Speaking at public hearings can be as daunting as public oral communication class itself, since many who step up to speak are not used to communicating in such forums. Ordinary citizens find themselves apprehensive about having to speak in front of large groups, perhaps with a microphone, to unfamiliar officials. They may face opponents or others who are hostile to their views. Sometimes, they must wait hours for their turns to speak. Those with jobs or small children face additional constraints because they must take time from work or find (and often pay) someone to watch their children. Plus, one can feel self-conscious about how one is perceived by the hearing board, depending on one's race/ethnicity, gender, class, educational background, and more. We elaborated on some of these cultural barriers to citizen participation in Chapter 10.

Due to the conditions that are typically imposed by crowded hearing rooms, limited time, volatile emotions, and long waiting times for speaking, some believe that public hearings are not an effective form of public participation. Daniels and Walker (2001) go further when they contend that some public lands management agencies such as the Forest Service exhibit a "Three-'I' Model . . . inform, invite, and ignore." For example, agency officials will inform the public about a proposed action, such as a timber sale, then "invite the public to a meeting to provide comments on that action, and ignore what members of the public say" (p. 9).

Results are never guaranteed in the public sphere, and public hearings reflect the diverse and messy norms of public life. Nevertheless, at their best, public hearings that invite wide participation by members of the public may generate comments and information that help agencies to shape or modify important decisions affecting the environment. Although they occasionally may be confrontational or ignored, public hearings provide many citizens their only opportunity to speak directly to government authority about matters of concern to them, their families, or their communities.

Mobilizing Public Comment: The Roadless Rule

A dramatic example of citizen mobilization occurred in the U.S. Forest Service's adoption of its **Roadless Rule** for U.S. national forests. The rule, adopted in 2001, prohibits road building and restricts commercial logging on nearly 60 million acres of national forest lands in 39 states. The final rule reflected a year and a half of public comment under NEPA. (Here, we must admit a personal interest, having participated as president of the Sierra Club in mobilizing individuals to participate in the public comment process. For a more critical review of NEPA's role in the Roadless Rule, see Walker, 2004.)

By the end of the process, the Forest Service had held more than 600 public meetings and received an unprecedented two million comments from members of the public, environmentalists, businesspeople, sports groups and motorized recreation associations, local residents, and state and local officials. As a result of the strong public support for protecting wild forests, the rule grew stronger, expanding the amount of protected (roadless) forest land. Forest Service Chief Mike Dombeck reflected, "In my entire career, this is the most extensive outreach of any policy I've observed" (Marston, 2001, p. 12, in Walker, 2004, p. 114).

Following its adoption, the Roadless Rule has been both praised and criticized for its public participation process. Its implementation initially was delayed by court challenges from logging interests and western state officials. In May 2005, the Bush administration dropped the Roadless Rule altogether after a hasty NEPA process. Environmental groups challenged this and the case continues to be argued in the federal courts. The Obama administration is defending the Roadless Rule as this book goes to press. (For the status of the Roadless Rule, see http://roadless.fs.fed.us.) Part of the controversy has been a debate over the meaning of public participation itself and the goals it is intended to serve. Walker (2004) asked, "Does the *number* of

European Parliament

Photo 12.3

The European Union, like many other places in the world, has been having a public debate about water privatization initiatives. Calling the first hearing on a Citizens' Initiative in 2014 "a milestone in the history of European democracy," Gerald Häfner (Greens/EFA, DE), of the Petitions Committee, said, "Today, we are switching to listening mode. The question now is how we can better legislate on an issue that is crucial. Water is a human right and should remain in public hands."

public meetings and *amount* [emphasis added] of comment letters received provide sufficient evidence of meaningful public participation?" (p. 115). We take up this question more generally in the next chapter by describing some of the criticisms of public comment in environmental decision making.

Overall, NEPA, and its guarantee of public comment, has proved to be one of the most empowering laws of the environmental history of the United States. In terms of its scope and involvement of the public, NEPA has been the cornerstone of the principle of direct participation in governance through the right of any citizen to comment directly to agencies responsible for decisions affecting the environment. Now, we want to turn to similar efforts that have been made globally.

Another Viewpoint: Is Private Money Shrinking Public Rights?

Public participation in democratic societies is based in an assumption that everyday people should have a voice in decisions impacting our lives. Today, some feel those voices are less heard by elected officials as a result of financial donations from corporations.

In a close 2010 U.S. Supreme Court ruling (5–4), *Citizens United v. Federal Election Commission*, political spending was to be protected as free speech. As a result, any person or institution may spend any amount of money to support a candidate any time up to an election. Further, that money could be pooled together in something called a "Super PAC" that would not require the names of the donors to be disclosed publicly.

Professor of law and political science Richard L. Hasen (2010) found in 1992 that the total amount of spending on elections was approximately US$1.5 million; in 2012, after the 2010 ruling, spending increased to over US$88 million. What rhetorical constraints do you think this trend in politics causes?

SOURCE: Hasen (2010).

SLAPP: Strategic Litigation Against Public Participation

A direct challenge to the right to comment sometimes occurs when some businesses act aggressively against their critics, whether they are local residents or environmental organizations. In this final section, we describe a particularly chilling strategy used to silence, discourage, or intimidate such critics' communication—a legal action known as "Strategic Litigation Against Public Participation" or a *SLAPP lawsuit*. Simply stated, a SLAPP occurs when someone criticizes, for example, a permit allowing a corporation to operate a hazardous waste inclinator, and is sued by the company for speaking against it. Professor of law George W. Pring and professor of sociology Penelope Canan (1996) defined a **SLAPP lawsuit** more formally as one involving "communications made to influence a governmental action or outcome, which, secondarily, resulted in (a) a civil complaint [lawsuit] . . . (b) filed against nongovernmental individuals or organizations . . . on (c) a substantive issue of some public interest or social significance" such as the environment (pp. 8–9).

In the end, the courts dismiss most SLAPP lawsuits because of citizens' First Amendment protections of rights to speak and petition government. But, the mere act of filing a lawsuit that alleges libel, slander, or interference with a business contract can be financially and emotionally crippling to the defendants. And this is part of the strategy—to discourage, chill, or silence critics. Let's look at some examples of SLAPPs being filed against individuals and how they responded.

Sued for Speaking Out

Consider, first, the case of Colleen Enk, who had publicly questioned a gravel mine proposed for the Salinas River near her neighborhood in San Luis Obispo County, California. Shortly after she had spoken out, Colleen found herself the target of a lawsuit by the developers for libel and defamation, among other charges. The lawsuit also sought financial damages. "It's pretty transparent why they did it," her attorney, Roy Ogden, said. "They wanted to shut her up" (Johnston, 2008, para. 12).

The developers had proposed to dig sand and gravel from the Salinas River over a 20-year period. After Colleen and her neighbors questioned the plans, a process server appeared at her door one evening in May 2008. She was not at home. The next day, Colleen voiced her concerns about the gravel mine again at a public meeting of the San Luis Obispo County Planning Commission. "My heart was pounding," she said (quoted in Johnston, 2008, para. 11). The next morning, the server caught her at home and served the summons to court.

Enk's attorney said, "Colleen has been sued for exercising the right of free speech in America. She's been stomped." Ultimately, the developer dropped the suit, but Colleen had been forced to pay several thousand dollars in attorney fees and court costs to defend her right to public comment.

More recently, a large cement corporation has proposed to build a coal-burning, cement-manufacturing plant near Wilmington, North Carolina. At a County Commissioners meeting, two local residents spoke against the plant. They cited the company's performance and the plant's potential health effects on children (Faulkner, 2011). In response, the corporation filed two lawsuits in federal court alleging the individuals slandered the company. One of the defendants, Kayne Darrell, reacted saying, "I'm kind of in shock right now. I'm just a mom and a housewife fighting a billion-dollar company trying to keep me from saying what I believe is right and true and from protecting my air and water" (quoted in Faulkner, 2011, para. 17). The Raleigh *News and Observer,* in an editorial, called the lawsuits "a blatant attempt at intimidation. If they're not outright SLAPP suits . . . they're a reasonable facsimile" ("Bullying Behavior," 2011, p. 9A). (Again, the cement company later dropped its lawsuit.)

The list of examples could go on. Every year, "thousands of Americans have been 'SLAPPed' for writing letters to the editor, circulating petitions, calling public officials, or speaking at public meetings, according to legal experts" (Johnston, 2008, para. 22). The goal of such lawsuits is to chill or silence the individual or organization that has spoken out in order to stifle debate in the public sphere. So, what can be done?

Response to SLAPPs

To figure out how to respond rhetorically to a SLAPP, it is important to assess what a SLAPP achieves. As Pring and Canan (1996) explain, businesses that file SLAPP suits

> seldom win a legal victory—the normal litigation goal—yet often achieve their goals in the real world. . . . Many [of those who are sued] are devastated, drop their political involvement, and swear never again to take part in American political life. (p. 29)

Even if the organization or individual citizen wins, he or she most likely "has paid large sums of money to cover court costs and has been thrown into the public eye for months or even years. This unlawful intimidation pushes people into becoming less active and outspoken on issues that matter" (California State Environmental Resource Center, 2004, para. 7). The goal, therefore, is not necessarily to win in court but to cost public advocates of environmental quality and health their time, resources, and sense of security. While not a direct legal challenge to standing in a court of law, it attempts to inhibit everyday people and environmental organizations from "standing up" or speaking against powerful interests in the public sphere.

The wicked nature of a SLAPP lawsuit has led many states to provide remedies for the individuals sued. In such cases, courts may agree to dismiss a lawsuit if it appears to be motivated by the unconstitutional purpose of silencing speech and to require the plaintiffs to pay court costs. Beyond this, two principal sources of defense have arisen in response to SLAPP lawsuits—one based in constitutional guarantees of democratic rights and the other from personal injury law. Pring and Canan (1996) refer to these two sources of defense as the "one-two punch" that has characterized successful defense against SLAPP lawsuits.

The first and most basic defense against a SLAPP is derived from the rights granted to citizens in the First Amendment to the U.S. Constitution, most importantly, freedom of speech and the right of the people to petition the government for redress of grievances. Often, a court will grant expedited hearings to dismiss a SLAPP if the citizen's criticism was part of a petition to the government. In such cases, the plaintiff (the party bringing the lawsuit) must show that the citizen's petition is a sham in order to proceed with the original lawsuit.

The second part of the defense involves what is known as a **SLAPP-back** suit against the corporation bringing the initial allegations against a citizen. Here, the defendant SLAPPs back by filing a countersuit alleging that the plaintiff infringed on the citizen's right to free speech or to petition the government. SLAPP-back lawsuits usually allow for recovery of attorneys' fees as well as punitive damages for violating constitutional rights and inflicting damage or injury on the defendant.

Indeed, environmentalists, labor, individual citizens, and others have won monetary awards in fighting SLAPP actions. Pring and Canan (1996) report that awards in SLAPP-backs occasionally have been large—jury verdicts of $5 million to a staggering $86 million were awarded against corporations that have brought SLAPP suits.

Increasingly, developers, polluters, and others have had to weigh the chances of a SLAPP-back before bringing a SLAPP action. Pring and Canan observed, "Even though SLAPP-backs are not a panacea, this risk of having to defend against them may prove to be the most effective SLAPP deterrent of all" (p. 169).

Growth of Public Participation Internationally

The expansion of rights of the public to participate actively in decisions about the environment has occurred not only in the United States. In the past decade, in fact, more and more nations have begun to guarantee public access to information and implement various forms of public participation in governmental decisions about the environment. Clearly, the European Union, the United Nations Economic Commission for Europe (UNECE), and many of the former nations of the Soviet Union have taken the lead in this area. For example, UNECE has successfully negotiated five environmental treaties, governing trans-boundaries environmental protections in Europe for air pollution, watercourses, and lakes and extending guarantees for environmental impact assessments.

Of these five treaties, the Convention on Access to Information, Public Participation, and Access to Justice in Environmental Matters—often called the *Aarhus Convention*—is unprecedented for its guarantees of public participation. Adopted in 1998 in the Danish city of Aarhus, the **Aarhus Convention** is a "new kind" of agreement, linking environmental rights and human rights (United Nations Economic Commission for Europe, 2008). Article 1 clearly announces its goal:

> In order to contribute to the protection of the right of every person of present and future generations to live in an environment adequate to his or her health and well-being, each party shall guarantee the rights of access to information, public participation in decision-making, and access to justice in environmental matters.

These three principles—access to information, public participation, and access to justice—are developed in detail, with concrete procedures for ensuring citizen access to these rights.

In many ways, the approach of the Aarhus Convention goes further than U.S. environmental law under NEPA by providing additional measures to help protect public interests. For example, the right of **access to justice** in Article 9 ensures that "any person who considers that his or her request for information . . . has been ignored, [or] wrongfully refused . . . has access to a review procedure before a court of law or another independent and impartial body" (United Nations Economic Commission for Europe, 2008, para. 13).

At their 2008 meeting in Latvia, European nations reaffirmed the goals of the Aarhus Convention, guaranteeing the rights of access to information and participation in decisions about the environment. Marek Belka, executive secretary of the UN

Economic Commission for Europe, observed that the convention's core principles "empower ordinary members of the public to hold governments accountable and to play a greater role in promoting more sustainable forms of development" (*Aarhus Parties*, 2008). Similar moves to implement or strengthen the role of the public in decisions affecting the environment are actively under way in China and central Asia, as well as in many parts of Africa and South America.

In addition to the Aarhus Convention, programs similar to the TRI have been established or are being implemented not only in Europe but Asia, Australia, Canada, and some countries in South America and Africa. These TRI-like initiatives range from emission inventories, which collect data on specific chemical releases, to a more comprehensive program known as **Pollutant Release and Transfer Register (PRTR)** that expands on this data collection. Like the TRI, PRTR programs require not only the collection of data but mandatory reporting of a facility's chemical releases as well as public access to such data. PRTR programs also differ in some nations. For example, Japan collects information on automobile emissions, while Mexico's PRTR is voluntary for industries.

Clearly, demands for public participation in environmental matters are increasing worldwide. For example, in early 2008, the Carter Center in Atlanta convened an International Conference on the Right to Public Information. More than 125 representatives from 40 nations gathered to "identify the necessary steps and measures to ensure the effective creation, implementation, and exercise of the right of access to public information" (Carter Center, 2008, para. 2). Elsewhere, new initiatives for public participation are emerging in Asia, Africa, and Latin America, with a vigorous movement for environmental protection and public access to information growing in China, particularly. (For updates on recent initiatives to strengthen public participation guarantees in China, see http://switchboard.nrdc.org/blogs/chinagreenlaw/; this blog is a joint project of the Natural Resources Defense Council and the China Environmental Culture Promotion Association.)

Public participation, as we have shown, is not always simply a matter of informing citizens and inviting comment. Instead, many environmental communication practitioners help foster public participation through focusing on the ways communities and environmental decision makers may be encouraged to become involved in a more interactive, participatory process. Communication scholar and international consultant Arvind Singhal (2001) defines **participatory communication** as "a dynamic, interactional, and transformative process of dialogue between people, groups, and institutions that enables people, both individually and collectively, to realize their full potential and be engaged in their own welfare" (p. 12). That is, he emphasizes how the process of people engaging each other shapes how we imagine our identities and agency in relation to decision-making processes. This approach involves determining the ways people decide to communicate with each other, as well as access to use or to own communication technologies. We elaborate on this more collaborative understanding of public participation in the next chapter.

SUMMARY

In this chapter, we identified some of the legal rights and civic forums that enable you and others to participate directly and publicly in decisions about the environment.

- In the introduction, we defined *public participation* as the ability of citizens and groups to influence environmental decisions through (1) the right to know, (2) the right to comment, and (3) the right of standing.
- Basic to effective public participation is a *right to know* or to have access to information to help one form an opinion. In the first section, we described two powerful tools for citizens' right to know: (1) the Freedom of Information Act (FOIA) and (2) the Toxic Release Inventory (TRI).
- Next, we explored the *right to comment* that is guaranteed by the National Environmental Policy Act (NEPA) and its requirements for (a) an Environmental Impact Statement (EIS) about a proposed action and (b) concrete procedures for public comment on this action.
- As a challenge to the right to comment, we described how some polluting industries, in particular, attempt to intimidate everyday people and environmental organizations through SLAPP suits, or Strategic Litigation Against Public Participation lawsuits. These are used to silence or intimidate individuals who criticize their environmental performance. Responses have included dismissal of charges under the First Amendment and a SLAPP-back, in which the aggrieved sue in return.
- Finally, initiatives guaranteeing public participation are expanding in Europe, Asia, and other continents. The most comprehensive of these rights-based approaches is Europe's Aarhus Convention, which seeks to implement rights of access to information, public comment on environmental matters, and review by the courts.

In the next chapter, we continue to discuss more informal forms of public participation, generally referred to as "collaboration." This kind of communication occurs when decision makers and everyday people sit down at the same table to make judgments about the environment together.

SUGGESTED RESOURCES

- Thomas Dietz and Paul C. Stern (Editors), *Public Participation in Environmental Assessment and Decision Making*. National Research Council. Washington, DC: National Academies Press, 2008.
- Michael E. Kraft, Mark Stephan, and Tory D. Abel. (2011). *Coming Clean: Information Disclosure and Environmental Performance*. Cambridge, MA: MIT Press, 2011.

- *Mother Jones* magazine analyzed data to create an interactive U.S. map so everyday people can search to see if there is a chemical plant that could cause a catastrophic release nearby. Since there are over 9,000 facilities that qualify nationally, chances are you can find one near you as well: http://www.motherjones.com/environment/2014/04/west-texas-hazardous-chemical-map.
- There are many websites offering tips for how to participate in a public hearing. Our Water West Virginia offers this succinct and helpful one on how to prepare and talking points: http://ourwaterwv.org/tips-for-public-hearing/.
- The U.S. EPA's public participation guide is available to download for free online in English, Spanish, French, and Chinese: http://www.epa.gov/oia/public-participation-guide/index.html.

KEY TERMS

Aarhus Convention   308

Access to justice   308

Accountability   390

Electronic Freedom of Information Amendments   292

Emergency Planning and Community Right to Know Act (EPCRA)   295

Environmental Impact Statement (EIS)   298

Environmental tort   293

Freedom of Information Act (FOIA)   292

National Environmental Policy Act (NEPA)   298

Notice of Intent (NOI)   300

Participatory communication   309

Pollutant Release and Transfer Register (PRTR)   309

Public comment   298

Public hearings   301

Public participation   290

Right to comment   290

Right to know   290

Roadless Rule   303

Scoping   300

Shock and shame response   296

SLAPP-back   307

SLAPP lawsuits   305

Sunshine laws   291

Superfund sites   297

Technical Assistance Grant (TAG) Program   297

Toxic Release Inventory (TRI)   295

Transparency   290

DISCUSSION QUESTIONS

1. Are there limits to what the public has a right to know? For example, should the U.S. government have a right to restrict the public's access to information about oil refineries, pipelines, nuclear plants, or other environmental facilities to prevent potential terrorists from exploiting vulnerabilities in these facilities?

2. Do public hearings merely allow angry citizens to blow off steam about controversial environmental actions? Do such forums serve an important role for public comment, or are they just window dressing?

3. Would the threat of a corporate SLAPP lawsuit discourage you from speaking about your concerns over an environmental or health concern at a public hearing? Be honest! Can you imagine or understand the psychological and emotional (and financial) fears that some might have in these circumstances?

4. We regularly see headlines today of international protests for democratic rights. Which countries do you believe are the most open to sharing information and listening to their citizens? Which are the least? Globally, do you feel we've steadily gained rights or do everyday people seem to be losing these rights?

5. Team up with three other students for 15 minutes to brainstorm how to act out one of the two rights (know and comment) as they relate to an environmental issue in the format of a two-minute TV political ad. You must use the name of the right and the corresponding democratic principle. Any reference to course materials is considered a plus. Be creative! Then, share all the ads in class and vote on your favorite one. Which right seemed most challenging to perform in this format? How did you decide who was most compelling?

The "historic agreement to protect British Columbia's Great Bear Rainforest—the world's largest remaining swath of coastal temperate rainforest, an area twice the size of Yellowstone—marks the culmination of 10 years of haggling, negotiating, and compromising among divergent agendas. . . . The plan, which places four and a half million acres off limits to logging and regulates logging practices on the remaining 10 million acres, is backed by environmentalists, industry, native people, and the provincial government" (Roser, 2006, para. 1).

CHAPTER 13

Managing Conflict

Collaboration and Environmental Disputes

In the long history of humankind (and animal kind, too) those who learned to collaborate and improvise most effectively have prevailed.

—Charles Darwin, biologist (1809–1882)

"This isn't a magic bullet," says Gerald Mueller of Missoula, Mont., who has been a mediator since 1988. . . . It takes hefty amounts of work and time. "It's not usually a lot of fun to do, because you're usually involved with people you may have had conflicts with and that you may not like."

— Journalist Lisa Jones (1996b, para. 3)

Frustrated with traditional forms of public participation, many environmentalists, businesses, and community leaders have turned to other ways to manage conflicts over environmental disputes—including, for example, disagreements over logging practices in Canada's Great Bear Rainforest, feuds between ranchers and conservationists over prairie dogs, opposition to dams that impede salmon runs, and more. The purpose of this chapter is to describe the adoption by many community leaders, environmentalists, natural resource industries, and others of some form of collaboration to manage environmental disputes.

Chapter Preview

- In the first section of this chapter, we describe the dissatisfaction with traditional forms of public participation, such as public hearings, and identify three alternatives for managing environmental conflicts—citizens' advisory committees, natural resource partnerships, and community-based collaboration.
- In the second section, we describe a form of communication called "collaboration," when this is appropriate, and the communication skills required for successful collaboration. To illustrate these concepts, we explore a successful case study of collaboration about the Great Bear Rainforest.
- Finally, in the third section, we consider criticisms of collaboration and identify some of the circumstances in which collaboration may not be appropriate for resolving environmental conflicts.

In the United States and globally, citizens, environmentalists, business leaders, and public officials are experimenting with new approaches to public participation in environmental disputes. They are talking with their opponents across the table, working through their differences, and in many cases resolving conflicts that have festered for years. These innovative forms of conflict management have been called by different names: community-based collaboration, citizen advisory boards, consensus decision making, and alternative dispute resolution models. Usually, they involve a form of communication called *collaboration*.

Collaboration is defined as "constructive, open, civil communication, generally as dialogue; a focus on the future; an emphasis on learning; and some degree of power sharing and leveling of the playing field" (Walker, 2004, p. 123). Occurring among multiple parties, collaboration is

> sometimes facilitated or mediated, and sometimes not; . . . [and] often strive[s] for specific implementable agreements. . . . But in all cases, the emphasis is on the discourse—the thoughtful process of deliberating on complex and often controversial issues. Listening and speaking is done as much to learn as to convince. (Daniels & Cheng, 2004, pp. 127–128)

For example, after years of conflict, ranchers, environmentalists, and off-road vehicle recreationalists (who often fought over public lands in the West) came together in an unprecedented agreement over Idaho's Owyhee Canyonlands; the agreement's designation of wilderness and wild rivers is one of the largest additions in recent years to the U.S. wild and scenic river system (Barker, 2010).

New Approaches to Environmental Disputes

Since passage of the National Environmental Policy Act (Chapter 12), the public's right to comment on government actions affecting the environment has been widely recognized, and forums for public involvement have proliferated. As we learned in the previous chapter, public comment on an environmental proposal typically takes the form of public hearings, citizen testimony, and written comments. Yet, some feel these processes produce more frustration and division than they do reasoned decision making. Officials and consultants often speak in technical jargon, using such phrases as *parts per billion* of chemical substances, and members of a community sometimes feel that their concerns don't matter, that their efforts to speak are dismissed by public officials and experts. In this section, we examine criticisms of public hearings and identify some of the emerging alternatives for public involvement.

Criticism of Public Hearings

Several years ago, public officials in a town near one of us announced a public hearing after they had decided informally to build a hazardous waste facility near residential homes and a hospital. Many of the town's residents and patients' advocates understandably were upset about this. The atmosphere at the public hearing was electric; many voiced their anger at officials who sat stone-faced in the front of the auditorium. While some calmly testified at the microphones provided to them, other members of the audience shouted at the officials. One young man rushed to the front of the auditorium and dumped a bag of garbage in front of the officials to dramatize his objection to hazardous waste. Area TV stations and news editorials denounced the "irrational" behavior of residents and the heavy reliance on emotion.

Are ordinary citizens really irrational? Or are public officials insensitive to the concerns of ordinary citizens, dismissing their fears because they lack technical expertise? Certainly, some officials feel that the behavior of the public is "overdramatized and hysterical" and that they must endure "the public gauntlet" of angry, shouting, sign-waving protesters (Senecah, 2004, pp. 17, 18). Yet, environmental communication scholar Susan L. Senecah (2004) poses the question differently: Are public hearings sometimes divisive or unproductive because of the way the public acts, or is there something wrong with the process itself? Senecah suggests that, in many local conflicts, a gap exists between what the public expects and their "actual experiences" from participating in these forums (p. 18). Although NEPA procedures require officials to solicit the views of the public, formal mechanisms for public participation are sometimes simply ritualistic processes that give members of the public little opportunity to influence decisions. It's no surprise, then, that ordinary citizens so often experience "frustration, disillusionment, skepticism, and anger" (Senecah, 2004, p. 18).

What has gone wrong? Stephen Depoe, director of the University of Cincinnati's former Center for Environmental Communication Studies, and John Delicath (Depoe & Delicath, 2004), now of the U.S. Government Accountability Office, surveyed the extensive research on traditional modes of public participation, such as written comments and public hearings. They identified five primary shortcomings:

1. Public participation typically operates on technocratic models of rationality, in which policy makers, administrative officials, and experts see their roles as educating and persuading the public of the legitimacy of their decisions.

2. Public participation often occurs too late in the decision-making process, sometimes even after decisions have already been made.

3. Public participation often follows an adversarial trajectory, especially when public participation processes are conducted in a decide–announce–defend mode on the part of officials.

4. Public participation often lacks adequate mechanisms and forums for informed dialogue among stakeholders.

5. Public participation often lacks adequate provisions to ensure that input gained through public participation makes a real impact on decisions' outcomes. (pp. 2–3)

Although formal mechanisms for citizens' involvement in influencing environmental decisions have been effective on some occasions, on others they have fallen far short of citizens' expectations. Too often, disputes over local land use or the cleanup of communities contaminated by chemical pollution linger for years. In many of these cases, citizens, businesses, government agencies, and environmentalists have turned to alternatives to public hearings to resolve conflicts over environmental problems.

Emergence of Alternative Forms of Public Participation

Starting in the 1990s, new forms of public involvement in environmental decisions began to emerge—from local, neighborhood initiatives to Environmental Protection Agency (EPA) collaborations with cities over new standards for safe drinking water. As citizens, public officials, businesses, and some environmentalists grew frustrated with traditional forms, they began to experiment with new ways of organizing public participation: scoping meetings, listening sessions, advisory committees, blue-ribbon commissions, citizen juries, negotiated rule making, consensus-building exercises, and professional facilitation, among others (Dietz & Stern, 2008). For example, the U.S. Institute for Environmental Conflict Resolution, an independent federal program, works with local groups and public officials to "find workable solutions to tough environmental conflicts" (www.erc.gov), while the EPA provides alternative dispute resolution (ADR) services to deal with environmental disputes and potential conflicts through its Conflict Prevention and Resolution Center (www.epa.gov/adr/cprc_adratepa.html).

At the heart of these experiments is some version of community or place-based collaboration among the relevant parties. Later in this chapter, we identify characteristics of collaboration that help to explain its success or failure. But first, let's look at three forms that collaboration about environmental conflicts can take: (1) citizens' advisory committees, (2) natural resource partnerships, and (3) community-based collaborations.

Citizens' Advisory Committees

One of the most widespread uses of collaboration is the **citizens' advisory committee (CAC)**. Also called citizens' advisory panels or boards (CAPs or CABs), these usually are groups that a government agency appoints to solicit input from diverse interests—local residents, business representatives, city planners, environmental scientists, and more—in a community about a project or problem. For example, the toxic landfill that many say sparked the environmental justice movement in Warren County, North Carolina (Chapter 10), was cleaned up after the efforts of a citizens' advisory committee (Pezzullo, 2001).

On a larger scale, the Department of Defense uses Restoration Advisory Boards (RABs) to advise military officials on the social, economic, and environmental impacts of military base closings and the restoration of military lands. The purpose of RABs, which were initiated in 1994, is to "achieve dialogue between the installation and affected stakeholders; provide a vehicle for two-way communication; and provide a mechanism for earlier public input" (Santos & Chess, 2003, p. 270). One successful RAB, composed of local interests and military officials, collaborated in helping the Department of Defense to convert the former Rocky Mountain Arsenal, a chemical weapons facility, into the Rocky Mountain Arsenal National Wildlife Refuge, where today, bison and other wildlife adapted to the high plains browse (www.fws.gov/refuge/rocky_mountain_arsenal).

The impetus for involving communities in the work of federal agencies is a result of the Federal Advisory Committee Act of 1972. The impact of this act can be seen in other agencies that also use citizen advisory panels. For example, the EPA uses citizen advisory panels to involve citizens in ongoing projects to clean up abandoned toxic waste sites. Similarly, the Department of Energy (DOE) relies on site-specific advisory boards to involve nearby residents during the cleanup of toxic waste at former energy sites, such as the nuclear weapons production facility in Fernald, Ohio.

For most citizen advisory committees, the government agency selects participants to represent various interests or points of view or to be "representative, that is, a microcosm of the socioeconomic characteristics and the issue orientation of the public in [a] particular area" (Beierle & Cayford, 2002, pp. 45–46). The committee's work normally takes place over time and may or may not assume that consensus will be achieved. Typically, the outcome of collaboration is a set of recommendations to the agency (Beierle & Cayford, 2002).

Natural Resource Partnerships

Particularly in western states, the idea of collaboration has taken off as diverse groups seek ways to manage differences over the uses of watersheds, public lands, and other natural features. As early as the 1990s, Colorado's *High Country News* observed that "coalitions of ranchers, environmentalists, county commissioners,

Joint Base Anacostia-Bolling/Flickr

Photo 13.1

U.S. Air Force photo by Senior Airman Steele C. G. Britton. Volunteers came together to pick up trash in support of the 24th Annual Potomac River Watershed Cleanup, April 21, 2012, at Joint Base Anacostia–Bolling. More than 160 volunteers came out to support the clean-up that was followed by a Blessing of the Fleet, a pass-in-review and chili cook-off at the Slip Inn Bar and Grill patio on base.

government officials, loggers, skiers, and jeepers are popping up as often as wood ticks across the Western landscape" (Jones, 1996a, p. 1). Sometimes called **natural resource partnerships**, these coalitions include private landowners, local officials, businesses, and environmentalists, as well as state and federal agencies. They are organized around an identifiable region—such as a watershed, forest, or rangeland—with natural resource concerns (for example, water quality, timber, agriculture, wildlife). Such partnerships operate collaboratively to integrate their differing values and approaches to the management of natural resource issues.

One of the earliest and longest running models of natural resource collaboration is the Applegate Partnership & Watershed Council, organized in 1992. It was formed after years of conflict among ranchers, local government, loggers, environmentalists, and the U.S. Bureau of Land Management (BLM) in the watershed of southwestern Oregon and northern California. Feuding parties finally decided to take a different approach. Local BLM official John Lloyd explained, "We got to the point where we just had to sit down and start talking" (Wondolleck & Yaffee, 2000, p. 7).

As they talked, it became apparent that conservationists, loggers, and community leaders all shared a love of the land and a concern for the sustainability of local communities. At its first meeting, the partnership agreed on a vision statement that foreshadowed a model later adopted by other communities in the West:

The Applegate Partnership is a community-based project involving industry, conservation groups, natural resource agencies, and residents cooperating to encourage and facilitate the use of natural resource principles that promote ecosystem health and diversity. Through community involvement and education, the partnership supports the management of all lands within the watershed in a manner that sustains natural resources and that will, in turn, contribute to economic and community stability within the Applegate Valley. (Wondolleck & Yaffee, 2000, pp. 140–141; for more information, see www.applegatepartnershipwc.org/about-us)

Similar natural resource partnerships addressing different issues have grown up across the United States. Among these are the following:

- Ponderosa Pine Forest Partnership in Southwest Colorado, a coalition of citizens, loggers, local colleges, and the U.S. Forest Service, working to enhance wildlife habitat and sustainable approaches to forestry (www.rlch.org/stories/ponderosa-pine-forest-partnership).
- Potomac Watershed Partnership is a restoration and stewardship project for the Potomac River and its surrounding lands in Maryland and Virginia. Its mission is "to create a collaborative effort among federal, state, and local partners to restore the health of the land and waters of the Potomac River Basin, thereby enhancing the quality of life and overall health of the Chesapeake Bay" (Potomac Watershed Partnership, 2009, para. 2).

Collaboration in natural resource partnerships differs somewhat from the agency-appointed citizens' advisory committee. Partnerships usually are voluntary, although officials from government agencies like the BLM or Forest Service may be among the partners. They focus on a geographical region and a wider range of ecological concerns. And, unlike a citizens' advisory committee, a natural resource partnership usually works on an ongoing basis to respond to new challenges and concerns in its region.

Community-Based Collaboration

Occasions for local disputes over the environment are numerous: the loss of green space, contamination of well water, tensions between automobile drivers and bicyclists, and so forth. Increasingly, local government and courts are encouraging the use of collaborative processes to avoid long, contentious conflicts that can drain resources, divide groups, and weaken the relationships in a community. **Community-based collaboration** involves individuals and representatives of affected groups, businesses, or other agencies in addressing a specific or short-term problem in a local community. Besides being court-appointed or agency-sponsored associations, these community-based groups may be voluntary associations without legal sanction or regulatory power.

Although they have some features in common with natural resource partnerships, community-based collaborations tend to focus on specific, local problems that involve shorter time frames. For example, in Sherman County, Oregon, a conflict arose over a proposal to locate a 24-megawatt wind farm in the community. With a farming population of 1,900, Sherman County lies directly in the path of relentless winds from the Pacific Ocean; for this reason, the area was proposed as a site for harvesting wind energy. In other communities, proposals for wind farms had generated considerable conflict—powerful, 200-foot-tall turbines can affect aviation, bird populations, cultural and historical sites, weed control, and other ecological matters (Policy Consensus Initiative, 2004).

In the face of potential controversy, Oregon's governor invited local farmers, citizens' groups, landowners, the Audubon Society, and representatives from local, state, and federal agencies, and business concerns to engage in a collaborative process to decide the fate of the proposed wind farm. Working together, the group identified possible wind-farm sites and related issues of concern. Their efforts eventually led to an agreement on a site that would have "minimal negative impacts on the community and environment" (Policy Consensus Initiative, 2004).

Each of these forms of participation—citizens' advisory committees, natural resource partnerships, and community-based collaboration—share certain characteristics that contribute to their eventual success (or failure). Therefore, in the next section of this chapter, we identify some of the conditions that must be in place for successful collaboration as well as the requirements for building trust among the participants and sustaining open, civil dialogue.

 FYI | **Case Studies of Successful Collaboration in Environmental Disputes**

- *The Fire Next Time:* www.pbs.org/pov

  A film about the way residents in the Flathead Valley, Montana, came together to resolve conflicts over loss of jobs and the environment that threatened to tear their community apart. For a copy of this film for local showing and discussion, see http://www.pbs.org/pov/thefirenexttime.

- *Cultivating Common Ground*: www.youtube.com

  The participants in the Lakeview Stewardship Group, a natural resource partnership, tell their story of successful collaboration to restore the 500,000-acre Lakeview Federal Stewardship Unit in the Fremont-Winema National Forest in Oregon in this brief YouTube video.

- *A River Reborn: The Restoration of Fossil Creek:* www.mpcer.nau.edu/riverreborn

  A film documenting environmental conflict and collaboration in an Arizona community as it struggled to protect Fossil Creek by removing a 100-year-old hydroelectric dam. The DVD, narrated by actor Ted Danson, is available at http://www.mpcer.nau.edu/riverreborn.

  For more case studies of successful resolution of environmental conflicts, see these online sites:

  1. The University of Michigan's Ecosystem Management Initiative site for collaboration, "Case Studies and Lessons Learned" at www.snre.umich.edu/ecomgt//collaboration.htm.

  2. The National Policy Consensus Center's "Policy Consensus Initiative" and its archive of case studies at http://www.policyconsensus.org/casestudies.

Collaborating to Resolve Environmental Conflicts

As we've seen, collaboration differs from the forms of public participation in Chapter 12. Collaboration is also sharply distinguished from more adversarial forms of managing environmental conflict, such as litigation or advocacy campaigns, although sometimes, these have led adversaries to put aside their differences and talk to each other. (We describe how this conflict–collaboration dynamic worked in the controversy over the Great Bear Rainforest later in this chapter.) One of the field's leading scholars in collaboration, Gregg Walker (2004) identifies eight attributes that distinguish collaboration from traditional forms of public participation:

1. Collaboration is less competitive.

2. Collaboration features mutual learning and fact finding.

3. Collaboration allows underlying value differences to be explored.

4. Collaboration resembles principled negotiation, focusing on interests rather than positions.

5. Collaboration allocates the responsibility for implementation across many parties.

6. Collaboration's conclusions are generated by participants through an interactive, iterative, and reflective process.

7. Collaboration is often an ongoing process.

8. Collaboration has the potential to build individual and community capacity in such areas as conflict management, leadership, decision making, and communications. (p. 124)

Walker's list helps us understand collaboration as a process that is distinct from more adversarial forms of public participation in environmental decisions.

In this section, we build on Walker's observations to describe some of the characteristics of successful collaboration. However, before going further, it's helpful to distinguish collaboration from two other, closely related forms of conflict resolution: arbitration and mediation. **Arbitration** is usually court ordered and involves the presentation of opposing views to a neutral, third-party individual or panel that, in turn, renders a judgment about the conflict. **Mediation** is a facilitated effort entered into voluntarily or at the suggestion of a court, counselors, or other institution. Most important, this form of conflict management involves an active mediator who helps the disputing parties find common ground and a solution on which they agree. Whereas collaboration may use a mediator on occasion, it requires active contributions from all participants.

With these distinctions in mind, let's look at the minimum, core conditions that are typically present when collaboration succeeds.

Requirements for Successful Collaboration

Most scholars and those who participate in effective collaborations cite a number of conditions and participant characteristics that must be present for collaboration to succeed.

Requirements of Successful Collaboration

1. Relevant stakeholders are at the table.

2. Participants adopt a problem-solving approach.

3. All participants have access to necessary resources and opportunities to participate in discussions.

4. Decisions are usually reached by consensus.

5. Relevant agencies are guided by the recommendations of the collaboration.

1. *Relevant stakeholders are at the table.* Collaboration begins when the relevant "stakeholders" agree to participate in "constructive, open, civil communication" (Walker, 2004, p. 123) to address a problem. **Stakeholders** are those parties to a dispute who have a real or discernible interest (a stake) in the outcome. Sometimes they're selected by a sponsoring agency to sit at the table, usually to represent certain interests or constituents, such as local businesses, residents, environmental groups, the timber industry, and so forth. In other cases, stakeholders self-identify and volunteer to participate. In most collaborations, stakeholders are place-based; that is, they live or work in the affected community or region. (For more information about the concept of the stakeholder in environmental decision making, see Dietz & Stern, 2008.)

2. *Participants adopt a problem-solving approach.* Communication among participants strives to solve problems instead of being adversarial or manipulative. Problem solving uses discussion, conversation, and information, seeking to define the concrete problem, the relevant concerns, the criteria for appropriate solutions, and finally a solution that addresses the concerns of all parties. Although conflict is expected in the discussion, collaboration keeps the focus on the interests or issues rather than on the people. It discourages adversarial or overtly persuasive stances and instead favors listening, learning, and trying to agree on workable solutions.

3. *All participants have access to necessary resources and opportunities to participate in discussions.* In a collaborative effort, solutions cannot be imposed. If agreement is to be reached by all parties, all participants must have an opportunity to be heard, to challenge others' views, to question, and to provide input to the solution. Each party must also have access to the same information as others do, including reports, expert briefings, and so on. Finally, the group must guard against the effects of different levels of power or privilege among the participants to ensure that all voices are respected and have opportunities to contribute to and influence the solution.

4. *Decisions usually are reached by consensus.* Most collaborative groups aim to reach decisions by **consensus**, which usually means *general agreement* or that discussions will not end until everyone has had a chance to share their differences and find

common ground. Consensus often means that all participants agree with the final decision. Daniels and Walker (2001) note, however, that some define consensus in a way that leaves room for some differences of opinion. In this use, consensus is an agreement that comes after the parties have attempted to identify the interests of all stakeholders and have crafted a decision addressing as many of these concerns as possible (p. 72).

Consensus can also be distinguished from *compromise,* another form of collaboration that groups use in reaching decisions. As interpersonal communication scholar Julia Wood (2009) observes, in a **compromise,** "members work out a solution that satisfies each person's minimum criteria but may not fully satisfy all members" (pp. 270, 271). In either case, a decision assumes some form of cooperation, or as Walker, Daniels, and Emborg put it, "appropriate collaboration"—a "fair, inclusive process; respectful interaction; . . . [and] mutual learning," among other traits (2015).

5. *Relevant agencies are guided by the recommendations of the collaboration.* The results of a collaborative effort usually are advisory to the agency that appointed the group, for example, the report of a citizens' advisory committee to the governmental agency handling the cleanup of a toxic waste site. Such agencies, however, are not always open to influence by citizen groups. Therefore, a successful collaboration assumes that there is an opportunity to influence the final decision, what Walker, Daniels, and Emborg (2015) call "decision space." **Decision space** refers to what decisions are open to the participants' influence, "what is 'on the table' for discussion and what matters are not?" Walker et al. give the example of local citizens' collaboration with the U.S. Forest Service about a national recreation area:

> As we designed and facilitated community workshops about this recreation area, we asked Forest Service officials to clarify what was "within" and "outside" the decision space. Recreation area curfews were within the decision space, while threatened and endangered species were not.

Obviously, the "greater the decision space, the greater the potential for meaningful" public involvement and collaboration (Walker et al., 2015).

Successful collaboration by parties with sharp disagreements is not always possible, particularly in environmental disputes where the stakes are high or the parties are too deeply divided by a history of discord or entrenched opposition. Guy Burgess and Heidi Burgess (1996), co-directors of the Conflict Research Consortium at the University of Colorado at Boulder, observed, "While consensus building can be very effective in low-stakes disputes . . . it does not work as well when the issues involve deep-rooted value differences, very high stakes, or irreducible win-lose confrontations" (p. 1). Those of us who have worked with environmental conflicts tend to feel that collaboration succeeds in these situations only when the adversaries come to feel something must change, and when they can identify a shared vision of the future.

From Conflict to Collaboration in the Great Bear Rainforest

It may be helpful to look at a case study of a difficult, but successful collaboration—the long-running conflict over logging in Canada's Great Bear Rainforest that ultimately became a model of collaboration among environmentalists, First Nations, the timber industry, and local communities. In exploring this conflict, we illustrate the importance of the minimum requirements research has found are necessary for effective collaboration.

Conflict in the Rainforest

The region known as the Great Bear Rainforest stretches 250 miles along Canada's Pacific Coast. It is the largest remaining, temperate rainforest in the world, encompassing more than 28,000 square miles. Mountains, streams, fjords, forests, and estuaries in the rainforest are home to tremendous biodiversity, including grizzly bears, the white Kermode bear (known as Spirit Bears), wolves, salmon, and millions of migratory birds, as well as 1,000-year-old trees. (See opening photo for this chapter.) This region is also home to First Nations—pre-European, indigenous people—whose communities often are accessible only by air or water (Armstrong, 2009; Smith, Sterritt, & Armstrong, 2007).

©Wolfgang Kaehler/Corbis

Photo 13.2 The protests and blockades of logging roads on Clayoquot Sound, British Columbia, were the start of 15 years of conflict, "the war in the woods," to preserve one of the planet's last, temperate rainforests.

The old-growth forests of the Great Bear Rainforest were also the scene of intense conflict among logging companies, First Nations, and environmentalists for many years. This conflict, nevertheless, evolved into a model of collaboration by unlikely groups, resulting in an unprecedented agreement to protect forests, indigenous rights, and local communities.

In the 1990s, the Great Bear Rainforest witnessed a sharp increase in logging and other extractive activities, often in First Nations' traditional territories. These activities also brought an era of conflict as environmental activists joined with First Nations' elders in protests and direct actions to protect the region's forests and watersheds. In 1993, in Clayoquot Sound on Vancouver Island, more than 900 people were arrested for nonviolent, direct-action protests, including blockades of logging roads, in "the largest mass arrest in Canadian history" (Smith et al., 2007, p. 2). (See Photo 13.2.) The protests were the start of 15 years of conflict and an international campaign, what newspapers called "the war in the woods" (p. 4), to preserve one of the planet's last temperate rainforests.

As logging blockades continued, environmental groups launched an international social media campaign, using blogs and social networking sites, to mobilize supporters. (See Chapter 9, "Digital Media and Environmental Activism.") For example, a spokesperson for Forest Ethics said, at the time, "The campaign has appeared on dozens of popular blogs and networking sites around the world. They include Facebook and Twitter. . . . Supporters have been sending about 100 e-mails per day to the [British Columbia] government through the group's website" (Johnson, 2008, paras. 9–10, 12–13). In its media efforts, Greenpeace, Forest Ethics, Rainforest Action Network, and other groups aimed to "tell the world's pulp and paper customers exactly what was happening to British Columbia's coastal rainforests and to urge wood product buyers to change their procurement practices" ("Conflict and Protest," n.d., para. 4). Many businesses began to cancel their contracts. Ultimately, over 80 companies, including well-known brands like Ikea, Home Depot, Staples, and IBM, responded, and "committed to stop selling wood and paper products" from the region ("Markets Campaign," n.d., para. 3).

As the markets campaign grew, other changes were starting in British Columbia. In 1999, senior industry leaders began a discussion about ways to resolve the controversy with environmentalists and First Nations. It was, they decided, "time to travel in new direction," including "sitting down face to face with environmental groups" (Armstrong, 2009, pp. 8–9). Old assumptions were "replaced by a recognition that the environment represents a core social value," that environmentalists were seen by the public "as both credible and influential," and that "customers expect their [forest industry] suppliers to resolve conflict, not simply rationalize it" (Smith et al., 2007, p. 4). The chief forester for Western Forests Products, Bill Dumont, put it more succinctly, "Customers don't want their two-by-fours with a protester attached to it. If we don't end it, they will buy their products elsewhere" (p. 4).

*Collaboration Among Loggers,
First Nations, Business, and Environmentalists*

The shift to dialogue received a significant boost in 2000 when the forest companies—joining together as the Coast Forest Conservation Initiative—agreed to halt logging in more than 100 watersheds in the Great Bear Rainforest. Smith, Sterritt, and Armstrong (2007) describe what happened next: "In return, ForestEthics, Greenpeace, and Rainforest Action Network modified their market campaigns," no longer asking customers to cancel their contracts. "This 'mutual standstill' created the conditions for a new beginning between the parties" (p. 5).

At the same time, the environmental groups, ForestEthics, Greenpeace, RAN, and Sierra Club of British Columbia, came together as the Rainforest Solution Project to develop shared proposals for protecting the Great Bear Rainforest. Shortly afterward, this group and industry's Coast Forest Conservation Initiative formed a coalition called the Joint Solutions Project. The new coalition would be "a structure for communications and negotiations" among the former opponents and would serve to facilitate "a broader dialogue with First Nations, the BC government, labour groups, and local communities" (Smith et al., 2007, p. 5). The way was opened finally for collaboration among all of the major stakeholders to begin crafting a far-reaching set of agreements for the Great Bear Rainforest.

Much work remained. By 2006, however, forest companies, First Nations, environmentalists, and the British Columbia government had negotiated a set of consensus recommendations that, if fully implemented, would protect key areas, manage other areas in a sustainable manner, and invest in local communities. On March 31, 2009, the parties announced the deadline had been met for implementation of these recommendations, including completion of a land use plan for an ecosystem-based management (EBM) of the Great Bear Rainforest. The agreements included the following:

1. *Protected Areas Network:* More than one third of the rainforest (8,150 square miles) would be protected in a new network of conservancies. These areas include old-growth forests, estuaries, wetlands, salmon streams, and habitat for key species, "representing the full diversity of habitat types," and quadrupling the total protected areas within the Great Bear Rainforest (Smith et al., 2007, p. 8).

2. *Ecosystem-Based Management:* Logging or other development occurring within the other areas of the Great Bear Rainforest would be subject to "ecosystem-based management (EBM) rules and guidance for governing resource use in the region" (Armstrong, 2009, p. 13). The EBM guidelines will attempt to ensure "high degrees of ecological integrity," including the goal of maintaining 70% of natural, old-growth forests across the region (p. 13).

3. *Coast Opportunities Fund:* An important aspect of the new agreement was economic investment in First Nations communities in the region. The newly created Coast Opportunities Fund, endowed by public and private sources with an

initial $120 million, "is designed to support conservation and environmentally responsible economic development initiatives for First Nations in the Great Bear Rainforest" (Armstrong, 2009, p. 15).

An agreement of this scale could not have happened without a number of key developments that sustained the Great Bear Rainforest collaboration. Importantly, the five minimum requirements for effective collaboration were clearly satisfied:

1. All of the relevant stakeholders were "at the table" (logging companies, environmentalists, First Nations, and the BC government).

2. When the parties finally sat down at the table, they agreed to use a problem-solving approach rather than advocacy and agreed early on ground rules for discussion.

3. As a result, the participants learned to work with one another and felt they had an equal opportunity to participate in discussions.

4. Consensus was reached on the key elements of the agreement—the Protected Areas Network, ecosystem-based management, and the Coast Opportunities Fund.

5. And the BC government and relevant agencies appear to be honoring their pledge to implement the recommendations from the collaboration, as this book goes to press.

Although core minimum requirements for collaboration were present, most of the participants acknowledged that the final discussions would not have occurred without other, key dynamics of change. Merran Smith of Forest Ethics, Art Sterritt from the Coastal First Nations, and Patrick Armstrong, a consultant who worked with companies in the Coast Forest Conservation Initiative reflected on several of these dynamics of change:

1. *Persistent Vision:* The vision of environmentalists and First Nations of "a protected, globally significant rainforest with healthy indigenous communities and a diverse economy . . . played an integral role in inspiring new people, keeping the process on track, and reminding participants of what they were trying to achieve" (Smith et al., 2007, p. 12).

2. *Power Shift:* As a result of the international market campaigns, the region's revenue from sales of forest products "hung in the balance"; Smith, Sterritt, and Armstrong (2007) noted that, "when forest companies and ultimately the BC government acknowledged this power, negotiations . . . experienced a fundamental shift" (p. 12). This power shift was "one of the most critical elements to achieving the eventual outcome in the Great Bear Rainforest" (p. 12). This is because, when all parties— environmentalists, First Nations, forest companies, and the BC government—held power that was needed to achieve a lasting solution, "all parties stayed at the table" (p. 12).

3. *Collaboration and Relationships:* The principal groups—environmentalists, First Nations, and forest companies—each formed an informal entity to resolve internal conflicts and represent their interests to the other groups. As a result, those in the Joint Solutions Project and in the broader communications with First Nations and the BC government "were empowered by their constituencies to move the dialogue forward" (p. 13).

4. *Leadership:* While some in each group resisted the idea of collaborating with their opponents, leaders within the forest industry, in First Nations, and among environmentalists arose to counter the resistance in their own ranks. As Smith, Sterritt, and Armstrong (2007) observed, "Leadership meant standing up to critics—and even to traditional allies—who disagreed with building solutions with one's opponents" (p. 14).

The Great Bear Rainforest agreements are steadily becoming a reality. Logging companies and environmentalists of the Joint Solutions Project are still working together and with the BC government and First Nations, as this book goes to press, to ensure that the two remaining goals of "high degrees of ecological integrity" in managing forests (70% of old-growth forests to be set-aside from logging) and the economic and environmental sustainability of First Nation communities would be achieved.

Act Locally!

Is Collaboration Always Possible?

Where a conflict has been resolved, there may be important lessons to be learned. One way to gain an appreciation for the challenges to successful collaboration is to investigate a local environmental conflict by interviewing some of its participants.

Identify a successful case of conflict resolution on your campus or locally, or a current conflict in which efforts are being made to bring the parties together. Invite one or more representatives from each side of the conflict to your class:

1. Ask the different participants to describe the nature of the conflict, their goals, and concerns.

2. In successful cases, ask the participants how they managed to come together and to collaborate, despite their differences. What kept them talking? How would they characterize their communication or relationships with each another?

3. In more difficult cases, or where the collaboration is ongoing, ask the different parties about what they perceive to be the obstacles to successful collaboration or to reaching consensus.

Alternatively, arrange for your class or group to screen one of the videos from the (above) "FYI: Case Studies of Successful Collaboration in Environmental Disputes." Were the basic requirements for effective collaboration present? What communication features helped to explain the willingness of the feuding parties to talk with one another and to reach an agreement?

Limits of Collaboration and Consensus

Not all attempts at collaboration are successful or even appropriate. In the following section, we describe a case that, at first, appeared to be a very successful experiment in bringing loggers, environmentalists, and local business and community leaders together, but an experiment that was criticized almost immediately. Before we look at this case, let's look at one other tool for evaluating collaborative efforts and for identifying the reasons that some efforts fail.

Evaluating Collaboration: The "Progress Triangle"

In *Working Through Environmental Conflict: The Collaborative Learning Approach*, Steven E. Daniels and Gregg B. Walker (2001) propose that environmental conflict management can best be thought of as "making progress," rather than *resolving* the conflict (p. 35). In understanding how multi-party stakeholders work through a conflict, they identify several key factors, visualized as **The Progress Triangle.** The triangle portrays conflict management as "three interrelated dimensions—substantive, procedural, and relationship" (p. 36). While these dimensions may be obvious, overlooking their importance can weaken or derail a collaborative effort. For example, poor procedures, lack of respect between stakeholders, or papering over disagreements can be costly. As we see below, procedures that excluded key stakeholders scuttled the substantive agreements the group had reach.

Let's use The Progress Triangle, as well as the five requirements for successful collaboration (Table 13.1), to evaluate one of the most highly touted examples of collaboration, an effort by a local community to develop a consensus approach for managing national forest lands in northern California. Though initially successful, the efforts of the **Quincy Library Group**, as the participants called themselves, ended in conflict and by moving in a different, more adversarial direction.

The Quincy Library Group: Conflict in the Sierra Nevada Mountains

The rural town of Quincy (fewer than 50,000 residents) is located 100 miles northeast of Sacramento, California. More important, it lies in the geographical center of the timber wars in the Plumas, Lassen, and Tahoe National Forests of the Sierra Nevada Mountains. Although logging increased from the 1960s through the 1980s in the three national forests around Quincy, the timber cut fell sharply in the 1990s due to shifting market demands and to Forest Service restrictions that protected old-growth trees and habitat for spotted owls and other endangered species.

As logging declined and local sawmills shut down, the area began to experience sharp conflicts between timber interests and environmentalists. For example, loggers and their families blamed the Forest Service for restricting the level of timber cuts and organized the Yellow Ribbon Coalition to lobby for their interests. Charges and countercharges also flew between the coalition and environmentalists over instances of tree spiking (see Chapter 2) and the use of nail clusters on forest roads to stop logging trucks (Wondolleck & Yaffee, 2000, p. 71). Plumas County supervisor Bill Coates expressed the fears of many in local communities: "Our small towns were already endangered. This [decline in logging] was going to wipe them out" (Wondolleck & Yaffee, 2000, p. 71).

Initial Success: Collaboration in Quincy

Despite the controversy, some in the community suggested that opponents might share a larger set of interests and values. Michael Jackson, an environmental attorney and member of Friends of Plumas Wilderness, was one of the earliest. In 1989, he wrote a letter to the local newspaper, the *Feather River Bulletin,* "arguing that environmentalists, loggers, and business needed to work together for 'our mutual future'" (quoted in Wondolleck & Yaffee, 2000, p. 71). In his letter, Jackson invited loggers to work with environmentalists toward a set of common goals:

> What do environmentalists believe we have in common with the Yellow Ribbon Coalition? We believe that we are all honest people who want to continue our way of life. We believe that we all love the area in which we live. We believe that we all enjoy beautiful views, hunting and fishing and living in a rural area. We believe that we are being misled by the Forest Service and by large timber, which controls the Forest Service, into believing that we are enemies when we are not. (quoted in Wondolleck & Yaffee, 2000, pp. 71–72)

By 1992, a few individuals in each of the warring camps—forest industry, community and business leaders, and environmentalists—began to talk about the effects of declining timber production on the community. Initially, three men agreed to talk among themselves: Bill Coates (Plumas County supervisor and a business owner who supported the timber industry), Tom Nelson (a forester for Sierra Pacific Industries), and Michael Jackson (the environmental attorney and a passionate environmentalist). The three "found more common ground than they had expected, and decided to try at least a truce, maybe even a full peace treaty, based upon that common ground" (Terhune & Terhune, 1998, para. 8).

Soon, other people joined the discussions of Coates, Nelson, and Jackson. Later observers recalled that these "early meetings had some very tense moments, and some participants were very uncomfortable at times" (Terhune & Terhune, 1998, p. 8). Meeting in the public library, they began calling themselves the Quincy Library Group (QLG). "Some only half-jokingly [noted] that meeting in a library would prevent participants from yelling at each other" (Wondolleck & Yaffee, 2000, p. 72).

By 1993, members of the QLG (www.qlg.org) had agreed among themselves on the Community Stability Plan, which the group hoped would guide management practices in the Plumas, Lassen, and Tahoe National Forests. Although this plan had no official status—the Forest Service was not involved in the discussions—it reflected the group's belief that "a healthy forest and a stable community are interdependent; we cannot have one without the other" (Terhune & Terhune, 1998, p. 11). The purpose of the Community Stability Plan was to integrate these values into a common vision: "to promote the objectives of forest health, ecological integrity, adequate timber supply, and local economic stability" (Wondolleck & Yaffee, 2000, p. 72). The group's plan set forth a series of recommendations to the Forest Service for implementing its vision:

> The plan would . . . prevent clear-cutting on Forest Service land or in wide protection zones around rivers and streams and would require group and single tree selection [logging] intended to produce an "all-age, multi-storied, fire-resistant forest approximating pre-settlement conditions." Under the plan, local timber mills would process all harvested logs. (Wondolleck & Yaffee, 2000, p. 72)

The Community Stability Proposal was the result of many meetings, difficult conversations, and the desire of all participants to reach consensus where possible. Their agreement was unusual among the contentious parties in Quincy and its surrounding communities. Nevertheless, the QLG soon confronted resistance that would shift it to a more adversarial process. Most important, the group failed to adequately assess the "decision space" (Walker, Daniels, & Emborg, 2015) needed to enable their recommendations to be implemented by the relevant agency. The Forest Service—which had not participated in the QGL collaboration—refused to entertain the group's Community Stability Proposal.

Frustrated by resistance from the Forest Service and criticized by other environmentalists, QLG members turned to the legislative process in Washington, DC. After successive lobbying trips in 1998, they persuaded Congress to enact a version of the Community Stability Proposal. In an unprecedented move, the Quincy Library Group Forest Recovery Act overrode the usual Forest Service decision-making process and directed it to include the QLG version in its management planning for the three national forests in the Quincy area.

Although we return (below) to some of the criticisms of the Quincy Library Group's experience, it is important to observe that, initially, the group received considerable praise for its collaborative work. Prompted by the feeling that something had to change, individuals in Quincy believed that conditions were ripe for some alternative mechanism for resolving the long-simmering dispute over logging in the area's national forests.

As they looked at their work, QLG members Pat and George Terhune (1998) stressed the importance of consensus in keeping the group together and focused.

> Votes are not taken until the group is pretty well convinced that the decision will be unanimous. If it isn't, then more discussion takes place, and if anybody is still opposed, the decision is either dropped or postponed for still more discussion. (p. 32)

Criticisms of the Quincy Library Group

Beyond the lack of sufficient "decision space," others also failed to share the QLG's vision for management of the national forests. "Although the group attracted widespread public participation at first, most outsiders who had offered new ideas said they got the cold shoulder and stopped attending. When the QLG chose to go the legislative route, others dropped out" (Red Lodge Clearinghouse, 2008). Others objected that the group had excluded key stakeholders—particularly environmental groups that were closely involved with U.S. National Forests. Although representatives of the timber industry (Sierra Pacific, for example) were at the table in Quincy, representatives from the Sierra Club and other forest advocacy groups were not. Some that had been excluded, for example, were upset that the QLG's proposal would double the levels of logging in the Lassen, Plumas, and Tahoe National Forests (Brower & Hanson, 1999).

As a consequence, environmentalists accused the Quincy Library Group of evading NEPA's requirement for an environmental impact statement and the opportunity for public comment (Chapter 12). Furthermore, they argued, by turning to the U.S Congress, the QLG enabled local interests to set new, national standards for managing wider, U.S. National Forests. This, in turn, led to lawsuits objecting to logging in many of the areas covered by the QLG's plan.

Indeed, by 2009, it had become clear that the QLG was frustrated in its aims. The excluded environmental groups continued to block implementation of many projects by filing lawsuits, as well as legal appeals within the Forest Service. (At the time of this controversy, the QLG developed a set of responses to the environmental concerns on its website; see "Concerns and Responses About the Quincy Library Group Bill," at http://www.qlg.org.)

Evaluating the QLG Collaboration

The Quincy Library Group's apparent success stalled almost immediately when the "relevant agency," the Forest Service, refused to implement the group's proposal for the three national forests. In other words, the QLG had limited *decision space* in which to deliberate. The Progress Triangle also suggests that the group's *procedures* led to a very limited view of *relationships* by excluding key interests from outside of the local community who were involved in decisions affecting the management of U.S. National Forests, such as environmental groups.

By excluding certain stakeholders, while allowing others (the timber industry) to sit at the table, the QLG also ensured that the substance of its proposals would be met with skepticism. Indeed, such exclusions soon led to mistrust of the collaboration's *substance*—its Community Stability Proposal. For example, environmentalists David Brower and Chad Hanson (1999) charged that the QLG procedures had allowed industry interests to capture the decision process: The Quincy plan, they argued, was "based on the premise of letting industry groups in rural timber towns dictate the fate of federally owned lands, essentially transferring decision-making power from the

American people and into the hands of extractive industries" (p. A25). Similar criticism came from other environmentalists, editorials, and scholars studying collaborative processes. For example, Wondolleck and Yaffee (2000) observed, instead of being "a model collaborative effort, the QLG suddenly became the focus of an acrimonious debate" (p. 265).

The Red Lodge Clearinghouse (2008), a project of the Natural Resources Law Center at the University of Colorado, summarized the QLG collaboration this way:

> It is as a collaborative that the Quincy Library Group is troubling. Many who have tried to participate have felt ostracized. And although it is developing policy for managing federal lands, the coalition has demonstrated little concern for involving the broader public in its process. . . . No one knows what would be happening now if QLG had stuck with consensus and insisted on trying to include everyone. Instead . . . the QLG took a top-down, federally mandated approach that has limited participation in the program.

The charge by Brower and Hanson that local groups can capture the decision process affecting U.S. public lands also illustrates a dilemma posed by so-called **place-based collaboration,** that is, a collaborative process involving stakeholders who live in, and have an interest or "stake" in, the place (local community or region) where a conflict occurs. To what extent do such place-based models provide mechanisms for managing contentious disputes with wider implications, and to what extent do they exclude key stakeholders and ignore national standards? The Quincy experience is not encouraging in this regard. (For example, see "Another Viewpoint: A Skeptic Looks at Collaboration.")

Another Viewpoint: A Skeptic Looks at Collaboration

In a well-publicized article in the western *High Country News,* the Sierra Club's former executive director, Michael McCloskey (1996), argued that collaborative processes like the Quincy Library Group give small local groups "an effective veto" over entire national forests. McCloskey cited two shortcomings of local or place-based collaboration:

1. Placed-based collaboration excludes key stakeholders. It ignores "the disparate geographical distribution of constituencies" (p. 7). That is, those who are sympathetic to environmental values often live in urban areas; therefore, they are not invited to participate in collaborative processes in the communities near the national forests where there is a dispute.

2. Placed-based collaboration undermines national standards for managing natural resources such as national forests. By transferring the power to decide the direction for public lands to small, local groups, local collaboration evades the need to hammer out "national rules to reflect majority rule in the nation" (p. 7).

As a result, McCloskey argued, such models are abdications of the role of government to represent the national (public) interest.

Common Criticisms of Collaboration

Although community-based collaborations have many advantages, Daniels and Walker (2001) have observed that they have not been universally accepted as a model for handling conflicts over natural resources. In closing, it may be useful to review some of the common criticisms of the use of collaboration and consensus decision making in environmental conflicts. Environmental scholars and facilitators who work with such disputes have found seven basic complaints or occasions for which collaboration may not be appropriate:

1. *Stakeholders may be unrepresentative of wider publics.* Some scholars have suggested that the more intensive modes of alternative participation, such as citizens' advisory councils and consensus-seeking groups, may be able to reach agreement, but they often do so only by excluding wider publics. For example, Beierle and Cayford (2002) report that "the exclusion of certain groups, the departure of dissenting parties, or the avoidance of issues ultimately made consensus possible—or at least easier—in 33%" of the cases they studied involving consensus-based efforts in which conflict was reported (p. 48). Environmental communication scholar William Kinsella (2004) also observes that highly involved individuals who serve on citizens' advisory boards "do not necessarily represent the larger public" (p. 90). In other situations, the questions of who is a stakeholder and who should set environmental policy lie at the heart of many local, national, and global environmental controversies.

2. *Place-based collaboration may encourage exceptionalism or a compromise of national standards.* As we witnessed with the Quincy Library Group, the exclusion of national environmental groups gave local interests greater control over the management of national resources. Daniels and Walker (2001) reported that such cases may "preclude meaningful opportunities for non-parties to review and comment on proposals" (p. 274), encouraging a kind of **exceptionalism,** or the view that, because a region has unique or distinctive features, it is exempt from the general rule. The concern of some critics is that, if place-based decisions reached at the local level in one area become a precedent for exempting other geographical areas, they may compromise more uniform, national standards for environmental policy.

3. *Power inequities may lead to co-optation.* One of the most common complaints about collaboration and consensus approaches is that power inequities among the participants may lead to the co-optation of environmental interests. The greater resources in training, information, and negotiation skills often brought to collaboration processes by industry representatives and government officials may make it harder for ordinary citizens and environmentalists to defend their interests. Environmentalists such as McCloskey (1996) are especially critical of such inequities in power and resources: "Industry thinks its odds are better in these forums [place-based collaboration]. . . . It believes it can dominate them over time and relieve itself of the burden of tough national rules" (p. 7).

4. *Pressure for consensus may lead to the lowest common denominator.* In cases of successful collaboration, groups striving for consensus may drop contentious issues or defer them until later. However, some critics fear that this tendency can go too far, that vocal minorities are given an effective veto over the process. "Any recalcitrant stakeholder can paralyze the process.... Only lowest common denominator ideas survive the process" (McCloskey, 1996, p. 7). Instead of a win-win solution, agreement on the least contentious parts is simply a deferral of the real sources of conflict to other forums or other times.

Conversely, a pressure for conformity among the group's members can lead to what psychologist Irving Janis (1977) called **groupthink,** that is, excessive cohesion that impedes critical or independent thinking. Indeed, in a broad survey, Robert S. Baron (2005) found that the symptoms of groupthink are widespread, with the result that groupthink is often an uninformed consensus.

5. *Consensus tends to delegitimize conflict and advocacy.* Conflict can be unpleasant. For many people, civil dialogue in forums where collaboration is the rule may be a safe harbor from controversy. The desire to avoid disagreement, however, is closely related to groupthink or to a premature compromise, thus postponing the search for long-term solutions. As a result, some charge that the desire for consensus "may serve to de-legitimize conflict and co-opt environmental advocates" (Daniels & Walker, 2001, p. 274).

On the other hand, successful instances of collaboration do not necessarily view all conflict as bad, nor do they avoid controversy in their deliberations. Indeed, what communication scholar Thomas Goodnight and others have called *dissensus* may serve an important communication role (Fritch, Palczewski, Farrell, & Short, 2006; Goodnight, 1991). **Dissensus** is a questioning of or disagreement with a claim or a premise of a speaker's argument. Rather than bringing a discussion to a halt, dissensus can be generative. If properly handled, it may invite more communication about the areas of disagreement between the differing parties. The attractiveness of collaboration and consensus models for managing environmental conflicts should not, therefore, overshadow the difficulties these processes may involve.

6. *Collaborative groups may lack authority to implement their decisions.* In the collaboration over the Great Bear Rainforest case, the BC provincial government and timber companies have worked to implement the group's consensus recommendations. But this is not always the case. Many citizens' advisory committees deliberate for extended periods without the assurance that their decisions will be accepted or implemented by federal agencies. The QLG, for example, ran into resistance from the Forest Service when it presented its proposal. The simple fact is that most collaborative groups are composed of nonelected citizens and other individuals whose authority—when present—is contingent upon the very governmental agency they are seeking to influence.

7. *Irreconcilable values may hinder agreement.* We suggested earlier that collaborative approaches do not work well when the issues involve deep-rooted value differences, high stakes, or irreducible, win–lose confrontations. Each of us has values that

we believe we cannot or should not compromise—for example, the health of our children, liberty, biodiversity, private property rights, or the right of people to be safe from industrial poisons. And many wilderness advocates believe that the natural environment has been compromised enough. For them, further compromises allowing logging or other development presumably are nonnegotiable.

SUMMARY

This chapter has explored an important alternative approach for managing environmental conflicts. At the heart of these approaches is the idea of *collaboration* and the search for consensus among the relevant parties.

- In the first section, we described the dissatisfaction of many with traditional forms of public participation, and we identified a range of alternatives for managing environmental conflicts: citizens' advisory committees, natural resource partnerships, and community-based collaboration.
- In the second section of this chapter, we identified five conditions for successful collaboration:
  - All relevant stakeholders are at the table
  - Participants adopt a problem-solving approach
  - All participants have equal access to resources and opportunities to participate in discussions
  - Decisions usually are reached by consensus
  - The relevant agencies are guided by the recommendations of the collaborating group

  And we illustrated these in a successful case of multi-party collaboration over logging in Canada's Great Bear Rainforest.
- Then, in section three, we turned to limitations of collaboration and introduced another tool for evaluating collaboration, The Progress Triangle; we illustrated—with the Quincy Library Group—the ways in which collaboration can fail when key requirements are overlooked.
- Finally, we described some of the common criticisms of collaboration and consensus approaches, ranging from a failure to include key stakeholders to pressure toward the lowest common denominator in order to reach consensus, and we noted that collaboration may not always be possible when a conflict involves deep differences over values or irreducible win–lose confrontations.

Neither collaboration nor more adversarial forms of public participation provide a magic answer to conflicts arising from complex human–environment relationships. In the end, disputes over deeply held values about the environment may require both conflict and conversation—collaboration with opponents and, at other times, advocacy of values that cannot or should not be compromised. Collaboration—like advocacy—has a place in managing environmental conflicts, but no single mode is always the most appropriate or effective path to a solution.

SUGGESTED RESOURCES

- *The Story of the Great Bear Rainforest*, a moving YouTube video of the struggle and successful collaboration by environmentalists, First Nations, and loggers to protect this large, temperate rainforest in western Canada. 3:02. Available at https://www.youtube.com/watch?v=OgN2PFAEtGM.
- Examining a contentious conflict—and the successful collaboration that resolved it—can be instructive. That's the goal of the film *Whose Home on the Range*: "Catron County, New Mexico—the 'toughest county in the West'—has been at the center of a struggle between ranchers, loggers, environmentalists, and the U.S. Forest Service over the management of federal land. The only physician in the county, concerned about the health of his community, began a process of dialogue among citizens" (Bullfrogfilms). *Whose Home on the Range* rental or sale: www.bullfrogfilms.com/catalog/whose.html; for the study guide go to http://bullfrogfilms.com/guides/whoseguideS.pdf.
- Resources and case studies at the U.S. Institute for Environmental Conflict Resolution (www.ecr.gov).
- *Towards an Environmental Justice Collaborative Model: An Evaluation of the Use of Partnerships to Address Environmental Justice Issues in Communities.* The U.S. EPA's Federal Interagency Working Group on Environmental Justice (2003) prepared this report on environmental justice collaboration, drawing from literature cited in this chapter. Available at http://www.epa.gov/evaluate/pdf/ej/towards-ej-collaborative-model-evaluation.pdf.
- For their 100th anniversary, the Rockefeller Foundation has launched a 100 Resilient Cities Centennial Challenge, funding local collaborative efforts to make cities more resilient with global resources and networks. Go to the Foundation's website to see how they are encouraging global and local collaboration: http://www.100resilientcities.org. See also the Resilience Alliance, with whom they are working in some locations: http://www.resalliance.org.

KEY TERMS

DISCUSSION QUESTIONS

1. Is conflict needed to force opponents to the negotiating table, or can collaboration be used at the start when an environmental concern is first identified?

2. Would you feel comfortable disagreeing with the majority in a collaborative process? Would you still support a group consensus even if your preferred solution was not adopted? Would you if you felt that the group had fairly considered your views before it reached its decision?

3. Some critics feel that the inequities in power and resources between representatives of industry and citizens make true consensus between these two groups impossible. "Industry thinks its odds are better in these forums. . . . It believes it can dominate them over time" (McCloskey, 1996, p. 7). Do you agree?

4. Should "outside" interests, like national environmental groups, be stakeholders in local, place-based collaborations about natural resources? How about a corporation, headquartered elsewhere, such as a timber or oil company? Can it use its local representatives to serve as its stakeholder? What about philanthropy organizations, such as the Rockefeller Foundation?

5. Is "compromise" a bad word? Is it possible in all environmental conflicts? How about cases such as oil or natural gas drilling in wilderness areas (like the Arctic National Wildlife Refuge) or near community water sources?

In 2014, the New York State Supreme Court ruled that towns may use local zoning ordinances to ban fracking or hydraulic fracturing. Dryden, New York, town supervisor Mary Ann Sumne said, "The oil and gas industry tried to bully us into backing down, but we took our fight all the way to New York's highest court." She added, "I hope our victory serves as an inspiration to people . . . elsewhere who are also trying to do what's right for their own communities" (quoted in Taylor & Kaplan, 2014, p. A15). Pictured above is scientist, writer, advocate, and New York resident Sandra Steingraber.

CHAPTER 14

Citizens' (and Nature's) Standing

Environmental Protection and the Law

> *The intent of granting an ecosystem legal status is to expand the vocabulary for expressing value.*
>
> —Clint Williams (2012, para. 7)

You may wonder what "law" has to do with a book about environmental communication. Actually, speaking in court was one of the first subjects studied in the educational curriculum in early Greece. Aristotle called this **forensic rhetoric**, the speaking of ordinary citizens who appeared before a judge to argue for their rights, as well as what was just or unjust, under Athenian law. Today, citizens in many countries, including the United States, have the right to speak—with the aid of legal counsel—on behalf of themselves and others in court. In the United States, for example, residents, environmentalists, and other members of the public have relied on environmental laws—the Clean Water Act, Food Quality Protection Act, Endangered Species Act, and others—to force the cleanup, for example, of polluted rivers, to ensure that our foods are free of pesticides, to save an endangered plant or animal, and to secure other environmental and health protections.

As such, courts and other legal forums are vital spaces to study when considering how we judge environmental controversies in the public sphere. In this chapter, we explore the ways in which ordinary citizens, as petitioners or plaintiffs, may be actively involved with the law in protecting both the natural environment and their community's health and well-being, as well as their own.

Chapter Preview

- The first section of this chapter describes the "right of standing" and citizen environmental lawsuits, that is, the ability of ordinary citizens to appear before courts of law to seek environmental protections.
- In the next two sections, we summarize three landmark cases that have affected the public's right of standing in environmental controversies, and we illustrate the importance of this right in a major ruling on global warming by the U.S. Supreme Court.
- Finally, we take up the question, "Do future generations, as well as nonhuman nature—dolphins, trees, and others—have a right of standing to protect their interests?"

When you've finished this final chapter, we hope you will have an appreciation of the dramatic changes in the past half-century giving ordinary citizens a right to stand in a court of law to argue for clean water in their community, or to require the government to protect the habitat for an endangered bird, and more. And, we hope, in the years ahead, you'll be part of the conversation about who else—future generations, as well as dolphins, whales, and other natural creatures—will have a similar right.

Right of Standing and Citizen Suits

Beyond the right to know and public comment (Chapter 12) is another route for citizen participation in legal environmental decision making: the right of standing. The **right of standing** is based on the presumption that an individual having a sufficient interest in a matter may stand before judicial authority to speak and seek protection of that interest in court. In both common law and provisions under U.S. environmental law, citizens—under specific conditions—may have standing to object to an agency's failure to enforce environmental standards or to hold a violator directly accountable.

Standing in a Court of Law

The right of standing in court developed originally from common law, where those who suffered what is called an **injury in fact** to a legally protected right could seek redress in court. The definition of *injury* under common law normally meant a concrete, particular injury that an individual had suffered due to the actions of another party. One of the earliest cases of standing in an environmental case took place in England and involved William Aldred, who in 1611 brought suit against his neighbor Thomas Benton. Benton had built a hog pen on an orchard near Aldred's house. Aldred complained that "the stench and unhealthy odors emanating from the pigs drifted onto [his] land and premises" and were so offensive that he and his family "could not come and go without being subjected to continuous annoyance" (in Steward & Krier, 1978, pp.117–118). Although Benton argued "one ought not have so delicate a nose, that he cannot bear the smell of hogs," the court sided with Aldred and ordered Benton to pay for the damage caused to Aldred's property.

Aldred was able to pursue his claim as a result of his and his family's "injury in fact" from the offensive odors. But, in the 20th century United States, this principle would be expanded to allow wider access to the courts by environmental advocates. Two developments modified the strict common-law requirement of concrete, particular injury, allowing a greater opening for citizens to sue on behalf of environmental values.

First, the 1946 **Administrative Procedure Act (APA)** broadened the right of judicial review for persons "suffering a legal wrong because of agency action, or adversely affected or aggrieved by agency action" (in Buck, 1996, p. 67). This was so because, under the APA, the courts generally have held that an agency must "weigh all information with fairness and not be 'arbitrary and capricious'" in adopting agency rules (Hays, 2000, p. 133). Thus, when an agency's actions depart from this standard, it is subject to citizen complaints under the APA; that is, because citizens have suffered from an "arbitrary and capricious" action, they have standing to seek protection in the courts. In succeeding years, this provision of the APA would be an important tool enabling environmental organizations to hold agencies accountable for their actions toward the environment.

A second development would expand the public's right of standing even more explicitly to environmental concerns.

Citizen Suits and the Environment

A major expansion of the public's right of standing came in the form of **citizen suits** in major environmental laws. The provision for such lawsuits enables citizens to go into a federal court to make an argument with the hope of persuading a judge to enforce an environmental law. For example, the Clean Water Act confers standing on any "persons having an interest which is or may be adversely affected" to challenge violations of clean water permits if the state or federal agency fails to enforce the statutory requirements (Clean Water Act, 2007).

Using the Clean Water Act's provision for the public's right of standing, for example, citizens in Appalachia have been filing lawsuits against the practice of **mountaintop removal coal mining**. In this type of mining, coal companies literally push the tops of mountains into nearby valleys, filling streams in their search for coal. This is a particularly destructive form of coal mining. An environmental group in West Virginia, for example, has just won a citizen lawsuit against a major coal company for allowing selenium, a chemical that harms aquatic life, to leak from a coal slurry impoundment into streams. As a result, the court will prescribe a remedy, as well as a financial penalty for the company (Quiñones, 2014). Other environmental laws that allow for citizen suits include the Endangered Species Act, the Clean Air Act, the Toxic Substances Control Act, and the Comprehensive Environmental Response Compensation and Liability Act (the Superfund law).

The purpose of a citizen suit is to challenge an agency's lack of enforcement of environmental standards. Local citizens and public interest groups are empowered to sue an agency directly to enforce the law. The rationale behind this is straightforward: When regulators "overlook local environmental deterioration or are compromised by interest group pressure, local groups in affected areas are empowered to trigger enforcement themselves" (Adler, 2000, para. 48). This is especially important in cases of **agency capture**, in which a regulated industry pressures or influences officials to ignore violations of a corporation's permit for environmental performance (for example, its air or water discharges).

The idea of citizen access to the courts—the right of standing—is increasingly recognized in other nations, as citizens seek to enforce environment laws. As we learned in Chapter 12, for example, the Aarhus Convention in Europe includes the right of "access to justice," or review by a court of law. And, recently, China has approved "sweeping new environmental protections . . . amid mounting concerns over pollution poisoning the nation's air, water and soil" (Leavenworth, 2014, p. 8A). Importantly, the law gives everyday citizens the right to seek legal action against influential polluters.

Landmark Cases on Environmental Standing

Citizens' rights of standing are subject not only to the provisions of specific laws (for example, the Clean Water Act) but also to judicial interpretations of the **cases and controversies clause** in Article III of the U.S. Constitution. Despite its arcane title, this clause serves an important purpose. The cases and controversies clause "ensures that lawsuits are heard *only if the parties are true adversaries, because only true adversaries will aggressively present to the courts all issues* [emphasis added]" (Van Tuyn, 2000, p. 42).

To determine if a party is a "true adversary," the U.S. Supreme Court, the highest federal court, uses three tests: (1) persons bringing a case must be able to prove they've been harmed, or suffered an "an injury in fact"; (2) this injury must be "fairly traceable" to an action of the defendant, and (3) the Court must be able to respond to the injury through a favorable ruling (Van Tuyn, 2000, p. 42). So, to be clear, even though some environmental laws provide for a citizen's right of standing, *citizens still must meet these three constitutional tests* before their case can proceed in a court.

The main question, then, in granting an individual standing is the meaning of *injury in fact*. What qualifies as an "injury" where individual citizens seek to enforce the provisions of an environmental law? The U.S. Supreme Court has worked out, at times, an uneven answer to this question in several landmark cases.

Sierra Club v. Morton (1972)

The U.S. Supreme Court's 1972 ruling in *Sierra Club v. Morton, Secretary of the Interior, et al. (Sierra Club v. Morton)* provided the first guidance for determining standing under the U.S. Constitution's cases and controversies clause in an environmental case. In this case, the environmental organization, the national Sierra Club, sued the federal government to compel it to block plans by Walt Disney Enterprises to build a resort in Mineral King Valley in California. Plans for the resort included the building of a road through Sequoia National Park with this region's thousand-year-old giant Sequoia trees.

In its suit, the Sierra Club argued that a road would "destroy or otherwise adversely affect the scenery, natural and historic objects, and wildlife of the park for future generations" (Lindstrom & Smith, 2001, p. 105). (See Photo 14.1.) Although the Supreme Court, by a narrow 4–3 decision, did find that such damage could constitute an injury in fact, it noted that the Sierra Club *did not allege that any of its members themselves had suffered any actual injury*, and therefore, they were not true adversaries. Instead, the Club had asserted a right of standing simply on the basis of its interest in protecting the environment. The Court rejected the organization's claim in the case, ruling that a long-standing *interest in a problem* was not enough to constitute an injury in fact (Lindstrom & Smith, 2001, p. 105).

Despite its ruling in *Sierra Club v. Morton*, the U.S. Supreme Court spelled out an expansive standard for a successful claim of standing. Justice Potter Stewart wrote the majority opinion, observing that, in the future, the Sierra Club would need only to allege an injury to *its members'* interests—for example, that particular members would no longer be able to enjoy an unspoiled wilderness or their normal recreational pursuits. The Sierra Club immediately and successfully amended its

| Photo 14.1 | View of Mineral King Valley, a part of Sequoia National Park since 1978. |

suit against Disney, arguing that such injury would occur to its members if the road through Sequoia National Park were to be built. (Subsequently, Mineral King Valley itself was added to Sequoia National Park, and Disney Enterprises withdrew its plans to build the resort.)

The Court's liberal interpretation of the test for injury in fact in *Sierra Club v. Morton,* along with the right of standing in many environmental laws, produced a 20-year burst of environmental litigation by citizens and environmental organizations. This trend continued until the Supreme Court issued a series of conservative rulings, later, that narrowed the basis for citizens' standing.

Lujan v. Defenders of Wildlife (1992)

In the 1990s, the U.S. Supreme Court handed down several rulings that severely limited citizen suits in environmental cases. In perhaps the most important case, **Lujan v. Defenders of Wildlife** (1992), the Court rejected a claim of standing by the environmental organization, Defenders of Wildlife, under the citizen suit provision of the Endangered Species Act. That law declares that "any person may commence a civil suit on his own behalf (A) to enjoin any person, including the United States and any other governmental instrumentality or agency . . . who is alleged to be in violation of any provision" of the act (Endangered Species Act, 1973, A71540 [g] [1]).

In its lawsuit, Defenders of Wildlife argued that the Secretary of the Interior (Lujan) had failed in his duties to ensure that U.S. funding of projects overseas—in this case, in Egypt—did not jeopardize the habitats of endangered species, as the law required (Stearns, 2000, p. 363).

Writing for the majority opinion, Justice Antonin Scalia stated that Defenders of Wildlife had failed to satisfy constitutional requirements for injury in fact that would grant standing under the ESA. He wrote that the Court rejected the view that the citizen suit provision of the statute conferred upon "*all* persons an abstract, self-contained, non-instrumental 'right' to have the Executive observe the procedures required by law" (*Lujan v. Defenders of Wildlife,* 1992, p. 573). Rather, he explained, the plaintiff must have suffered a tangible and particular harm not unlike the common-law requirement (Adler, 2000, p. 52). This ruling seriously constrained *Sierra Club v. Morton,* in which Sierra Club members needed only to prove injury to their interests—that is, that they couldn't enjoy their recreational pursuits or experience of wilderness.

Consequently, courts began to sharply limit citizen claims of standing under citizen suit provisions of environmental statutes. Writing in the *New York Times,* Glaberson (1999) reported that the Court's rulings in the 1990s were one of the most "profound setbacks for the environmental movement in decades" (p. A1).

Friends of the Earth, Inc. v. Laidlaw Environmental Services, Inc. (2000)

In a later case, the Supreme Court modified its strict *Lujan* doctrine, holding that the knowledge of a possible threat to a legally recognized interest (clean water) was enough to establish a "sufficient stake" by a plaintiff in enforcing the law (Adler, 2000, p. 52).

In 1992, the national environmental organization Friends of the Earth, and CLEAN, a smaller, locally based environmental group, sued Laidlaw Environmental Services in Roebuck, South Carolina, using the Clean Water Act citizen suit provision. Their lawsuit, *Friends of the Earth, Inc. v. Laidlaw Environmental Services, Inc.* (2000), alleged that Laidlaw had repeatedly violated its permit limiting the discharge of pollutants (including mercury, a highly toxic substance) into the nearby North Tyger River. Residents of the area who had lived by or used the river for boating and fishing testified that they were "concerned that the water contained harmful pollutants" (Stearns, 2000, p. 382).

The Supreme Court ruled that Friends of the Earth and CLEAN did not need to prove an actual (particular) harm to residents. Writing for the majority opinion, Justice Ruth Bader Ginsburg stated that injury to the plaintiff came from lessening the "aesthetic and recreational values of the area" for residents and users of the river due to their knowledge of Laidlaw's repeated violations of its clean water permit (Adler, 2000, p. 56). In this case, the plaintiffs were not required to prove that Laidlaw's violations of its water permit had contributed to actual deterioration in water quality. It was sufficient that they showed that *residents' knowledge of these violations* had discouraged their normal use of the river.

A cautionary note: Although the right of standing has been expanded recently, courts may still find that a petitioner sometimes fails to meet one of the three requirements under the Supreme Court's interpretation of the cases and controversies clause of the Constitution. As this book goes to press, a federal district judge has ruled a physician in Pennsylvania, for example, lacked standing to sue the state's Department of Environmental Protection "because he failed to meet certain criteria to show he suffered an 'injury in fact' that 'must be concrete in both a qualitative and temporal sense'" (Colaneri, 2014, para. 2). The physician, a nephrologist who specializes in the treatment of renal diseases and other illnesses, had argued that "he was unable to obtain critical information about the quality of local water needed on a daily basis for his practice from [natural] gas drillers" who were fracking in the area (para. 2).

Global Warming and the Right of Standing

The public's right of standing arose again as a pivotal issue in *Massachusetts, et al., Petitioners v. Environmental Protection Agency et al.* (*Massachusetts v. EPA*), in 2007, in the U.S. Supreme Court's first-ever ruling on global warming. Twelve states, several cities, and American Samoa, along with the Center for Biological Diversity, Sierra Club, Greenpeace, the Union of Concerned Scientists, and a number of other groups petitioned the Supreme Court to direct the EPA to regulate motor vehicles' tailpipe emissions of greenhouse gases, invoking the Clean Air Act. In a 5–4 ruling, the Court sharply rebuked the Bush administration's claim that the EPA lacked this authority or, if it had the authority, could choose not to exercise it.

The central issue in this case was whether carbon dioxide (CO_2) and other "greenhouse" gases were an "air pollutant," under the definition of this term in the Clean Air Act. That law identified "any air pollution agent ... [as] including any physical, chemical . . . substance or matter which is emitted into or otherwise enters the ambient air" (Clean Air Act Amendments, 1990, 7602[g]). While the majority opinion ultimately agreed that carbon dioxide did meet this definition, the justices first had to decide the question of the plaintiffs' standing to argue before the Court.

In a strategic move, the petitioners decided to list the coastal state of Massachusetts first, or as the lead **plaintiff** in the case, or the one who brings a civil action in a court of law (hence, the name, *Massachusetts v. EPA*). This proved to be important, because only one plaintiff is required to be a true adversary for the case to proceed on its merits, and Massachusetts could meet the requirement because, it argued, rising sea levels threatened the economic interests of its citizens. The state also argued that EPA's failure to set greenhouse gas standards was "arbitrary and capricious" (an action prohibited by the EPA).

In ruling on the question of Massachusetts' right of standing, Justice John Paul Stevens delivered the majority opinion of the Court:

> EPA's steadfast refusal to regulate greenhouse gas emissions presents a risk of harm to Massachusetts that is both "actual" and "imminent," *Lujan*, 504 U.S. . . . and there is a "substantial likelihood that the judicial relief requested" will prompt EPA to take steps to reduce that risk. (*Massachusetts et al. v. Environmental Protection Agency et al.*, 2007, p. 3)

Finally, responding to the Bush administration's objection that a ruling for the plaintiffs would not solve the problem, Justice Stevens wrote, "While regulating motor-vehicle emissions may not by itself *reverse* global warming, it does not follow that the Court lacks jurisdiction to decide whether EPA has a duty to take steps to *slow* or *reduce* it" (*Massachusetts et al. v. Environmental Protection Agency et al.*, 2007, p. 4).

With these rulings, the Supreme Court reaffirmed the rationale that citizens, as well as states, have an interest in the enforcement of environmental quality under the provisions of specific laws such as the Clean Water Act or the Clean Air Act. However, disagreement over the criteria for standing in environmental cases is likely to continue. (See "FYI: Do Nations Threatened With Rising Sea Levels From Global Warming Have a Right of Standing?") At stake are differing interpretations of injury in fact and the rights of citizens to compel the government to enforce environmental laws.

☞ FYI   Do Nations Threatened With Rising Sea Levels From Global Warming Have a Right of Standing?

Do Pacific island nations like Kiribati, Vanuatu, and Tuvalu or low-lying countries such as Bangladesh have standing to sue the United States, China, and other nations that are the main sources of greenhouse gases that contribute to climate change and rising sea levels, which threaten their existence? A study by the Foundation for International Environmental Law and Development says "small island nations and other threatened countries have the right and likely the procedural means to pursue an inter-state case before the United Nations' International Court of Justice."

"Some of these countries are getting increasingly desperate," Chrisoph Schwarte, the paper's lead author, said. . . . Many leaders are looking for ways to make the United States and others understand the threats they face from rising sea levels, droughts and storm surges. . . . If nothing significant happens within the next two or three years, I really wouldn't be surprised if countries go to court."

Suing to force a country to reduce its greenhouse gas emissions is a tricky proposition. If a person is harmed or a livelihood threatened, he or she can take the offender to court for the damages incurred. But what if the victim is an entire country, and the damages—thousands forced to relocate or the loss of tourism dollars because of coral bleaching—are expected but not yet seen? More complicated still is pinpointing the perpetrator. A smokestack in China? The 24–7 air conditioner blasting from a shopping mall in Iowa? Emissions-belching SUVs in Sydney, Australia?

SOURCE: Adapted from Friedman, L. (2010, October 4).

Kyodo via AP Images

Photo 14.2 Dangerously high tides on the Pacific island nation of Kiribati, due to rising sea levels have forced some residents to seek refugee status in New Zealand, claiming it is too dangerous to remain on the island. The president of Kiribati has bought land on another island in Fiji, anticipating national relocation. Establishing the right of standing is vital to nations when international rulings impact their existence.

As this book goes to press, the Supreme Court, relying on *Massachusetts v. EPA*, has upheld a major initiative by U.S. president Obama's EPA that aims to cut greenhouse-gas emissions such as CO_2 emissions from coal-burning power plants and factories (Barnes, 2014).

Do Future Generations and Nonhuman Nature Have a Right of Standing?

As we noted in Chapter 11, an important element in the Brundtland Commission Report (1987) was a definition of "sustainable development" that "meets the needs of the present *without compromising the ability of future generations to meet their own needs* [emphasis added]" (section 1). A concern for future generations is the central theme in the idea of **intergenerational justice.** This is the principle of fairness, or justice, in relationships, not only between children and adults, but between the present and future (unborn) generations.

Meanwhile, previously in this chapter, we discussed the significance of the *Sierra Club v. Morton* ruling; yet, we have not addressed the dissenting opinion, which opened a legal door for considering the standing of nonhuman nature. In this final section, we explore these concerns by asking more specifically, "Should future generations and nonhuman nature have a right of standing?"

The Standing of Future Generations

Long before the Brundtland Commission was formed, ethical concerns for nature and future generations had been compelling themes in indigenous cultures. (Democracy in the United States, in fact, is greatly indebted to indigenous thought.) One such ethical concern was in the Great Law of Peace of the Iroquois Nation that took into account intergenerational rights in actions toward the environment:

> The thickness of your skin shall be seven spans. . . . Look and listen for the welfare of the whole people and have always in view not only the present but also the coming generations, even those whose faces are yet beneath the surface of the ground—the unborn of the future Nation. (Constitution of the Iroquois Nations, Section 28)[1]

This desire to make decisions based on how they will impact not one generation, but seven, is known today as the **seventh generation principle**. Popularized by the corporate brand of the same name that sells everything from cleaning products to toilet paper, it is worth exploring a bit more in depth what this principle might entail in the legal sphere.

Intergenerational Justice and Climate Change

The idea of intergenerational justice lies at the heart of environmental ethics, as well as many international, environmental agreements. Article 3 of the United

Nations Framework Convention on Climate Change, for example, embodies this principle in recognizing that climate change is a concern for future generations. It declares, "The Parties should protect the climate system for the benefit of present and future generations of humankind, on the basis of equity and in accordance with their . . . responsibilities and respective capabilities" (United Nations Framework Convention on Climate Change, 1992). Similar language in other global agreements led UN secretary-general Ban Ki-moon recently to conclude that "concern for future generations has developed as a guiding principle of international norms" (Report of the United Nations Secretary-General, 2013, p. 24).

Despite the references (usually in the preambles) to "future generations" in many environmental agreements, there is, nevertheless, "*no legally binding instrument* [emphasis added] at international level that commits States to the protection of future generations" (Mary Robinson Foundation—Climate Justice, 2013, p. 1). In other words, citizens (for the most part) lack standing to appear on behalf of "future generations" in a court of law to force a nation to take action to mitigate climate change.

We said, "for the most part." Actually, as this book goes to press, three countries— Tunisia, Ecuador, and the Dominican Republic—have embedded a concern for future generations in their constitutions, giving their citizens, in theory, standing to appear before judicial authority. Tunisia's constitution, for example, obliges that country to "contribute to the protection of the climate . . . for future generations" (quoted in King, 2014, para. 3). In the United States, the Supreme Court recognized in *Massachusetts vs. EPA* that *states* (not future generations), may have a right of standing to require the EPA to regulate greenhouse gases. Recent initiatives, however, are attempting to extend future individuals' legal right of standing to address climate change. Let's look at some of these efforts.

Children and the Public Trust Doctrine

An innovative legal strategy to secure a right of standing in cases addressing climate change has been the attempted use of the **public trust doctrine**. Derived from common law, this is the "principle that certain natural and cultural resources are preserved for public use, and that the government owns and must protect and maintain these resources for the public's use" (Nolo, 2014, para. 1). Traditionally, the global atmosphere, and climate, have not fallen under the scope of the public trust doctrine; however, that may be changing.

For several years, two nonprofit foundations, Kids vs. Global Warming and Our Children's Trust, have pursued a legal campaign "in which children sue states and the federal government seeking to invoke the common law 'public trust doctrine' for climate change" (Jacobs, 2014, para. 4). Under their theory of this doctrine, "citizens have standing to ask courts . . . [to] order the legislature and executive branches to carry out their duty of protecting the natural resources everyone relies on" (Ebersole, 2014, para. 7).

Consider, for example, the case of *Alec L. v. Gina McCarthy*, coordinated by Our Children's Trust. In this case, young people from California, as plaintiffs, sued the EPA and other government agencies under the public trust doctrine:

> Their theory is that the atmosphere qualifies as part of the public trust to be protected for future generations. The children contended that by failing to take greater action to limit carbon emissions, the U.S. EPA and the Departments of Agriculture, Energy and Interior are violating the doctrine. (Jacobs, 2014, para. 4)

In their brief, the youth argued that, "Failure to rapidly reduce CO_2 emissions and protect and restore the balance of the atmosphere is a violation of Youth's constitutionally protected rights"; it is also something, they argued, that is "redressable" by the courts, that is, courts can, in principle, order Congress to undertake certain actions to lessen greenhouse gases (Davis-Cohen & *StudentNation*, 2014, para. 1). (See Photo 14.3.)

As this book goes to press, the U.S. Court of Appeals for the District of Columbia has "bluntly dismissed" the case, ruling that "the public trust doctrine 'remains a matter of state law'" (Jacobs, 2014, para. 1, 5). The court basically said that the children (plaintiffs) had failed to satisfy a key requirement for standing; that the question was one that could properly be addressed by a federal court. To date, this has generally been the fate of similar attempts to invoke the public trust doctrine.

iMatter/kids vs global warming

Photo 14.3	Our Children's Trust plaintiffs in a federal lawsuit invoking the public trust doctrine to compel action on climate change stand outside a courtroom with images and text on posters to help publicize the legal debates inside the courtroom for broader support in the public sphere.

Nevertheless, with help from Our Children's Trust, young Americans have continued to press their campaign. Invoking the public trust doctrine, they have filed suits in all 50 states. Some of these cases have been denied; others are being decided as this book goes to press. (See "Act Locally! What Can Children Do About the Future?")

Act Locally!

What Can Children Do About the Future?

As this book goes to press, an appeals court in Oregon has given youth plaintiffs a second chance to use the public trust doctrine to gain standing in cases involving climate change:

> In a nationally significant decision in the case *Chernaik v. Kitzhaber,* the Oregon Court of Appeals ruled a trial court must decide whether the atmosphere is a public trust resource that the state of Oregon, as a trustee, has a duty to protect. Two youth plaintiffs were initially told they could not bring the case by the Lane County Circuit Court. The trial court had ruled that climate change should be left only to the legislative and executive branches. Today, the Oregon Court of Appeals overturned that decision. (Western Environmental Law Center, 2014, para. 1)

> In previous court tests of this theory, appeals courts had blocked use of the public trust doctrine in addressing climate change. One of the two teenagers who filed this case, Kelsey Juliana, said afterward: "This decision makes me feel very proud to be an Oregonian. It validates the younger generation's voice and lets us know that we are listened to and considered, and that our future matters" (para. 3).

> *Act Locally!* Track the results of this Oregon case: What was its outcome when it returned to trial court? What other state cases are ongoing, as you read this chapter? Are any cases succeeding at the appeals level?

SOURCE: Western Environmental Law Center. (2014, June 11). Youth Win Reversal in Critical Climate Recovery Case. [Press release]. Retrieved from http://www.westernlaw.org/article/youth-win-reversal-critical-climate-recovery-case-press-release-61114

Nonhuman Nature: Should Trees, Dolphins, and Rivers Have Standing?

Recently, some environmental communication scholars have begun to suggest that we pay attention to the ways in which nature or nonhuman species "speak," engage, and influence both humans and their own environments. Tema Milstein (2012), for example, calls for us to consider "nature as co-present, active, and [a] dynamic force in human-nature relationships" (p. 171). Such a perspective encourages us to view nature—rivers, orcas, wolves, sequoias, watersheds, and ecosystems—as active "subjects," that is, having agency to act or make known their interests. Similarly, Richard A. Rogers (1998) proposes a "rehearsal of ways of listening to non-dominant voices and extra-human agents and their inclusion in the production of meaning" (p. 268).

In the United States, ethical concern for nonhuman nature has been a theme in conservation literature as well as environmental ethics. In his classic plea for a "**land ethic**," the pioneering ecologist Aldo Leopold, for example, lamented that "the farmer who clears the woods off a 75 per cent slope, turns his cows into the clearing, and dumps its rainfall, rocks, and soil into the community creek, is still (if otherwise decent) a respected member of society" (Leopold, 1949/1966, p. 245). The problem, he wrote, is that, "Obligations have no meaning without conscience, and the problem we face is *the extension of the social conscience from people to land* [emphasis added]" (p. 246).

The value ascribed to nature also has been a theme in the global environmental movement. The **Deep Ecology movement**, for example, affirms that "the well-being and flourishing of human and nonhuman life on Earth have value in themselves (synonyms: inherent worth, intrinsic value, inherent value). [Furthermore] These values are independent of the usefulness of the nonhuman world for human purposes" (Foundation for Deep Ecology, 2012, para. 1).

The belief that nature has intrinsic value raises an important question for us, in this chapter: Does our valuing of nature also imply that nature has a *legal* right to standing, that is, a right to "speak" for its interests in courts of law? Is a view of non-human nature as a "subject," and, therefore, having a right to speak in courts or before governmental agencies, even possible?

The question of standing for nature famously arose in the U.S. Supreme Court's deliberations in *Sierra Club v. Morton* in 1972 (described above). In the close 4–3 ruling against the Sierra Club, three Justices chose to write *dissenting opinions*. One of these dissents, by Justice William O. Douglas, has become a classic statement of the right of standing for nature.

Justice Douglas argued that even trees and rivers should have standing.

> The critical question of "standing," would be simplified . . . if we fashioned a . . . rule that allowed environmental issues to be litigated before federal agencies or federal courts in the name of the inanimate object about to be despoiled, defaced, or invaded by roads and bulldozers and where injury is the subject of public outrage. (*Sierra Club v. Morton*, 1972, p. 405 U. S. 741)

In arguing for the right of standing of natural objects, Douglas pointed out that even inanimate objects, such as ships and corporations, are sometimes parties in litigation. This is a precedent, he wrote, for natural objects:

> So it should be as respects valleys, alpine meadows, rivers, lakes, estuaries, beaches, ridges, [and] groves of trees. . . . The river, for example, is the living symbol of all the life it sustains or nourishes. . . . The river as plaintiff speaks for the ecological unit of life that is part of it. Those people who have a meaningful relation to that body of water—whether it be a fisherman, a canoeist, a zoologist, or a logger—must be able to speak for the values which the river represents and which are threatened with destruction. (*Sierra Club v. Morton*, 1972, p. 405 U.S. 743)

Justice Douglas's argument for a right of standing for alpine meadows, estuaries, and "groves of trees," however, was a minority opinion, that is, the Supreme's Court majority (official) ruling did not recognize such a right, nor have any U.S. courts since recognized a right of standing for nature in cases where certain parties—developers or chemical companies, for example—have despoiled the land or polluted the air or water. Instead, the Supreme Court has made clear over the decades that the right of standing is limited to humans who are injured in fact:

> In *Friends of the Earth, Inc. v. Laidlaw Environmental Services (TOC), Inc.* . . . the Court explained, "[t]he relevant showing for purposes of Article III standing . . . is not injury to the environment but injury to the plaintiff." . . . Even if the relevant statute prohibits activity that results in injury to a river, a species, or an ecosystem, the courts will only hear the case if the *human* plaintiff can demonstrate that *she* has suffered a particularized, concrete injury (or injury-in-fact) as a result of (i.e. causation) the prohibited activity. Further, it is the injury to the *human* plaintiff that must be likely to be redressed by a favorable ruling. (Benzoni, 2008, p. 349)

In other words, only a *human* can satisfy the U.S. Supreme Court's three requirements that govern the right of standing—that a plaintiff show (a) an injury, (b) that is clearly traceable to the offending party, and (c) is something that a favorable ruling of a court can redress. (For some of the practical problems facing the case for legal standing for natural objects, see, "Another Viewpoint: Who Speaks for Nature, in Court?")

Another Viewpoint: Who Speaks for Nature, in Court?

"Conferring a sort of legal personhood to [a] river is . . . is impractical in the American legal system," says one environmental attorney.

"What has frustrated environmentalists is that harm to the environment has been defined as harm to the users," says [attorney] Andrew S. Levine. . . . Damage to a river, for example, would be defined by losses by trout fishermen or businesses serving anglers. Damage to a forest would be defined by losses to hunters or loggers.

"How do you put a cost on a view-scape? On the clarity of the water? On the diversity of species?" Levine says.

The intent of granting an ecosystem legal status is to expand the vocabulary for expressing value. "The downside is, where does it end?" says Levine. . . .

"Who speaks for the trees," Levine asks. "In the real world, there would be 20 people who claim to speak for the trees."

SOURCE: Williams, C. (2012). "Should Nature Have a Legal Standing?" Mother Nature Network, at www .mnn.com/earth-matters/wilderness-resources/stories/should-nature-have-a-legal-standing#ixzz 35SuXl532

Several other nations, however, have granted certain species and/or eco-systems rights of protection. In a dramatic move, India, for example, has declared that dolphins and other cetaceans such as whales and porpoises should be perceived as "nonhuman persons" due to their intelligence and self-awareness. As a result, the country banned the use of these creatures for public entertainment and "forbid them from being held captive anywhere in India" (Ketler, 2013, para. 1). India's action reflected a call by the Helsinki Group, an international organization of scientists who adopted a *Declaration of Rights for Cetaceans: Whales and Dolphins* on May 22, 2010, in Helsinki, Finland, which affirms that "all cetaceans as persons have the right to life, liberty and wellbeing." (See "FYI: Declaration of Rights for Cetaceans: Whales and Dolphins.")

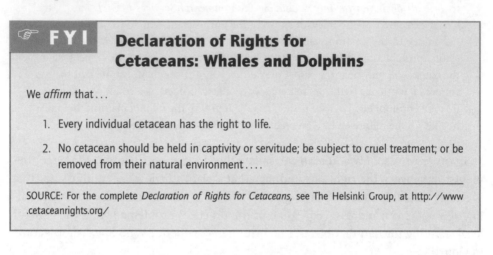

☞ **FYI** **Declaration of Rights for Cetaceans: Whales and Dolphins**

We *affirm* that . . .

1. Every individual cetacean has the right to life.

2. No cetacean should be held in captivity or servitude; be subject to cruel treatment; or be removed from their natural environment . . . .

SOURCE: For the complete *Declaration of Rights for Cetaceans,* see The Helsinki Group, at http://www.cetaceanrights.org/

Similarly, New Zealand recently granted legal standing to the Whanganui River, the country's third-largest river (Williams, 2012). The river "will be recognized as a person when it comes to the law—much like a corporation," under agreement signed between the Whanganui River iwi, an indigenous people, and the New Zealand government. Under the agreement, "the river is given legal status under the name Te Awa Tupua, and two guardians, one from the Crown and one from a Whanganui River iwi, will be given the role of protecting [and therefore, speaking for] the river" (Williams, 2012, para. 2). Given these expanding efforts to consider nonhuman rights to standing, it will be interesting to see whether corporations or rivers perform more persuasively in court.

SUMMARY

- As you've seen by this point, the right of ordinary citizens—often with the legal aid of an environmental organization—to stand before a court to argue for environmental protections has been a struggle over the past half century. Given the intertwined history of legal theory, political ideology, and the U.S. Supreme Court's rulings, an expansive right of standing—for citizens, future generations, or nature—has never been fully guaranteed. That is why we feel that, going forward, analyzing the law on rights of standing will be important in your study of environmental communication.

- In our first section, we described this basic right of standing, as well as citizen environmental lawsuits, that is, the ability of ordinary citizens to appear before courts of law to speak on behalf of environmental protections.

- We then summarized three landmark cases that affected the right of standing of public advocates in environmental controversies, and we illustrated the importance of this right in a major Supreme Court ruling on global warming, *Massachusetts vs. EPA* (2007).
- Finally, we posed, and tried to answer, the question, "Should future generations, as well as nonhuman nature—such as trees, dolphins, and rivers—have right of standing?"
- As you finish this chapter, you may have noticed that the question of who (and what) has standing in environmental cases is very much an open question. We hope, therefore, that in the years ahead you will be a participant in these conversations—whether or not future generations or the nonhuman natural world—orcas, manatees, estuaries, African elephants, alpine meadows, and other species and natural communities—will have a similar right.

SUGGESTED RESOURCES

- *Topix: Environmental Law* updates news about environmental law from thousands of sources on the Web, at http://www.topix.com/law/environmental.
- For a reading of the U.S. Supreme Court's ruling on global warming, in *Massachusetts et al. v. Environmental Protection Agency et al.* (2007), see http://www.law.cornell.edu/supct/html/05–1120.ZS.html.
- Our Children's Trust has been at the forefront of youth who are using the public trust doctrine to pursue more forceful action on climate change. Watch their compelling video at http://www.youtube.com/watch?v=89xZ7gOaCu4.
- Mary Christina Wood, *Nature's Trust: Environmental Law for a New Ecological Age.* (Cambridge University Press, 2013). The author critiques environmental law for its failure to address ecosystems' collapse and the impact of climate change, arguing for a new approach based on the public trust doctrine.
- Christopher D. Stone, *Should Trees Have Standing?* (Oxford University Press, 3rd ed., 2010). Originally published in 1972, Stone's book is probably the best-known argument for the legal right of standing for nature.

KEY TERMS

Administrative Procedure Act (APA)   345

Agency capture   346

Cases and controversies clause   346

Citizen suits   345

Deep Ecology movement   356

Forensic rhetoric   343

Friends of the Earth, Inc., v. Laidlaw Environmental Services, Inc.   349

Injury in fact   344

Intergenerational justice   352

Land ethic   355

DISCUSSION QUESTIONS

1. Can you imagine yourself being a party to an environmental law case, having "standing" before a governmental agency or court of law, to speak about your interests? Under what circumstance might this happen?

2. In *Lujan v. Defenders of Wildlife*, Justice Antonin Scalia stated that the U.S. Supreme Court *rejected* the view that the citizen suit provision of the Endangered Species Act conferred upon "*all* persons an abstract, self-contained, non-instrumental 'right' to have the Executive observe the procedures required by law" (1992, p. 573). Why not? If the law itself gives citizens the right to bring lawsuits to enforce the Act, why can the Court stop them?

3. Do trees have standing? How about streams, plants, and animals? Future generations? What about Pacific Island nations and residents of low-lying villages, threatened with a rise in ocean levels? Do they have standing to sue the United States, China, and other nations that emit greenhouse gases for contributing to global warming?

4. Are Pacific Island nations *true adversaries,* that is, (a) do they suffer an injury in fact, (b) can this injury be *fairly traceable* to an action of the defendant, and (c) is a court is able to redress the injury through a favorable ruling?

NOTE

1. In 1987, the United States Senate acknowledged that the Great Law of Peace of the Iroquois Nations served as a model for the Constitution of the United States (U.S. S. Con. Res. 76, 2 Dec. 1987).

Glossary

Aarhus Convention: Adopted in 1998 in the Danish city of Aarhus, this is an environmental agreement of the United Nations Economic Commission for Europe (UNECE). It addresses three areas: access to information, public participation in decision making, and access to justice in environmental matters (similar to a right of standing).

Acceptable risk: From a technical perspective, a judgment based on the numerical estimate of deaths or injuries expected annually from exposure to a hazard. From a cultural perspective, a judgment of what dangers society is willing or unwilling to accept and who is subject to this risk; such a judgment inevitably involves values.

Access to justice: One of three core principles of the Aarhus Convention; Article 9 ensures that "any person who considers that his or her request for information . . . has been ignored, [or] wrongfully refused . . . has access to a review procedure before a court of law or another independent and impartial body."

Accountability: The requirement that political authority meet agreed-upon norms and standards.

Acute single-dose tests: Occurs when risks are assessed by considering an individual hazard in isolation with a short-term exposure at a relatively high level.

Administrative Procedure Act (APA): Enacted in 1946, this law laid out new standards for the operation of U.S. government agencies; it required that proposed actions be published in the *Federal Register* and that the public be given an opportunity to respond; it also broadened the right of judicial review for persons "suffering a legal wrong" resulting from "arbitrary and capricious" actions on the part of agencies.

Advocacy: Persuasion or argument in support of a cause, policy, idea, or set of values.

Advocacy campaign: A strategic course of action involving communication that is undertaken for a specific purpose.

Aesthetics: The role of art or taste. Applied aesthetics of popular culture include but are not limited to: the position of the camera, the use of light, the role of sound, design of costume and sets, the pace of a moving picture, and the application of special effects.

Affordances: Drawing on Jerome Gibson's classic *The Ecological Approach to Visual Perception,* affordances describe what are enabled by various media and what are not.

Agency capture: The pressuring or influencing of officials by a regulated industry to ignore violations of a corporation's permit for environmental performance (for example, its air or water discharges).

Agenda setting: The ability of media to affect the public's perception of the salience or importance of issues; in other words, news reporting may not succeed in telling people what to think, but it succeeds in telling them what to think about.

Androcentric bias: Much of technical risk assessment is based on the assumption that the person being exposed to a danger is an average adult male.

Antagonism: Recognition of the limit of an idea, a widely shared viewpoint, or an ideology that allows an opposing idea or belief system to be voiced.

Apocalyptic narrative: A literary style used by some environmental writers to warn of impending and severe ecological crises; evokes a sense of the end of the world as a result of the overwhelming desire to control nature.

App-centric: Mobile applications (on smartphones, iPads, etc.) as central portals for searching online.

Arbitration: The presentation of opposing views to a neutral, third-party individual or panel that, in turn, renders a judgment about the conflict; usually court ordered.

Attitude–behavior gap: Although individuals may have favorable attitudes or beliefs about environmental issues, they may not take any action; their behavior, therefore, is disconnected from their attitudes.

Audience: One of the three terms of the "rhetorical situation": the people being addressed, their beliefs, actions, and larger cultural understandings.

Bali Principles of Climate Justice: One of the first declarations redefining climate change from the perspective of environmental justice and human rights, crafted by a coalition of international nongovernmental organizations (NGOs) in Bali, Indonesia, in June 2002.

Base: A campaign's core supporters.

Bias of communication: The ways any given medium creates and limits conditions of possibility across space and time in a particular culture.

Black swan events: Unexpected, high-magnitude events that are beyond what modern society can usually predict.

Bottom-up sites: These provide online tools that enable users to start petitions on platforms like Facebook and Twitter.

Brundtland Commission Report: A UN report defining "sustainable development" as: "aims to promote harmony among human beings and between humanity and nature . . . that meets the needs of the present without compromising the ability of future generations to meet their own needs."

Business as usual (BAU): The continued growth of carbon-based economies. *Carbon based* refers to the energy sources—primarily fossil fuels or the burning of oil, coal, and natural gas—used to produce electricity, fuel transportation, and heat and power other dimensions of modern life.

Buycott (girlcott or pro-cott): A "concerted effort to make a point of spending money—as well as to convince others to make a point of spending money—on a product or service in the hopes of affirming specific condition(s) or practice(s) of an institution" (Pezzullo, 2011, p. 125).

Cases and controversies clause: The portion of Article III of the U.S. Constitution that ensures that lawsuits are heard by true adversaries in a dispute on the assumption that only true adversaries will represent to the courts the issues in a case; an important test of true adversary status is whether persons bringing the action are able to prove an injury in fact.

Circuit of culture: A flexible framework that reminds us of three characteristics of media: (1) culture always is changing and moving as part of broader networks or contexts; (2) to study it, one must choose which elements will or will not be the focus of one's analysis; and (3) people involved may or may not be involved in more than one element (Pezzullo, 2010).

Citizen environmental journalism: The sharing with others by ordinary citizens of information about an environmental event or condition they have witnessed, including news organizations, scientists conducting studies, or friends.

Citizen scientists: Those who gather and interpret knowledge from their own experiences and observation, as well as collecting relevant secondhand technical data.

Citizen suits: Action brought by citizens in federal court asking that provisions of an environmental law be enforced; the right to bring such suits is a provision of major environmental laws.

Citizens' advisory committee: Also called a citizens' advisory panel or board; a group appointed by a government agency to solicit input from diverse interests in a community—for example, citizens, businesses, and environmentalists—about a project or problem.

Civil disobedience: A peaceful form of protest that violates laws and accepts legal consequences (such as arrest) in order to point out ongoing injustices.

Clicktivism: The taking of action simply by clicking on a response link online.

Climate: Refers to average atmosphere changes over a long period of time.

Climate justice: Views the environmental and human impacts of climate change from the frame of social justice, human rights, and concern for indigenous peoples. The movement for climate justice asserts that global warming not only disproportionately impacts the most vulnerable regions and peoples of the planet but that these peoples and nations often are excluded from participation in the forums addressing this problem.

Collaboration: "Constructive, open, civil communication, generally as dialogue; a focus on the future; an emphasis on learning; and some degree of power sharing and leveling of the playing field" (Walker, 2004, p. 123).

Commons: The resources accessible to all people and not privately owned, such as air, water, and the Earth.

Communication: The symbolic mode of interaction that we use in constructing environmental problems and in negotiating society's different responses to them.

Communication tasks (of a campaign): (a) To create support or demand for the campaign's objectives, (b) to mobilize this support from relevant constituencies (audiences) to demand accountability, and (c) to develop a strategy to influence decision makers to deliver on their objectives.

Community-based collaboration: An approach to problem solving that involves individuals and representatives of affected groups, businesses, and other agencies in addressing a specific or short-term problem defined by the local community. Like natural resource partnerships, collaborative groups are usually voluntary associations without legal sanction or regulatory powers.

Compromise: An approach to problem solving in which participants work out a solution that satisfies each person's minimum criteria but may not fully satisfy all.

Condensation symbol: Graber (1976) defined a condensation symbol as a word or phrase that "stirs vivid impressions involving the listener's most basic values" (p. 289); political scientist Murray Edelman (1964) stressed the ability of such symbols to "condense into one symbolic event or sign" powerful emotions, memories, or anxieties (p. 6).

Consensus: Usually means *general agreement* or that discussions will not end until everyone has had a chance to share their differences and find common ground.

Conservation: The term used by early 20th-century forester Gifford Pinchot to mean the wise and efficient use of natural resources.

Constitutive: Communication about nature also helps us construct or compose representations of nature and of environmental problems as subjects for our understanding. Such communication invites a particular perspective, evokes certain values (and not others), and thus creates conscious referents for our attention and understanding.

Constraints: the cultural limitations and possibilities of the context, one of the three terms of the rhetorical situation.

Content flood: The sheer volume of text, images, data, and other information produced via social media and the Internet.

Cost-benefit analysis: A type of economic evaluation of technical risk in which both the disadvantages (costs) and advantages (benefits) of an investment, purchase, or process are weighed in financial, supposedly neutral, mathematical ways.

Crisis discipline: Term used to characterize the new discipline of conservation biology; coined by biologist Michael Soulé (1985) to refer to the duty of scientists, in the face of a looming biodiversity crisis, to offer recommendations to address this worsening situation, even with imperfect knowledge.

Critical discourse: Modes of representation that challenge society's taken-for-granted assumptions and offer alternatives to prevailing discourses.

Critical rhetoric: The questioning or criticism of a behavior, policy, societal value, or ideology; may also include the articulation of an alternative policy, vision, or ideology.

Cruel irony: Those who have contributed the least to the energy policies and demands that cause climate change and have profited the least from these industries have been and will continue to be impacted the most by climate change.

Cultivation analysis: Associated with the work of media scholar George Gerbner (1990), the theory that repeated exposure to a set of messages tends to produce, in an audience, agreement with the views contained in those messages.

Cultivation in reverse: The media's cultivation of an anti-environmental attitude through the persistent lack of environmental images or by directing the attention of viewers and readers to other, non-environmental stories.

Cultural model of risk communication: An approach that involves the affected public in assessing risk and in designing risk communication campaigns and that recognizes cultural knowledge and the experience of local communities.

Cultural rationality: In Plough and Krimsky's (1987) view, a basis for risk evaluation that includes personal, familiar, and social concerns; a source of judgment that arises when the social context and experience of those exposed to environmental dangers enter definitions of risk.

Cultural theory of risk assessment: Rejects an individualist, rationalist notion of risk in favor of one that believes our perceptions of risk are informed by cultural values.

Data visualizations: Illustrates numerical data through an image.

Decision space: An opportunity for collaboration to influence the final decision; refers to what decisions are open to the participants' influence, what is "on the table" for discussion and what is not.

Decorum: One of the virtues of style in the classical Greek and Latin rhetorical handbooks; usually translated as *propriety* or *that which is fitting* for the particular audience and occasion.

Deep Ecology movement: Affirms that "the well-being and flourishing of human and nonhuman life on Earth have value in themselves (synonyms: inherent worth, intrinsic value, inherent value). These values are independent of the usefulness of the nonhuman world for human purposes" (Foundation for Deep Ecology, 2012, para. 1).

Delhi Climate Justice Declaration: Final declaration of Climate Justice Summit, New Delhi, 2002, which declared, "Climate change is a human-rights issue"; also resolved "to actively build a movement from the communities" to address climate change from a social justice perspective.

Dialogic communication: Two-way communication.

Digital technologies: Communication media that are based on computational devices (blogs, social media, mobile devices, etc.).

Digitally-mediated social networks: "[I]nteractive and self-configurable" sites that enable individuals and groups to share their "outrage and hope" (Castells, 2012), initiating projects and coordinating with others to create change.

Direct action: Physical acts of protest such as road blockades, sit-ins, and tree spiking.

Discourse: A pattern of knowledge and power communicated through linguistic and nonlinguistic human expression; discourse functions to circulate a coherent set of meanings about an important topic.

Discourse of the free market: Usually, the absence of governmental restriction on business or commercial activity; the belief that the private marketplace is self-regulating and ultimately promotes social good.

Disparate impact: Term used to denote the discrimination resulting from environmental hazards in minority communities; adopted from the 1964 Civil Rights Act, which used it to recognize forms of discrimination that result from the disproportionate burdens experienced by some groups regardless of the conscious intention of others in their decisions or behaviors.

Dissensus: Term coined by communication scholar Thomas Goodnight (1991), meaning a questioning of, refusal of, or disagreement with a claim or a premise of a speaker's argument.

Dominant discourse: A discourse that has gained broad or taken-for-granted status in a culture; for example, the belief that growth is good for the economy; its meanings help to legitimize certain policies or practices.

Dominant Social Paradigm (DSP): A dominant discursive tradition of several centuries that has sustained attitudes of human dominance over nature. The DSP affirms society's belief in economic growth and its faith in technology, limited government, and private property.

Earth Day (1970): Twenty million people took part in protests, teach-ins, and festivals throughout the country, in one of the largest demonstrations in American history.

Eco-label certification programs: Certified products reflect an independent group's assurance to consumers that the product is environmentally friendly or produced in a manner that did not harm the environment.

Ecology: From the Greek ("house"), ecology is the scientific study of interactions among biological organisms and their environments; in the 1970s, the term was use by the media, students, and others commonly to refer to environmental concerns generally.

Economic blackmail: A false choice presented between financial worth and environmental protection; it deflects attention away from the fact that jobs can be provided while meeting basic health and environmental standards.

Electronic Freedom of Information Amendments: Amendments to the Freedom of Information Act (FOIA) that require federal agencies to provide public access to information in electronic form. This is done typically by posting a guide for making a freedom of information request on the agency's website.

Emergency Planning and Community Right to Know Act (Right to Know Act): Enacted in 1986, this act requires industries to report to local and state emergency planners the use and location of specified chemicals at their facilities, as well as emergency notification procedures.

Encoding and decoding: Any medium is both encoding, or created with a message, and decoding, or interpreted by audiences or receivers.

Environmental art: Usually references two types of artists: (1) artists who foreground particular materials in their work—such as sticks, stones, flowers, and mud—and is displayed in the outdoors; and (2) and artists who aspire to communicate about environment problems through their work.

Environmental communication: The pragmatic and constitutive vehicle for our understanding of the environment as well as our relationships to the natural world; the symbolic medium that we use in constructing environmental problems and in negotiating society's different responses to them.

Environmental impact statement (EIS): Required by the National Environmental Policy Act (NEPA) for proposed federal legislation or actions significantly affecting the quality of the environment; an EIS must describe (a) the environmental impact of the proposed action, (b) any adverse environmental effects that could not be avoided should the proposal be implemented, and (c) alternatives to the proposed action.

Environmental justice: As used by community activists and scholars studying the environmental justice movement, the term refers to (a) calls to recognize and halt the disproportionate burdens imposed on poor and minority communities by environmentally harmful conditions, (b) more inclusive opportunities for those who are most affected to be heard in the decisions made by public agencies and the wider environmental movement, and (c) a vision of environmentally healthy, economically sustainable, and culturally thriving communities.

Environmental melodrama: A genre used to clarify issues of power and the ways advocates "moralize" an environmental conflict. As a genre, melodrama "generates stark, polarizing distinctions between social actors and infuses those distinctions with moral gravity and pathos," and is therefore "a powerful resource for rhetorical invention" (Schwarze, 2006, p. 239).

Environmental news services (ENSs): Online platforms offering access to both working journalists and readers looking for more in-depth environmental news and timely information.

Environmental racism (and more broadly, environmental injustice): Refers not only to threats to communities' health from hazardous waste landfills, incinerators, agricultural pesticides, sweatshops, and polluting factories but also the disproportionate burden that these practices placed on people of color and the workers and residents of low-income communities. See also **Disparate impact**.

Environmental skepticism: An attitude that disputes the seriousness of environmental problems and questions the credibility of environmental science.

Environmental tort: A legal claim for injury or a lawsuit.

Exceptionalism: The view that, because a region has unique or distinctive features, it is exempt from the general rule. Some critics are concerned that place-based decisions reached at the local level in one area can become a precedent for exempting other geographical areas and thus compromise more uniform, national standards for environmental policy.

Executive Order 12898 on Environmental Justice: Issued by President Clinton in 1994, Executive Order 12898, titled "Federal Actions to Address Environmental Justice in Minority Populations and Low-Income Populations," instructed each federal agency "to make achieving environmental justice part of its mission by identifying and addressing . . . disproportionately high and adverse human health or environmental effects of its programs, policies, and activities on minority populations and low-income populations in the United States" (Clinton, 1994, p. 7629).

Exigency: a set of conditions that have been constituted as a "problem," grievance, or crisis that becomes marked by a sense of urgency; one of the three terms of the rhetorical situation.

Experiential knowledge: That which has been learned through direct experience.

Factory farming: The practice of large-scale, industrial agriculture known for severely restricting the movements or space in which chickens, pigs, and cows are kept and/or reproduce.

First National People of Color Environmental Leadership Summit: A key moment in the new movement for environmental justice, when delegates from local communities and national leaders from social justice, religious, environmental, and civil rights groups met in Washington, DC, in October 1991.

Food sovereignty: the right of everyday people, farmworker unions, and sovereign countries to public participation in agricultural and food policy.

Forensic rhetoric: Aristotle's name for the practice of ordinary citizens' speaking before a judge to argue for their rights under Athenian law.

Four-step procedure for risk assessment: Procedure used by agencies to evaluate risk in a technical sense; the four steps are (1) hazard identification, (2) assessment of human exposure, (3) modeling of the dose responses, and (4) a characterization of the overall risk.

Fracking: See **Hydraulic fracturing.**

Frames: First defined by Erving Goffman (1974) to refer to the cognitive maps or patterns of interpretation that people use to organize their understanding of reality. See also **Media frames.**

Freedom of Information Act (FOIA): Enacted in 1966, this act provides that any person has the right to see documents and records of any federal agency (except the judiciary or Congress).

Friends of the Earth, Inc., v. Laidlaw Environmental Services, Inc.: A 2000 case in which the Supreme Court reversed its strict *Lujan* doctrine, ruling that plaintiffs did not need to prove an actual (particular) harm; rather, the knowledge of a possible threat to a legally recognized interest (clean water) was enough to establish a sufficient stake in enforcing the law.

Gamification: The encouragement of something through play, often involving point scoring, competition, and rules of play.

Gatekeeping: The role of editors and media managers in deciding to cover or not cover certain news stories; a metaphor used to suggest that individuals in newsrooms decide what gets in and what stays out.

Goal (of a campaign): Describes a long-term vision or value, such as protection of old-growth forests, reduction of arsenic in drinking water, or making economic globalization more democratic.

Green consumerism: Marketing that encourages the belief that, by buying allegedly environmentally friendly products, consumers can do their part to protect the planet.

Green marketing: A corporation's attempt to associate its products, services, or identity with environmental values and images; generally used for (a) product promotion (sales), (b) image enhancement, or (c) image repair. Recently defined to include communication about environmentally beneficial product modifications.

Green product advertising: The attempt to sell commodities "that are presumed to be environmentally safe" for retail, "designed to minimize negative effects on the physical environment or to improve its quality," and an effort "to produce, promote, package, and reclaim products in a manner that is sensitive or responsive to ecological concerns" (American Marketing Association, 2014).

Greenwashing: An attempt to promote the appearances of products and commodity consumption as environmental or "green," while deliberately disavowing environmental impacts.

Groupthink: Term coined by psychologist Irving Janis (1977) referring to an excessive cohesion in groups that impedes critical or independent thinking, often resulting in uninformed consensus.

Hazard: In Sandman's (1987) model of risk, or what experts mean by risk (that is, expected annual mortality). See **Outrage**.

Hydraulic fracturing or **fracking:** A method used in drilling for natural gas that involves injecting large volumes of water and chemicals under high pressure into rock or shale strata to create fissures, which releases the trapped gas.

Hypermediacy: A heterogeneous space in which representations of events "open on to other representations or other media" in which mediations themselves are multiplied (Bolter & Grusin, 1999).

Image enhancement: The use of PR (public relations) to improve the brand or ethos of the corporation itself by associating it with positive environmental messages, practices, and products.

Image events: Actions by environmentalists that stage visual events to take advantage of news media's desire for pictures, particularly images of conflict.

Image repair: (also called *crisis management*): The use of public relations (PR) to restore a company's credibility after an environmental harm or accident.

Indecorous voice: One is deemed as inappropriate or unqualified for speaking in official forums, which is based in the assumption that ordinary people may be too emotional or ignorant to testify about chemical pollution or other environmental issues.

Infographics: Visual interpretations of information or data such as a table or chart that explain something (like a report, concept, or set of connections).

Information deficit model: An assumption in early science communication that providing more information or facts will change attitudes.

Injury in fact: Under common law, this normally meant a concrete, particular injury that an individual had suffered due to the actions of another party. Currently, it is one of three tests used by U.S. courts to determine a plaintiff's standing or right to seek redress in court for a harm to a legally protected right; criteria for defining injury in fact have varied from the denial of enjoyment or use of the environment to a concrete, tangible harm to the plaintiff.

Interactive Maps: Two-way electronic communications that synthesize complex data through a cartographic representation and respond to user activity.

Intergenerational justice: The principle of fairness, or justice, in relationships, not only between children and adults, but between the present and future (unborn) generations.

Intermediation: How various media interact with each other and people.

Irony: The use of language that is the opposite of one's belief, often for humor.

Irreparable: A forewarning or opportunity to act before it is too late to preserve what is unique or rare before it is lost forever. Cox (1982, 2001a) identified the four characteristics of an appeal to the irreparable nature of a decision or its consequences: A speaker establishes that (a) the decision threatens something unique or rare and thus of great value, (b) the existence of what is threatened is precarious and uncertain, (c) its loss or destruction cannot be reversed, and (d) action to protect it is therefore timely or urgent.

Jeremiad: Originally named for the lamentations of the Hebrew prophet Jeremiah, jeremiad refers to speech or writing that laments or denounces the behavior of a people or society and warns of future consequences if society does not change its ways.

Land ethic: The pioneering ecologist Aldo Leopold's principle of respect for the natural world. The problem, he wrote, is that,

"Obligations have no meaning without conscience, and the problem we face is *the extension of the social conscience from people to land*" (Leopold, 1949/1966, p. 246; emph. added).

Legitimizers: Sources such as official spokespersons and experts who, presumably, bring authority or credibility to news about risk.

Leverage: From Archimedes' famous claim, "Give me a place to stand and a lever long enough, and I will move the world," that is, the application of a certain kind of action (assuming "a place to stand") produces a dynamic that can move—or leverage—a much larger force.

Life-cycle-assessment (LCA): A way of studying how a product's "life" develops from cradle to grave. A LCA might include raw material extraction, production, distribution, use, and disposal.

Lujan v. Defenders of Wildlife: A 1992 case in which the Supreme Court rejected a claim of standing by the group Defenders of Wildlife under the citizen suit provision of the Endangered Species Act (ESA), ruling that the Defenders had failed to satisfy constitutional requirements for injury in fact because plaintiffs had not suffered a tangible and particular harm. It overturned the more liberal standard established in *Sierra Club v. Morton.*

Mainstreaming: An alleged effect in consistent viewers of media whereby differences are narrowed toward cultural norms represented in media programs.

Massachusetts et al. v. Environmental Protection Agency et al.: The U.S. Supreme Court's first-ever ruling on global warming in which a state (Massachusetts) had "standing," and could successfully argue that the EPA had the authority, under the Clean Air Act, to regulate carbon dioxide and other greenhouse gases as "air pollutants."

Material rhetoric: The interpretation of how humans and the physical world (including buildings and bodies) constitute meaning.

Mean world syndrome: A view of society as a dangerous place, peopled by others who want to harm us.

Media effects: The influence of different media content, frequency, and forms of communication on audiences' attitudes, perceptions, and behaviors.

Media frames: The central organizing themes that connect different semantic elements of a news story (headlines, quotes, leads, visual representations, and narrative structure) into a coherent whole to suggest what is at issue. See also **Frames.**

Media political economy: The influence on news content of ownership and economic interests of the owners of news stations and television networks.

Mediation: A facilitated effort, entered into voluntarily or at the suggestion of a court, counselors, or other institution, that involves an active mediator who helps the disputing parties find common ground and a solution upon which they can agree.

Meme: An idea, phrase, or image that is widely shared in a culture.

Message: A phrase or sentence that concisely expresses a campaign's objective and the values at stake in the decision of the primary audience. Although campaigns develop considerable information and arguments, the message itself is usually short, compelling, and memorable and accompanies all of a campaign's communication materials.

Metaphor: One of the major tropes; *Mother Nature, Spaceship Earth, population bomb,* and the *web of life* are just a few examples. A metaphor's function is to invite a comparison by talking about one thing in terms of another.

Modern society: A historical period that can be defined by the rise of industrial capitalism, the increased significance of the nation-state, and the scientific revolution.

Mountaintop removal coal mining: The removal of the tops of mountains in the Appalachians to expose seams of coal buried in the mountain; a particularly environmentally destructive form of mining.

Multimodality: The multiple and intersecting forms of networked relationships that are developed in sustaining a movement.

Naming: The mode by which we socially represent objects or people and know the world, including the natural world.

Narrative framing: Media's organization of phenomena through stories to aid audiences' understanding.

National Environmental Justice Advisory Council (NEJAC): A federal advisory committee in the Environmental Protection Agency (EPA) that is intended to provide the EPA administrator with independent advice, consultation, and recommendations related to environmental justice.

National Environmental Policy Act (NEPA): Requires every federal agency to prepare an environmental impact statement and invite public comment on any project that would affect the environment. Signed into law by President Richard M. Nixon on January 1, 1970, NEPA is the cornerstone of modern environmental law.

Natural resource partnerships: Informal working groups organized around regions with natural resource concerns such as the uses of rangelands and forests as well as protection of wildlife and watersheds. Partnerships operate collaboratively to integrate their differing values and approaches to the management of natural resource issues.

Nature: The physical world that generally exceeds human creation (trees, birds, bears, clouds, rainbows, oceans, seashells, and so forth).

Neurotoxin: A chemical substance that can cause adverse effects in the brain and nervous system.

News hole: The amount of space that is available for a news story relative to other demands for the same space.

Newsworthiness: The ability of news stories to attract readers or viewers; often defined by such criteria for selecting and reporting environmental news as prominence, timeliness, proximity, impact, magnitude, conflict, oddity, and emotional impact.

Notice of Intent (NOI): A statement of an agency's intent to prepare an EIS for a proposed action. The NOI is published in the

Federal Register and provides a brief description of the proposed action and possible alternatives.

Objective (of a campaign): A specific action or decision that moves a group closer to a broader goal; a concrete and time-limited decision or action.

Objectivity and balance: Norms of journalism for almost a century, the commitment to which is made by news media to provide information that is accurate and without reporter bias and, where there is uncertainty or controversy, to balance news stories with statements from all sides of the issue.

Opinion leaders: Those persons whose statements are influential with the media and members of the primary audience.

Opponents: Those potential audience members who strongly disagree and are unlikely to be persuaded by an advocacy campaign.

Outrage: In Sandman's (1987) model, a term for factors the public considers in assessing the acceptability of their exposure to a hazard. See **Hazard**.

Participatory communication: Defined by Arvind Singhal (2001) as, "a dynamic, interactional, and transformative process of dialogue between people, groups, and institutions that enables people, both individually and collectively, to realize their full potential and be engaged in their own welfare."

Persuadables: Members of the public who are undecided but potentially sympathetic to a campaign's objectives; they often become primary targets in mobilizing support.

Place-based collaboration: A collaborative process involving stakeholders *who live in, and have an interest or "stake" in, the place* (local community or region) where a conflict occurs.

Plaintiff: One who brings a civil action in a court of law.

Plan EJ 2014: A set of strategies of the U.S. EPA to recommit and to reinvigorate environmental justice efforts through legal, scientific, information, and resource development and communication.

Pollutant Release and Transfer Register (PRTR): Like the U.S. Toxic Release Inventory (TRI), international PRTR programs require mandatory reporting of specific chemical releases into the air, water, or land and public access to these data.

Pragmatic: Instrumental; a characteristic of environmental communication whereby it educates, alerts, persuades, mobilizes, and helps to solve environmental problems.

Praxis: The study of efficient action or the best means to achieve an objective. In communication studies, we further emphasize the importance of critical theory to this study and define *praxis* as an ongoing process of critical theoretical reflection and embodied action in which the two inform each other in order to provide further insight and practice that may improve the world within specific contexts.

Precautionary principle: As defined by the 1998 Wingspread conference, "When an activity raises threats of harm to human health or the environment, precautionary measures should be taken even if some cause and effect relationships are not fully established scientifically. In this context the proponent of an activity, rather than the public, should bear the burden of proof" (Science and Environmental Health Network, 1998, para. 5).

Preservation: The movement to protect from harm and to maintain certain places as wilderness.

Primary audience: Decision makers who have the authority to act or implement the objectives of a campaign.

Principles of Environmental Justice: Principles adopted by delegates at the First National People of Color Environmental Leadership Summit in 1991 that enumerated a series of rights, including "the fundamental right to political, economic, cultural, and environmental self-determination of all peoples."

Productivist discourse: Discourse supporting "an expansionistic, growth-oriented ethic" (Smith, 1998).

Progressive ideal: Put forth by the 1920s' and 1930s' Progressive movement, the concept of a neutral, science-based policy as the best approach to government regulation of industry.

Public comment: Required of federal agencies under the NEPA, public input must be solicited by a federal agency on any proposal significantly affecting the environment; usually takes place at public hearings and in written reports, letters, e-mails, or faxes to the agency.

Public hearing: The common mode of participation by ordinary citizens in environmental decision making at both the federal and state levels; a forum for public comment to an agency before the agency takes any action that might significantly impact the environment.

Public participation: The ability of individual citizens and groups to influence environmental decisions through (a) the right to know or access to relevant information, (b) public comments to the agency that is responsible for a decision, and (c) the right, through the courts, to hold public agencies and businesses accountable for their environmental decisions and behaviors.

Public sphere: The forums and interactions in which different individuals engage each other in conversation, argument, debate, and questions about subjects of shared concern or that affect a wider community.

Public trust doctrine: Derived from common law, this is the "principle that certain natural and cultural resources are preserved for public use, and that the government owns and must protect and maintain these resources for the public's use" (Nolo, 2014, para. 1).

Public will: The desire of many people for some specific change; once identified and marshaled, the public will can be a powerful, rhetorical force in society for legitimizing or infusing an action with symbolic legitimacy.

Public will campaigns: "Organized, strategic initiatives designed to legitimize and garner [mobilize] public support . . . as a mechanism of achieving . . . change" (Salmon et al., 200, p. 4).

Quincy Library Group (QLG): A high-profile effort by a local community to develop a consensus approach for managing national forest lands in northern California. The effort ended by moving in a different, more adversarial direction, a move that appears to have undercut its initial goals.

Remediation: A process where older media (such as a 35 mm photo or film) are re-presented or refashioned in a new medium (for example, in a web video).

Resilience: An organism's ability to adapt and to persist at the same time.

Rhetoric: The faculty (power) of discovering the available means of persuasion in the particular case.

Rhetorical genres: Distinct forms or types of speech that share characteristics distinguishing them from other types of speech.

Rhetorical perspective: A focus on purposeful and consequential efforts to influence society's attitudes and ways of behaving through communication, which includes public debate, protests, news stories, advertising, and other modes of symbolic action.

Rhetorical situation: Defined by three terms: (1) *exigency,* a set of conditions that have been constituted as a "problem," grievance, or crisis that becomes marked by a sense of urgency; (2) *audience,* the people being addressed, their beliefs, actions, and larger cultural understandings; and (3) *constraints,* the cultural limitations and possibilities of the context.

Right to comment: The opportunity to publicly address the agency or entity that is responsible for a decision about actual or potential harms and benefits one perceives as a result of that decision.

Right to know: The public's ability to have access to relevant information.

Right of standing: The legal status accorded a citizen who has a sufficient interest in a matter and who may speak before judicial authority (courts) to protect that interest.

Risk assessment: The evaluation of the degree of harm or danger from some condition such as exposure to a toxic chemical.

Risk communication (general): The symbolic mode of interaction that we use in identifying, defining, assessing, and negotiating environmental and public health dangers.

Risk communication (technical): The translation of technical data about environmental or health risks for public understanding, with the goal of educating a target audience.

Risk society: Term coined by German sociologist Ulrich Beck (1992) to characterize today's society according to the large-scale nature of risks and the threat of irreversible effects on human life from modernization.

Roadless Rule: Adopted by the U.S. Forest Service in 2001, the rule prohibits road building and restricts commercial logging on nearly 60 million acres of national forest lands in 39 states.

Sacrifice zones: Term coined by sociologist Robert Bullard (1993) to denote communities that share two characteristics: "(1) They already have more than their share of environmental problems and polluting industries, and (2) they are still attracting new polluters" (p. 12).

Scoping: A preliminary stage in an agency's development of a proposed rule or action, including any meetings and how the public can get involved; it involves canvassing interested members of the public to determine what the concerns of the affected parties might be.

Second National People of Color Environmental Leadership Summit: A gathering subsequent to the first Summit, was held in Washington, DC, from October 23 to 26, 2002; highlighting women's roles as leaders in the movement, the second event was even larger.

Secondary audience: Segments of the public, coalition partners, opinion leaders, and the media whose support is useful in holding decision makers accountable for the campaign's objectives; also called *public audiences.*

Self-initiating: Refers to the ability of individuals, through what are often called *bottom-up* online sites, to initiate social change actions via social media that engage others.

Seventh Generation principle: Derived from the Constitution of the Iroquois Nations, this principle refers to the desire to make decisions based on how they will impact not one generation, but seven.

Shannon–Weaver model of communication: A linear model that defines human communication as the transmission of information from a source to a receiver.

Shock and shame response: If community members found out that a local factory was emitting high levels of pollution, their shock could push the community into action. In some cases, the polluting facility itself may feel shame from disclosure of its poor performance.

Sierra Club v. Morton: A 1972 case that established the first guidance for determining standing under the Constitution's cases and controversies clause in an environmental case; the Supreme Court held that the Sierra Club need only allege an injury to its members' interests—for example, that its members could not enjoy an unspoiled wilderness or their normal recreational pursuits.

SLAPP-back: A lawsuit against the corporation bringing an initial SLAPP suit against a citizen. Defendants SLAPP-back by filing a countersuit alleging that plaintiffs infringed on a citizen's right to free speech or to petition government; a SLAPP-back suit allows for recovery of attorneys' fees as well as punitive damages for violating constitutional rights and inflicting damage or injury on the defendant (malicious prosecution).

SLAPP lawsuits: Strategic Litigation Against Public Participation—as defined by Pring and Canan (1996), a SLAPP is a lawsuit involving "communications made to influence a governmental action or outcome, which secondarily, resulted in (a) a civil complaint [lawsuit] . . . (b) filed against nongovernmental individuals or organizations . . . on (c) a substantive issue of some public interest or social significance" such as the environment (pp. 8–9).

Stakeholders: Those parties to a dispute who have a real or discernible interest (a stake) in the outcome.

Standing: See **Right of standing.**

Strategy: A critical source of influence or leverage to bring about a desired change.

Strip-mining: Removal of surface land to expose the underlying mineral seams.

Sublime: An aesthetic category that associates God's influence with the feelings of awe and exultation that some experience in the presence of wilderness.

Sublime response: Term used to denote (a) the immediate awareness of a sublime object (such as Yosemite Valley), (b) a sense of overwhelming personal insignificance and awe in its presence, and (c) ultimately, a feeling of spiritual exaltation.

Sunshine laws: Laws intended to shine the light of public scrutiny on the workings of government, requiring open meetings of most governmental bodies.

Superfund: Legislation enacted in 1980 authorizing the EPA to clean up toxic sites and hold the responsible parties accountable for the costs.

Superfund sites: Abandoned chemical waste sites that have qualified for federal funds for their cleanup under the Comprehensive Environmental Response Compensation and Liability Act (commonly called the Superfund law).

Sustainability: The capacity to negotiate environmental, social, and economic needs and desires for current and future generations.

Sustainability curriculum: Distinguished from an environmental curriculum by engaging all three facets of sustainability—the Three Es (Environmental protection, Economic health, and Equity) or the Three Ps (People, Prosperity, and the Planet).

Sustainability gap: "In almost all areas of sustainability, we know scientifically what we need to do and how to do it; but we are just not doing it" (Agyeman, 2005, p. 40).

Sustainable infrastructure: The physical elements needed to allow an organization to operate (including everything from the chair you sit on in class to the light bulbs used in the stairwells to the energy source used to heat water in dorms).

Sustainable self-representation: Ability of digital media to create and sustain an identity that avoids "both the fickleness of changing news agendas, the vicissitudes of reporting and editorial practices, and the contending corporate interests of large-scale news conglomerates" (p. 591).

Symbolic action: The property of language and other acts to do something as well as literally to say something; to create meaning and orient us consciously to the world.

Symbolic annihilation: Media's erasure of the importance of a theme by the indirect or passive de-emphasizing of that theme.

Symbolic legitimacy: The perceived authority or credibility of a source of knowledge, such as scientists.

Symbolic perspective: Focuses on the cultural sources that construct our perceptions of the world.

Synecdoche: the part standing for the whole, as in references simply to melting glaciers to signal the wider impacts caused by global warming.

Tactics: Specific actions—alerts, meetings, protests, and so forth—that implement a broader strategy.

Technical approach to risk assessment: The expected annual mortality (or other severity) that results from some exposure.

Technical Assistance Grant (TAG) Program: A program initiated in 1986 to help communities at Superfund sites by providing funds for citizen groups to hire consultants who can help them understand and comment on information provided by the EPA and the industries responsible for cleaning these sites.

Technical model of risk communication: The translation of technical data about environmental or human health risks for public understanding, with the goal of educating a target audience.

Technocracy: John Dewey's term denoting a government ruled by experts.

Terministic screens: The means whereby language orients us to see certain things—some aspects of the world and not others. Defined by literary theorist Kenneth Burke (1966) to mean "if any given terminology is a *reflection* of reality; by its very nature as a terminology it must be a *selection* of reality; and to this extent it must function also as a *deflection* of reality" (p. 45).

The Progress Triangle: Portrays environmental conflict management as involving three interrelated dimensions—substantive, procedural, and relationship.

Three Es and Three Ps: Sustainability is often described as the **Three Es** of Environmental protection, Economic health, and Equity (social justice) or the **Three Ps:** People, Prosperity, and the Planet.

Toxic Release Inventory (TRI): An information-reporting tool established under the Emergency Planning and Community Right to Know Act (1986) that enables the EPA to collect data annually on any releases of toxic materials into the air and water by designated industries and to make this information easily available to the public.

Toxic tours: "Non-commercial expeditions organized and facilitated by people who reside in areas that are polluted by toxics, places that Bullard (1993) has named 'human sacrifice zones' . . . Residents of these areas guide outsiders, or tourists, through where they [residents] live, work, and play in order to witness their struggle" (Pezzullo, 2004, p. 236). See also **Sacrifice zones.**

Traditional news media: Newspapers, news magazines, and television and radio news programs, as opposed to online and other digital media.

Transcendentalism: Belief that a correspondence exists between a higher realm of spiritual truth and a lower one of material objects, including nature.

Transparency: Openness in government; citizens' right to know information that is important to their lives. In regard to the environment, the United Nations has declared that the principle of transparency "requires the recognition of the rights of participation and access to information and the right to be informed. . . . Everyone has the right of access to information on the environment with no obligation to prove a particular interest" (Declaration of Bizkaia on the Right to the Environment, 1999).

Tree spiking: The practice of driving metal or plastic spikes or nails into trees in an area that is scheduled to be logged to discourage the cutting of the trees.

Trope of uncertainty: An appeal that functions to nurture doubt in the public's perception of scientific claims and thereby to delay calls for action; in rhetorical terms, the trope of uncertainty turns, or alters, the public's understanding of what is at stake, suggesting there is a danger in acting prematurely, a risk of making the wrong decision.

Tropes: Words that turn a meaning from its original sense in a new direction.

Utilitarianism: Theory that the aim of action should be the greatest good for the greatest number.

Viral media: A metaphor that suggests how communication spreads digitally in random patterns and exponentially.

Visual rhetoric: Visual images function both pragmatically—to persuade—and constitutively, to construct or challenge a particular "seeing" of nature or what constitutes an "environmental problem."

Voices of the "side effects": Term used by Beck (1992) to refer to those individuals (or their children) who suffer the side effects of the risk society, such as asthma and other illnesses from air pollutants, chemical contamination, and so forth.

Wilderness: A "wilderness, in contrast with those areas where man [sic] and his own works dominate the landscape, is hereby recognized as an area where the earth and its community of life are untrammeled by man, where man himself is a visitor who does not remain" (1964 Wilderness Act).

Wireframe: A way to visualize the structure, function, and navigational steps of a website.

Wise Use groups: Groups that organize individuals who oppose restrictions on the use of their own (private) property for purposes such as protection of wetlands or habitat for endangered species; also called *property rights groups.*

Witnessing: Traditionally defined as an act of hearing and seeing oral and written evidence through firsthand experience.

References

Aarhus Parties commit to strengthening environmental democracy in the UNECE region and beyond. (2008, June 11–13). UN Economic Commission for Europe. Retrieved from http://www.unece.org

Abraham, J. (2014, April 9). Years of living dangerously—a global warming blockbuster. *The Guardian*. Retrieved from http://www.theguardian.com/environment/climate-consensus-97-per-cent/2014/apr/09/years-of-living-dangerously-global-warming-blockbuster

Abramson, R. (1992, December 2). Ice cores may hold clues to weather 200,000 years ago. *Los Angeles Times*, p. A1.

ACCCE. (2008, April 16). *I believe*. [TV ad]. Retrieved from http://www.youtube.com

Adam, D., Walker, P., & Benjamin, A. (2007, September 18). Grim outlook for poor countries in climate report. Retrieved from http://www.guardian.co.uk/

Ader, C. R. (1995). A longitudinal study of agenda setting for the issue of environmental pollution. *Journalism and Mass Communication Quarterly, 72,* 300–311.

Adler, J. H. (2000, March 2–3). *Stand or deliver: Citizen suits, standing, and environmental protection*. Paper presented at the Duke University Law and Policy Forum Symposium on Citizen Suits and the Future of Standing in the 21st Century. Retrieved from http://www.law.duke.edu/journals

Aerial wolf hunting on Kenai Peninsula put on hold. (2012, May 7). *Alaska Dispatch*. Retrieved from http://www.alaskadispatch.com/article/aerial-wolf-hunting-kenai-peninsula-put-hold

After cold winter, fewer Americans believe in global warming. (2013, May 9). Retrieved from http:// www.upi.com/Science_News/2013/05/09/After-cold-winter-fewer-Americans-believe-in-global-warming/UPI-43441368125124

Agyeman, J. (2005). *Sustainable communities and the challenge of environmental justice*. Cambridge, MA: MIT Press.

Agyeman, J., Doppelt, B., & Lynn, K. (2007). The climate-justice link: Communicating risk with low-income and minority audiences. In S. C. Moser & L. Dilling (Eds.), *Communicating a climate for change: Communicating climate change and facilitating social change* (pp. 119–138). Cambridge, UK: Cambridge University Press.

Allan, S., Adam, B., & Carter, C. (Eds.). (2000). *Environmental risks and the media*. London, England, & New York, NY: Routledge.

Allan, S., & Ewart, J. (2015). Citizen science/ citizen journalism: New forms of environmental reporting. In A. Hansen & R. Cox (Eds.). *Routledge handbook of environment and communication*. London, England: Routledge.

Alston, D. (1990). *We speak for ourselves: Social justice, race, and environment.* Washington, DC: Panos Institute.

Alter, C. (2014, September 21). Hundreds of thousands converge on New York to demand climate-change action. *Time.com.* Retrieved from http://time.com/3415162/peoples-climate-march-new-york-manhattan-demonstration/

American College Personnel Association. (n.d.). ACPA sustainability task force student learning outcomes assessment materials guidebook. Washington, DC: National Center for Higher Education. Retrieved from http://www.acpa.nche.edu/sites/default/files/ACPA_Sustainability.pdf

American Marketing Association. (2014). *Dictionary.* Retrieved from https://www.ama.org/resources/Pages/Dictionary.aspx?dLetter=G

America's Power. (2009). *Ad archive.* Retrieved from http://www.americaspower.org

Anderson, A. (2015). News organization(s) and the production of environmental news. In A. Hansen & R. Cox (Eds.). *Routledge handbook of environment and communication.* London, England: Routledge.

Andrews, R. N. L. (2006). *Managing the environment, managing ourselves: A history of American environmental policy* (2nd ed.). New Haven, CT: Yale University Press.

Antibiotics and agriculture. (2010, June 30). *New York Times,* p. A24.

Antonetta, S. (2012, March/April). Beehive collective: Metaphor crafters. *Orion Magazine.* Retrieved from http://www.orionmagazine.org/index.php/articles/article/6695

Archibald, R. (2014, Winter). Photojournalism upheaval heralds multimedia's rise. *SEJournal,* pp. 20–21.

Armstrong, P. (2009, July 30). Conflict resolution and British Columbia's Great Bear Rainforest: Lessons learned 1995–2009. Retrieved from http://www.coastforestconservationinitiative.com/

As arctic sea ice melts, experts expect new low. (2008, August 28). *New York Times,* p. A16.

Atwood, M. (2013, November 12). We must tackle climate change together. *Huffington Post.* Retrieved from http://www.huffingtonpost.ca/margaret-atwood/atwood-climate-change_b_4256145.html

Augustine, R. M. (Speaker). (1991). *Documentary highlights of the First National People of Color Environmental Leadership Conference* [Videotape]. Washington, DC: United Church of Christ Commission for Racial Justice.

Augustine, R. M. (Speaker). (1993, October 21–24). *Environmental justice: Continuing the dialogue* [Cassette recording]. Recorded at the Third Annual Meeting of the Society of Environmental Journalists, Durham, NC.

Austin, A. (2002). Advancing accumulation and managing its discontents: The US anti-environmental movement. *Sociological spectrum, 22,* 71–105.

Avaaz.org. (2014, October). Saving the Maasai's land. Retrieved from http://www.avaaz.org/en/highlights.php

Babbitt, B. (1995, December 13). *Between the flood and the rainbow.* Retrieved from http://www.fs.fed.us/eco/eco-watch

Bailey, A., Giangola, L., & Boykoff, M. T. (2014). How grammatical choice shapes media representations of climate (un)certainty. *Environmental Communication, 8*(2), 197–215.

Bailey, R. (2002, August 14). *Starvation a by-product of looming trade war.* Cato Institute. Retrieved from http://www.cato.org

Bali Principles of Climate Justice. (2002). Retrieved from http://www.ejnet.org

Banerjee, S. (2003). *Arctic National Wildlife Refuge: Seasons of life and land.* Seattle, WA: Mountaineer Books.

Barker, R. (2010, May 2). Idaho at forefront of collaboration on public land use. *Idaho Statesman.* Retrieved from http://www.idahostatesman.com/

Barnes, R. (2014, June 23). Supreme Court: EPA can regulate greenhouse gas emissions, with some limits. *Washington Post*. Retrieved from http://www.washingtonpost.com/politics/supreme-court-limits-epas-ability-to-regulate-greenhouse-gas-emissions/2014/06/23/c56fc194-f1b1-11e3-914c-1fbd0614e2d4_story.html

Baron, R. S. (2005). So right it's wrong: Groupthink and the ubiquitous nature of polarized group decision making. In M. P. Zanna (Ed.), *Advances in experimental social psychology* (Vol. 37, pp. 219–253). San Diego, CA. Elsevier Academic.

Bartlett, P. F., & Chase, G. W. (2004). Introduction. In P. F. Bartlett & G. W. Chase (Eds.), *Sustainability on campus: Stories and strategies for change* (pp.1-28). Cambridge, MA: MIT Press.

Baum, L. (2012). It's not easy being green ... or is it? A content analysis of environmental claims in magazine advertisements from the United States and United Kingdom. *Environmental Communication: A Journal of Nature and Culture, 6*(4), 423–440.

Beck, G. (2011, February 1). *Al Gore blames blizzards on ... global warming?* Retrieved from http://www.glennbeck.com

Beck, U. (1992). *Risk society: Towards a new modernity*. Newbury Park, CA: Sage.

Beck, U. (1998). Politics of risk society. In J. Franklin (Ed.), *The politics of risk society* (pp. 9–22). London: Polity.

Beck, U. (2009). *World at risk*. Cambridge, UK: Polity.

Beder, S. (2002). *Global spin: The corporate assault on environmentalism* (Rev. ed.). White River Junction, VT: Chelsea Green.

Begley, S. (2010, March 29). Their own worst enemies: Why scientists are losing the PR wars. *Newsweek*, p. 20.

Beierle, T. C., & Cayford, J. (2002). *Democracy in practice: Public participation in environmental decisions*. Washington, DC: Resources for the Future.

Belsky, J. (2011a). Stop featuring shark as food. *Change.org*. Retrieved from http://www.change.org/p/food-network-stop-featuring-shark-as-food

Belsky, J. (2011b). Victory. *Change.org*. Retrieved from http://www.change.org/p/food-network-stop-featuring-shark-as-food

Benzoni, F. (2008). Environmental standing: Who determines the value of other life? *Duke University Law & Policy Forum, 18*, 347–370.

Bercovitch, S. (1978). *The American jeremiad*. Madison: University of Wisconsin Press.

Berg, P., Baltimore, D., Boyer, H. W., Cohen, S. N., & Davis, R. W. (1974). Potential biohazards of recombinant DNA molecules. *Science, 185*, 303.

Besley, J. C., & Shanahan, J. (2004). Skepticism about media effects concerning the environment: Examining Lomborg's hypothesis. *Society and Natural Resources, 17*(10), 861–880.

Biggers, J. (2014, May 2). Rising tide on campuses: 7 Wash U students arrested over Peabody Coal trustee. *Huffington Post*. Retrieved from http://www.huffingtonpost.com/jeff-biggers/rising-tide-on-campuses-7_b_5255200.html

Bitzer, L. (1968). The rhetorical situation. *Philosophy and Rhetoric, 1*, 1–14.

Blair, C. (1999). Challenges and openings in rethinking rhetoric: Contemporary U.S. memorial sites as exemplars of rhetoric's materiality. In J. Selzer & S. Crowley (Eds.), *Rhetorical bodies: Toward a material rhetoric* (pp. 16-57). Madison: University of Wisconsin Press.

Bolter, J. D., & Grusin, R. (1999; 2000). *Remediation: Understanding new media*. Cambridge, MA: MIT Press.

Booker, C. (2009, November 28). Climate change: This is the worst scientific scandal of our generation. *The Telegraph*. Retrieved from http://www.telegraph.co.uk

Boulding, K. E. (1965, May 10). Earth as a spaceship. Address at Washington State

University. Retrieved from http://www.colorado.edu/

Bowen, M. (2008). *Censoring science: Inside the political attack on Dr. James Hansen and the truth of global warming.* New York, NY: Dutton.

Boykoff, M. T. (2007). Flogging a dead norm? Newspaper coverage of anthropogenic climate change in the United States and United Kingdom from 2003 to 2006. *Area, 39*(4), 470–481.

Boykoff, M. T., & Boykoff, J. M. (2004). Bias as balance: Global warming and the US prestige press. *Global Environmental Change, 14*(2), 125–136.

Boykoff, M. T., McNatt, M. M., & Goodman, M. K. (2015). Communicating in the anthropocene: The cultural politics of climate change news. In A. Hansen & R. Cox (Eds.). *Routledge handbook of environment and communication.* London, England: Routledge.

Boykoff, M. T., & Yulsman, T. (2013). Political economy, media, and climate change: Sinews of modern life. *Wiley Interdisciplinary News: Climate Change, 4*(5), 359–371.

BP. (2009). *Alternative energy.* Retrieved from http://www.bp.com

Brainard, C. (2008a, February 19). Dispatches from AAAS: A few thoughts on meeting's media-oriented panels. *Columbia Journalism Review.* Retrieved December 9, 2008, from http://www.cjr.org

Brainard, C. (2008b, August 27). Public opinion and climate: Part II. *Columbia Journalism Review.* Retrieved from http://www.cjr.org

Brainard, C. (2015). The changing ecology of news and news organisations: Implications for environmental news. In A. Hansen & R. Cox (Eds.). *Routledge handbook of environment and communication.* London, England: Routledge.

Braude, A. (2012, October 10). Film industry heads to solar with Avatar 2. Retrieved,from http://www.ases.org/2012/10/film-industry-heads-to-solar-with-avatar-2/

Brower, D., & Hanson, C. (1999, September 1). Logging plan deceptively marketed, sold. *San Francisco Chronicle*, p. A25.

Brown, L. R. (2000, July 25). The rise and fall of the Global Climate Coalition. *Earth Policy Institute.* Retrieved from http://www.earth-policy.org/plan_b_updates/2000/alert6

Brown, P., & Mikkelsen, E. J. (1990). *No safe place: Toxic waste, leukemia, and community action.* Berkeley: University of California Press.

Brulle, R. J., & Jenkins, J. C. (2006, March). Spinning our way to sustainability? *Organization & Environment, 19*(1), 82–87.

Brundtland Commission (1987). *Report of the World Commission on Environment and Development.* United Nations.

Brune, M. (2013, February 13). Doing the right thing. *Sierra Club blog.* Retrieved from http://sierraclub.typepad.com/michaelbrune/2013/02/keystone-xl-civil-disobedience-sierra-club-whitehouse.html

Buck, S. J. (1996). *Understanding environmental administration and law.* Washington, DC: Island Press.

Bullard, R., & Wright, B. H. (1987). Environmentalism and the politics of equity: Emergent trends in the black community. *Midwestern Review of Sociology, 12*, 21–37.

Bullard, R. D. (1993). Introduction. In R. D. Bullard (Ed.), *Confronting environmental racism: Voices from the grassroots* (pp. 7–13). Boston, MA: South End Press.

Bullard, R. D. (Ed.). (1994). *Unequal protection: Environmental justice and communities of color.* San Francisco, CA: Sierra Club Books.

Bullard, R. D. (2014). Retrieved from http://drrobertbullard.com/2014/01/01/state-of-the-environmental-justice-executive-order-after-20-years-3/

Bullard, R. D., Mohai, P., Saha, R., & Wright, B. (2007, March). *Toxic wastes and race at twenty, 1987–2007*. Cleveland, OH: United Church of Christ. Retrieved from http://www.ejrc.cau.edu

Bullying behavior. (2011, March 14). *Raleigh News & Observer*, p. 9A.

Burgess, G., & Burgess, H. (1996). *Consensus building for environmental advocates*. (Working Paper #96–1). Boulder, CO: University of Colorado Conflict Research Consortium.

Burke, K. (1966). *Language as symbolic action: Essays on life, literature, and method*. Berkeley: University of California Press.

Business Wire. (2008, November 6). *New poll data reveals 70 percent public opinion approval for coal-fueled electricity*. Retrieved from http://biz.yahoo.com

California State Environmental Resource Center. (2004). *"Eco-SLAPPs" are a frequent occurrence*. Retrieved from http://www.serconline.org

Caldwell, L. K. (1988). Environmental impact analysis (EIA): Origins, evolution, and future directions. *Policy Studies Review, 8*, 75–83.

Cameron, J. (Director). (2009). *Avatar* [Motion Picture]. United States: 20th Century Fox.

Campbell, K. K., & Huxman, S. S. (2008). *The rhetorical act* (4th ed.). Belmont, CA: Thomson Wadsworth.

Can Wal-Mart be sustainable? (2009, August 6). *New York Times*. Retrieved from http://www.nytimes.com.

Cantrill, J. G., & Oravec, C. L. (1996). Introduction. In J. G. Cantrill & C. L. Oravec (Eds.), *The symbolic earth: Discourse and our creation of the environment* (pp. 1–8). Lexington: University of Kentucky Press.

Carson, R. (1962). *Silent spring*. Boston, MA: Houghton Mifflin.

Carter Center. (2008, February 27–29). *International Conference on the Right to Public Information*. Retrieved from http://www.cartercenter.org

Carus, F. (2011, March 30). Greenpeace targets Facebook employees in clean energy campaign. Retrieved from http://www.guardian.co.uk

Carvalho, A. (2007). Ideological cultures and media discourses on scientific knowledge: Re-reading news on climate change. *Public Understanding of Science, 16*, 223–243.

Carvalho, A. (2009, September 16). Environmental communication in Europe. *Indications*. Message posted to http://indications.wordpress.com/

Castells, M. (2012). *Networks of outrage and hope: Social movements in the Internet age*. Cambridge, UK: Polity.

Center for Environmental Education. (n.d.). *Will the whales survive?* [Brochure]. Author.

Center for Health, Environment, and Justice. (2003). *Love Canal: The journey continues*. Retrieved from http://www.chej.org/

Chase, G. W., & Rowland, P. (2013). The Ponderosa project: Infusing sustainability curriculum. In P. F. Bartlett & G. W. Chase (Eds.), *Sustainability on campus: Stories and strategies for change* (pp. 91–105). Cambridge, MA: MIT Press.

Chavis, B. F., & Lee, C. (1987). *Toxic wastes and race in the United States: A national report on the racial and socio-economic characteristics of communities with hazardous waste sites*. New York, NY: Commission for Racial Justice, United Church of Christ.

Cherney, D. (1990, May 1). Freedom riders needed to save the forest: Mississippi summer in the California redwoods. *Earth First! Journal, 1*, 6.

Cicero, M. T. (1962). *Orator* (Rev. ed.). (H. M. Hubell & G. L. Hendrickson, Trans.). Cambridge, MA: Harvard University Press.

Clarida, M. Q. (2014, May 2). Day after arrest, divest protesters renew calls for open meeting with corporation. *The Harvard Crimson*. Retrieved from http://www.thecrimson.com/article/2014/5/2/divest-harvard-petition-rally/

Clark, G., Halloran, M., & Woodford, A. (1996). Thomas Cole's vision of "nature" and the conquest theme in American culture. In C. G. Herndl & S. C. Brown (Eds.), *Green culture: Environmental rhetoric in contemporary America.* (pp. 261–280). Madison: University of Wisconsin Press.

Clark, J., & Slyke, T. V. (2011). How journalists must operate in a new networked media environment. In R. W. McChesney & V. Pickard (Eds.), *Will the last reporter please turn out the lights* (pp. 238–248). New York, NY, & London, England: The New Press.

Clean Air Act Amendments. (1990). U.S.C. Title 42. Chapter 85. Subsection III. 7602 [g]. Retrieved from http://www.law.cornell.edu/uscode

Clean Water Act—Citizen Suits. (2007, July 30). Washington, DC: U.S. Environmental Protection Agency. Retrieved from http://www.epa.gov

Climate Central. Retrieved from http:www.ClimateCentral.org

Climate change protestors get through tight security to disrupt Stephen Harper during event. (2014, January 6). *National Post.* Retrieved from http://news.nationalpost.com/2014/01/06/climate-change-protesters-get-through-tight-security-to-disrupt-stephen-harper-during-event/

Climate Outreach Network. (2014). Uncertainty and the IPCC. Retrieved from http://talkingclimate.org/wp-content/uploads/2011/10/Uncertainty-the-IPCC.pdf

Climate Science Rapid Response Team. (2014). Retrieved from http://climaterapidresponse.org

Clinton, W. J. (1994, February 16). Federal actions to address environmental justice in minority populations and low-income populations. Executive Order 12898 of February 14, 1994. *Federal Register, 59,* 7629.

CNN Wire Staff. (2010, June 16). Oil estimate raised to 35,000–60,000 barrels a day. Retrieved from http://edition.cnn.com/

Cohen, B. C. (1963). *The press and foreign policy.* Princeton, NJ: Princeton University Press.

Cohen, K. (2013, November 22). Fracking safe, says Obama administration. *ExxonMobile Perspectives.* Retrieved from http://www.exxonmobilperspectives.com/2013/11/22/fracking-safe-says-obama-administration/

Cohen-Cruz, J. (1998). *Radical street performance: An international anthology.* London, England: Routledge.

Colaneri, K. (2014, July 1). Federal court dismisses doctor's lawsuit over Act 13 "gag rule." *National Public Radio.* Retrieved from http://stateimpact.npr.org/pennsylvania/2014/07/01/federal-court-dismisses-doctors-lawsuit-over-act-13-gag-rule/

Cole, L. W., & Foster, S. R. (2001). *From the ground up: Environmental racism and the rise of the environmental justice movement.* New York, NY: New York University Press.

Conflict and protest. (n.d.). *The Rainforest Solutions Project.* Retrieved from http://www.savethegreatbear.org/

Connor, S. (2007, January 18). Hawking warns: We must recognise the catastrophic dangers of climate change. *The Independent.* Retrieved from http://www.independent.co.uk

Constitution of the Iroquois Nations. (n.d.). *Indigenous peoples literature.* Retrieved from http://www.indigenouspeople.net/iroqcon.htm

Consumers beware: What labels mean. (2006, November 19). *Raleigh News and Observer,* pp. 23A–24A.

Corbett J. B. (2006). *Communicating nature: How we create and understand environmental messages.* Washington, DC: Island Press.

Corburn, J. (2005). *Street science: Community knowledge and environmental justice.* Cambridge, MA: MIT Press.

Core values for the practice of public participation. (2008). *The International Association for Public Participation.* Retrieved from http://www.iap2.org

The costs of fracking. (2012, September 20). Environment America Research and Policy Center. Retrieved from http://www.environmentamerica.org/reports/ame/costs-fracking

Cottle, S. (2000). TV news, lay voices and the visualization of environmental risks. In S. Allan, B. Adam, & C. Carter (Eds.), *Environmental risks and the media* (pp. 29–44). London, England: Routledge.

Council on Environmental Quality. (1997, January). *The National Environmental Policy Act: A study of its effectiveness after twenty-five years.* Washington, DC: Council on Environmental Quality, Executive Office of the President. Retrieved from http://ceq.eh.doe.gov

Council on Environmental Quality (CEQ). (2007, December). *A citizen's guide to the NEPA: Having your voice heard.* Retrieved from http://ceq.hss.doe.gov

Cox, J. R. (2001a). The irreparable. In T. O. Sloane (Ed.), *Encyclopedia of rhetoric* (pp. 406–409). Oxford, England, & New York, NY: Oxford University Press.

Cox, J. R. (2001b). Reclaiming the "indecorous" voice: Public participation by low-income communities in environmental decision making. In C. B. Short & D. Hardy-Short (Eds.), *Proceedings of the Fifth Biennial Conference on Communication and Environment* (pp. 21–31). Flagstaff: Northern Arizona University School of Communication.

Cox, J. R. (2010). Beyond frames: Recovering the strategic in climate communication. *Environmental Communication: A Journal of Nature and Culture 4*(1), 122-133.

Cozen, B. (2013). Mobilizing artists: Green patriot posters, visual metaphors, and climate change activism. *Environmental Communication, 7*(2), 297–314.

Cress, M. (2013, August 20). Glass straw: The original glass straw. *BeStrawFree.org.* Retrieved from http://glassdharma.blogspot.com/2013/08/from-milo-cress-best rawfreeorg.html

Crompton, T. (2008). *Weathercocks and signposts: The environment movement at a crossroads. World Wildlife Fund-UK.* Retrieved from http://wwf.org.uk/strategiesforchange

Cunningham, B. (2003). Re-thinking objectivity. *Columbia Journalism Review, 42,* 24–32.

Cushman, J. H. (1998, April 26). Industrial group plans to battle climate treaty. *New York Times,* p. A1.

Dahlstrom, M. F., & Scheufele, D. A. (2010). Diversity of television exposure and its association with the cultivation of concern for environmental risks. *Environmental Communication: A Journal of Nature and Culture, 4*(1), 54–65.

The Daily Show with Jon Stewart. (2014, January 13). Coal miner's water—A terrorist plot? Available at http://thedailyshow.cc.com/videos/qq6s5s/coal-miner-s-water—a-terrorist-plot-

Daley, J. (2010, January 7). Why the decline and rebirth of environmental journalism matters. *Yale Forum on Climate Change and the Media.* Retrieved from http://www.yaleclimatemediaforum.org/

Daniels, S. E., & Cheng, A. S. (2004). Collaborative resource management: Discourse-based approaches and the evolution of TechnoReg. In M. L. Manfredo, J. J. Vaske, B. L. Bruyere, D. R. Field, & P. J. Brown (Eds.), *Society and natural resources: A summary of knowledge* (pp. 127–136). Jefferson, MO: Modern Litho.

Daniels, S. E., & Walker, G. B. (2001). *Working through environmental conflict: The collaborative learning approach.* Westport, CT: Praeger.

Darwin, C. (n.d.). *BrainyQuote.com.* Retrieved from http://www.brainyquote.com/quotes/quotes/c/charlesdar393305.html

Davis-Cohen, S., & StudentNation. (2014, April 25). Youth are taking the government to court over its failure to address climate change. *The Nation.* Retrieved from http://www.thenation.com/blog/ 179533/can-ancient-doctrine-force -government-act-climate-change#

Declaration of Bizkaia on the Right to the Environment. (1999, February 10–13). Retrieved from http://unesdoc.unesco .org

Delhi Climate Justice Declaration. (2002). India Climate Justice Forum. Delhi, India: India Resource Center. Retrieved from http://www.indiaresource.org

DeLuca, K. M. (1999). *Image politics: The new rhetoric of environmental activism.* New York, NY: Guilford.

DeLuca, K. M. (2001). Trains in the wilderness: The corporate roots of environmentalism. *Rhetoric & Public Affairs,* 4(4), 633–652.

DeLuca, K. M. (2005). *Image politics: The new rhetoric of environmental activism.* London, England: Routledge.

DeLuca, K. M., & Demo, A. T. (2000). Imaging nature: Watkins, Yosemite, and the birth of environmentalism. *Critical Studies in Media Communication, 17*(3), 241–260.

DeLuca, K. M., & Peeples, J. (2002). From public sphere to public screen: Democracy, activism, and the "violence" of Seattle. *Critical Studies in Media Communication, 19*(2), pp. 125–151.

DeLuca, K. M., Sun, Y., & Peeples, J. (2011). Wild public screens and image events from Seattle to China. In S. Cottle & L. Lester (Eds.), *Transnational protests and the media* (pp. 143–158). New York, NY: Peter Lang.

Democracy Now! (2014). "Exclusive: 4 years after BP disaster, ousted drilling chief warns U.S. at risk of another oil spill," April 21. 2014. Available at http://www .democracynow.org/2014/4/21/exclu sive_4_years_after_bp_disaster

de Onís, K. M. (2012). "Looking both ways": Metaphor and the rhetorical alignment of intersectional climate justice and reproductive justice concerns. *Environmental Communication: A Journal of Nature and Culture, 6*(3), 308–327.

Depoe, S. P. (1991). Good food from the good earth: McDonald's and the commodification of the environment. In D. W. Parson (Ed.), *Argument in controversy: Proceedings from the 7th SCA/ AFA Conference on Argumentation* (pp. 334–341). Annandale, VA: Speech Communication Association.

Depoe, S. P. (2006). Preface. In S. P. Depoe (Ed.), *The environmental communication yearbook* (Vol. 3, pp. vii–ix). London, England: Routledge.

Depoe, S. P., & Delicath, J. W. (2004). Introduction. In S. P. Depoe, J. W. Delicath, & M-F. A. Elsenbeer (Eds.), *Communication and public participation in environmental decision making* (pp. 1–10). Albany: State University of New York Press.

Dewey, J. (1927). *The public and its problems.* New York, NY: Henry Holt.

Di Chiro, G. (1996). Nature as community: The convergence of environment and social justice. In W. Cronon (Ed.), *Uncommon ground: Rethinking the human place in nature* (pp. 298–320). New York, NY: Norton.

Di Chiro, G. (1998). Environmental justice from the grassroots: Reflections on history, gender, and expertise. In D. Faber (Ed.), *The struggle for ecological democracy: Environmental justice movements in the United States* (pp. 104–136). New York, NY: Guilford Press.

Dietz, T., & Stern, P. C. (2008). *Public participation in environmental assessment and decision making.* Washington, DC: National Academies Press.

DiFrancesco, D. A., & Young, N. (2010). Seeing climate change: The visual construction of global warming in Canadian national print media. *Cultural geographies, 18*(4), 517–536.

Dirikx, A., & Gelders, D. (2009). Global warming through the same lens: An exploratory framing study in Dutch and French newspapers. In T. Boyce & J. Lewis (Eds.), *Media and climate* change (pp. 200–210). Oxford, UK: Peter Lang.

Dizard, J. E. (1994). *Going wild: Hunting, animal rights, and the contested meaning of nature.* Amherst: University of Massachusetts Press.

Dlouhy, J. (2011, June 28). *EPA officials: "Sound science" will guide hydraulic fracturing study.* Retrieved from http://fuelfix.com/blog/2011

Dlouhy, J. A. (2014, March 6). *FracFocus falls short, report to feds concludes.* Retrieved from http://fuelfix.com/blog/2014/03/06/report-to-feds-fracfocus-falls-short/

Dobrin, S. I., & Morey, S. (Eds.). (2009). *Ecosee: Image, rhetoric, nature.* Albany: State University of New York Press.

Douglas, M., & Wildavsky, A. B. (1982). *Risk and culture: An essay on the selection of technical and environmental dangers.* Berkeley: University of California Press.

Doyle, A. (2008, February 4). "Tipping point" on horizon for Greenland ice. *Reuters.* Retrieved from http://www.reuters.com

Drajem, M. (2014, January 17). New health warning raises concern in West Virginia over chemical spill. *Insurance Journal.* Bloomberg.

Dunlap, R. E., & Van Liere, K. D. (1978). The "new environmental paradigm": A proposed instrument and preliminary analysis. *Journal of Environmental Education, 9,* 10–19.

Durban Declaration on Carbon Trading. (2004, October 10). Glenmore Centre, Durban, South Africa. Retrieved from http://www.carbontrade watch.org/

Earth First! (2014). No compromise in defense of Mother Earth. *Earth First! Journal.* Retrieved from http://www.earthfirstjournal.org

Ebersole, J. (2014, April 25). The public trust doctrine: Spurring government action on the environment. *Georgetown International Law Review* [Blog]. Retrieved from http://gielr.wordpress.com/

Eckholm, E. (2010, September 15). U.S. zeroes in on pork producers' antibiotics use. *New York Times,* pp. A13–19.

Edelman, M. (1964). *The symbolic uses of politics.* Urbana: University Illinois Press.

Eder, K. (1996a). The institutionalization of environmentalism: Ecological discourse and the second transformation of the public sphere. In S. Lash, B. Szerszynski, & B. Wynne (Eds.), *Risk, environment, and modernity: Towards a new ecology* (pp. 203–223). London, England: Sage.

Eder, K. (1996b). *The social construction of nature.* London, England: Sage.

Edwards, A. R. (2005). *The sustainability revolution: Portrait of a paradigm shift.* Gabriola Island, British Columbia, Canada: New Society.

Egan, T. (2003, May 3). Smithsonian is no safe haven for exhibit on Arctic Wildlife Refuge. *New York Times,* p. A20.

Ehrlich, P. R. (1968). *The population bomb.* San Francisco, CA: Sierra Club Books.

Ehrlich, P. R., & Ehrlich, A. H. (1996). *Betrayal of science and reason: How anti-environmental rhetoric threatens our future.* Washington, DC: Island Press/Shearwater Books.

Eiglad, E. (2010). Foreword. In B. Tokar, *Toward climate justice: Perspectives on the climate crisis and social change* (pp. 7–12). Porsgrunn, Norway: ommunalism Press.

Eilperin, J. (2006, January 29). Debate on climate shifts to issue of irreparable change. *Washington Post,* p. A1.

Entman, R. M. (1993). Framing: Toward clarification of a fractured paradigm. *Journal of Communication, 43*(4), 51–58.

Environmental Justice Foundation. (2014). *The gathering storm: Climate change, security and conflict.* London, England. Retrieved from http://ejfoundation.org/sites/default/files/public/EJF_climate_conflict_report_web-ok.pdf

Environmental Working Group. (2010). Chromium-6 is widespread in US tap water. Retrieved from http://static.ewg .org/

Ericson, R. V., Baranek, P. M., & Chan, J. B. L. (1989). *Negotiating control: A study of news sources.* Milton Keynes, England: Open University Press.

Evernden, N. (1992). *The social creation of nature.* Baltimore, MD: Johns Hopkins University Press.

Eyal, C. H., Winter, J. P., & DeGeorge, W. F. (1981). The concept of time frame in agenda setting. In G. C. Wilhoit (Ed.), *Mass communication yearbook* (pp. 212–218). Beverly Hills, CA: Sage.

Farrell, T. B., & Goodnight, G. T. (1981). Accidental rhetoric: The root metaphors of Three Mile Island. *Communication Monographs, 48,* 271–300.

Farrior, M. (2005, February). *Breakthrough strategies for engaging the public: Emerging trends in communications and social science for biodiversity project.* Retrieved from http://www.biodiversityproject.org

Faulkner, W. (2011, March 4). Titan suing two residents for slander over comments made at commissioners meeting. *Star News Online.* Retrieved from http://www.starnewsonline.com

Federal Trade Commission. (n.d.). Guides for the use of environmental marketing claims. Section 260.7. *Environmental marketing claims.* Retrieved from http:// www.ftc.gov

Federal Trade Commission. (2010, October). *Proposed revisions to the green guides.* Retrieved from http://www.ftc.gov

Fermino, J. (2010, September 11). Camera captures BP's PR disaster for all to see. *New York Post.* Retrieved from http:// www.nypost.com/

Ferris, D. (Speaker). (1993, October 21–24). *Environmental justice: Continuing the dialogue* [Cassette recording]. Recorded at the Third Annual Meeting of the Society of Environmental Journalists, Durham, NC.

Fiorino, D. J. (1989). Technical and democratic values in risk analysis. *Risk Analysis, 9,* 293–299.

Fischer, F. (2000). *Citizens, experts, and the environment: The politics of local knowledge.* Durham, NC: Duke University Press.

Fish, S. (2008, August 3). *Think again: I am therefore I pollute.* Retrieved from http://fish.blogs.nytimes.com

Fiske, J. (1987). *Television culture.* London, England: Methuen.

Flaccus, G. (2011, June 17). Sewage pile, illegal dump on Calif toxic tour list. *CNSNews.com.* Retrieved from http:// www.cnsnews.com

Flannery, T. (2005). *The weather makers: How man is changing the climate and what it means for life on earth.* New York, NY: Grove Press.

Flatt, V. B. (2010, June 5). What if drilling goes really wrong? *The News & Observer,* p. 7A.

Foderaro, L. (2014, September 22). At march, clarion call for action on climate. *New York Times,* pp. A1, 17.

Foer, J. S. (2010). *Eating animals.* New York, NY: Back Bay Books.

Food & Water Watch. (2013, September 24). *The social costs of fracking.* Retrieved from http://www.foodandwaterwatch .org/water/fracking/

Foucault, M. (1970). *The archaeology of knowledge (and the discourse on language).* New York, NY: Vintage.

Foundation for Deep Ecology. (2012). The Deep Ecology platform. Retrieved from http://www.deepecology.org/platform .htm

Fountain, H. (2014, March 12). Ohio looks at whether fracking led to 2 quakes. *New York Times,* p. A20.

Frederick, H. H. (1992). Computer communications in cross-border coalition-building: North American NGO networking against NAFTA. *International Communication Gazette, 50*(2–3), pp. 217–241.

Friedman, L. (2010, October 4). Developing countries could sue for climate action—study. *New York Times.* Retrieved from http://www.nytimes.com

Friedman, L. (2014, September 22). Marchers jam New York in massive call for climate action. *E&E News Climate Wire*. Retrieved from http://www.eenews.net/climatewire/2014/09/22/stories/1060006200

Friedman, S. M. (2004). And the beat goes on: The third decade of environmental journalism. In S. Senecah (Ed.), *The environmental communication yearbook* (Vol. 1, pp. 175–187). Mahwah, NJ: Erlbaum.

Friedman, S. M. (2015). The changing face of environmental journalism in the United States. In A. Hansen & R. Cox (Eds.). *Routledge handbook of environment and communication*. London, England: Routledge.

Friends of the Earth, Inc. v. Laidlaw Environmental Services, Inc. (TOC), Inc.—528 U.S. 167. (2000). Justia U.S. Supreme Court. Retrieved from http://supreme.justia.com/cases/federal/us/528/167/case.html

Fritch, J., Palczewski, C. H., Farrell, J., & Short, E. (2006). Disingenuous controversy: Responses to Ward Churchill's 9/11 essay. *Argumentation and Advocacy, 42*(4), 190–205.

Froomkin, D. (2010, June 1). Gulf oil spill: Markey demands BP broadcast live video feed from the source. *Huffpost Social News*. Retrieved from http://www.huffingtonpost.com

Fuller, B. (1963). *Operating manual for Spaceship Earth*. New York, NY: Dutton.

Fuller, M. C., Kunkel, C., Zimring, M., Hoffman, I., Soroye, K. L., & Goldman, C. (2010, September). *Driving demand for home energy improvements*. Environmental Energies Technology Division, Lawrence Berkeley National Laboratory. Retrieved from http://eetd.lbl.gov/EAP/EMP/reports/lbnl-3960e-web.pdf

Gamson, W. A., & Modigliani, A. (1989). Media discourse and public opinion on nuclear power: A constructionist approach. *American Journal of Sociology, 95*, 1–37.

Gardiner, B. (2014, January 31). Air of revolution: How activists and social media scrutinize city pollution. *The Guardian*. Retrieved from http://www.theguardian.com/cities/2014/jan/31/air-activists-social-media-pollution-city

Gendlin, F. (1982). A talk with Mike McCloskey: Executive director of the Sierra Club. *Sierra, 67*, 36–41.

Gerbner, G. (1990). Advancing on the path to righteousness, maybe. In N. Signorielli & M. Morgan (Eds.), *Cultivation analysis: New directions in research* (pp. 249–262). Newbury Park, CA: Sage.

Gerbner, G., Gross, L., Morgan, M., & Signorielli, N. (1986). Living with television: The dynamics of the cultivation process. In J. Bryant & D. Zillmann (Eds.), *Perspectives on media effects* (pp. 17–40). Hillsdale, NJ: Erlbaum.

Gerrard, M. B., & Foster, S. R. (Eds.). (2008). *The law of environmental justice* (2nd ed.). Chicago, IL: American Bar Association.

Gibbs, L. (1994). Risk assessments from a community perspective. *Environmental Impact Assessment Review, 14*, 327–335.

Gibson, J. J. (1986). *The ecological approach to visual perception*. Hillsdale, NJ: Erlbaum.

Gidez, C. (2010). Six lessons from the Chevron hoax: What marketers should take away from the hijacking of marketer's public-relations push. *Advertising Age*. Retrieved from http://adage.com/article/guest-columnists/marketing-lessons-chevron-hoax/146658/

Gillis, J. (2013, October 10). By 2047, coldest years may be warmer than hottest in past, scientists say. *New York Times*, p. A9.

Gillis, J. (2014, April 29). What does today owe tomorrow? *New York Times* Retrieved from http://www.nytimes.com/2014/04/29/science/what-does-today-owe-tomorrow.html?_r=0

Gillis, J., & Chang, K. (2014, May 12). Scientists warn of rising oceans from polar melt. *New York Times*. Retrieved from http://www.nytimes.com/2014/05/13/science/earth/collapse-of-parts-of-west-antarctica-ice-sheet-has-begun-scientists-say.html?hpw&rref=world&_r=0

Glaberson, W. (1999, June 5). Novel antipollution tool is being upset by courts. *New York Times*. Retrieved from http://query.nytimes.com

Gladwell, M. (2010, October 4). Small change: Why the revolution will not be tweeted. *New Yorker*. Retrieved from http://www.newyorker.com

Glanz, J. (2004, February 19). Scientist says administration distorts facts. *New York Times*, p. A21.

Goffman, E. (1974). *Frame analysis: An essay on the organization of experience*. Cambridge, MA: Harvard University Press.

Goldenberg, R. (2014, May 9). Why is the green movement so dominated by white dudes? *Mother Jones*. Retrieved from http://m.motherjones.com/environment/2014/05/white-men-run-big-environmental-organizations

Goodnight, G. T. (1982). The personal, technical, and public spheres of argument: A speculative inquiry into the art of public deliberation. *Journal of the American Forensic Association, 18*, 214–227.

Goodnight, G. T. (1991). Controversy. In D. Parson (Ed.), *Argument in controversy* (pp. 1–12). Annandale, VA: Speech Communication Association.

Gore, A. (2007). *The assault on reason*. New York, NY: Penguin Press.

Gottlieb, R. (1993). *Forcing the spring: The transformation of the American environmental movement*. Washington, DC: Island Press.

Gottlieb, R. (2002). *Environmentalism unbound: Exploring new pathways for change*. Cambridge, MA: MIT Press.

Gottlieb, R. (2003). Reconstructing environmentalism: Complex movements, diverse roots. In L. S. Warren (Ed.), *American environmental history* (pp. 245–256). Malden, MA: Blackwell.

Graber, D. A. (1976). *Verbal behavior and politics*. Urbana: University of Illinois Press.

Grace Communication Foundation. (2015). About the Meatrix, The Meatrix (website). Available at http://www.themeatrix.com/about

Grady, D. (2010, September 7). In feast of data on BPA plastic, no final answer. *New York Times*, pp. D1, D4.

Gray, S. (2010, June 15). New Orleans' cuisine crisis. *Time*. Retrieved from http://www.time.com/

Green marketing insight from GFK Roper's Green Gauge® at green marketing conference. (2011, February 24). *PR Web*. Retrieved from http://www.prweb.com

Greenberg, M., Sachsman, D. B., Sandman, P., & Salomone, K. L. (1989). Risk, drama and geography in coverage of environmental risk by network TV. *Journalism Quarterly, 66*(2), 267–276.

Greenpeace. (1990, April). *Ordinary people, doing extraordinary things* [Videotape]. Public service announcement broadcast on VH-1 Channel.

Greenpeace. (2008, December 22). *BP wins coveted "Emerald Paintbrush" award for worst greenwash of 2008*. Retrieved from http://weblog.greenpeace.org

Green Retail Decisions. (2014, April 30). Leading CEOs make supply chain pledge at Walmart's inaugural sustainable product expo. *Stagnito Media*. Retrieved from http://www.greenretaildecisions.com/news/2014/04/30/leading-ceos-make-supply-chain-pledge-at-walmarts-inaugural-sustainable-product-expo-

Greider, W. (2003, August 5). Victory at McDonald's. *The Nation*, pp. 8, 10, 36.

Grossman, K. (1994). The People of Color Environmental Summit. In R. D. Bullard (Ed.), *Unequal protection: Environmental justice and communities of color*. San Francisco, CA: Sierra Club Books.

Gulledge, J. (2011, March 1). Sixth independent investigation clears "Climategate" scientists. Retrieved from Pew Center on Global Climate Change. Web site: http://www.pewclimate.org

Habermas, J. (1974). The public sphere: An encyclopedia article (1964). *New German Critique, 1*(3), 49–55.

Haeckel, E. (1904). *The wonders of life: A popular study of biological philosophy.* New York, NY, & London, England: Harper & Brothers.

Hall, J. (2001, May/June). How the environmental beat got its groove back. *Columbia Journalism Review.* Retrieved from http://backissues.cjrarchives.org

Hall, S. (1973). Encoding and decoding in the television discourse (pp. 507–517). University of Birmingham: Birmingham Centre for Cultural Studies.

Hall, S. (1984). Encoding and decoding. *Culture, media, language.* London, England: Hutchinson.

The Halliburton loophole. (2009, November 2). *New York Times.* Retrieved from http://www.nytimes.com/2009/11/03/opinion/03tue3.html?_r=0

Hamilton, A. (1925). *Industrial poisons in the United States.* New York, NY: Macmillan.

Hanna, J. (2014, May 13). Ice melt in part of Antarctica "appears unstoppable," NASA says. *CNN.* Retrieved from http://www.cnn.com/2014/05/12/us/nasa-antarctica-ice-melt/index.html

Hansen, A. (2010). *Environment, media, and communication.* New York, NY: Routledge.

Hansen, A. (2011, February). Towards reconnecting research on the production, content and social implications of environmental communication. *International Communication Gazette, 73*(1–2), 7–25.

Hansen, A. (2015). News coverage of the environment: A longitudinal perspective. In A. Hansen & R. Cox (Eds.). *Routledge handbook of environment and communication.* London, England: Routledge.

Haraway, D. (1991). *Simians, cyborgs, and women: The reinvention of nature.* New York, NY: Routledge.

Harlow, T. (2001, March 16). Message posted on Infoterra LISTSERV. Retrieved from http://www.peer.org

Harman, W. (1998). *Global mind change: The promise of the twenty-first century.* San Francisco, CA: Berrett-Koehler.

Harris, S. (1977). *What's so funny about science?* Los Altos, CA: Wm. Kaufmann.

Harvey, D. (1996). *Justice, nature, and the geography of difference.* Malden, MA: Blackwell.

Hasen, R. L. (2010, March 9). The numbers don't lie. *Slate.com.* Retrieved from http://www.slate.com/articles/news_and_politics/politics/2012/03/the_supreme_court_s_citizens_united_decision_has_led_to_an_explosion_of_campaign_spending_.html

Hassett, K. A. (2013, April 4). Benefits of hydraulic fracking. *American Enterprise Institute.* Retrieved from http://www.aei.org/article/economics/benefits-of-hydraulic-fracking

Hawken, P. (2007). *Blessed unrest: How the largest social movement in history is restoring grace, justice, and beauty to the world.* New York, NY: Penguin Books.

Hayes, K. (2010, September 20). U.S. says BP permanently "kills" Gulf of Mexico well. *Reuters.* Retrieved from http://www.reuters.com/

Hays, S. P. (1989). *Beauty, health, and permanence: Environmental politics in the United States, 1955–1985.* Cambridge, England: Cambridge University Press.

Hays, S. P. (2000). *A history of environmental politics since 1945.* Pittsburgh, PA: University of Pittsburgh Press.

Headapohl, J. (2011, May 7). Obama touts clean energy jobs in weekly address. Retrieved from http://www.mlive.com/

Hecht, S. (2011, May 11). Anti-coal satire (with My First Inhaler) punks Peabody Energy. Retrieved from http://legalplanet.wordpress.com

Heffter, E. (2013, April 11). Sinclair known for conservative political tilt. *Seattle Times.*

Retrieved from http://seattletimes.com/html/businesstechnology/2020756844_fishersinclairxml.html

Heller, N. (2011). Are scientists confusing the public about global warming? Retrieved from http://www.climatecentral.com

Helvarg, D. (2004). *The war against the greens: The "wise-use" movement, the new right, and the browning of America.* Boulder, CO: Johnson Books.

Herndl, C. G., & Brown, S. C. (1996). Introduction. In C. G. Herndl & S. C. Brown (Eds.), *Green culture: Environmental rhetoric in contemporary America* (pp. 3–20). Madison: University of Wisconsin Press.

Herrick, J. A. (2009). *The history and theory of rhetoric: An introduction.* Boston, MA: Pearson.

Hogue, C. (2010, August 16). Recasting TSCA. *Chemical and Engineering News, 88*(33), 35–37. doi:1021/CEN081010131116

Holcomb, J., & Mitchell, A. (2014). The revenue picture for American journalism and how it is changing. *Pew Research Journalism Project.* Retrieved from http://www.journalism.org/2014/03/26/the-revenue-picture-for-american-journalism-and-how-it-is-changing

Holmes, D. (2014, February 20). "Cli-fi": Could a literary genre help save the planet? *The Conversation.* Retrieved from http://theconversation.com/cli-fi-could-a-literary-genre-help-save-the-planet-23478

Hope, D. S. (2009). Reporting the future: A visual parable of environmental ethics in Robert and Shana Parke Harrison's The architect's brother. *Visual Communication Quarterly, 16,* 32–49.

Howell, R. A. (2013). It's not (just) "the environment, stupid!" Values, motivations, and routes to engagement of people adopting lower-carbon lifestyles. *Global Environmental Change, 23,* 281–290.

Hsieh, S. (2014, April 22). People of color are already getting hit the hardest by climate change. *The Nation.* Retrieved from http://www.thenation.com/blog/179407/people-color-are-already-getting-hit-hardest-climate-change

Humes, E. (2011a). *Force of nature: The unlikely story of Walmart's green revolution.* New York, NY: HarperCollins.

Humes, E. (2011b, May 31). Wal-Mart's green hat. *Los Angeles Times.* Retrieved from http://articles.latimes.com

Hyman, Mark. (2011, February 3). Global warming hoax. Beyond the headlines with Mark Hyman. Commentary. Retrieved from http://www.behindtheheadlines.net/sections/videos/vid_8.shtml

Innis, H. A. (2008). *The bias of communication* (2nd Ed.). Toronto, Ontario, Canada: University of Toronto Press.

Institute for Energy Research. (2010, June 30). New study: Kerry-Lieberman to destroy up to 5.1 million jobs, cost families $1,042 per year, wealthiest Americans to benefit. Retrieved from http://www.instituteforenergyresearch.org

Intergovernmental Panel on Climate Change. (2007). *Climate change 2007: Synthesis report.* United Nations Environment Program. Retrieved from http://www.ipcc.ch/

Intergovernmental Panel on Climate Change. (2013). *Climate change 2013. The physical science basis. Summary for policymakers.* Geneva, Switzerland. Author. Available at ipcc.org

Intergovernmental Panel on Climate Change. (2014, January 30). IPCC publishes full report *Climate change 2013: The physical science basis.* IPCC press release. Retrieved from http://www.ipcc.ch/pdf/press/press_release_wg1_full_report.pdf

Iredale, W. (2005, December 18). Polar bears drown as ice shelf melts. *Sunday Times* [UK]. Retrieved from http://www.timesoline.co.uk

Irvine, S. (1989). *Beyond green consumerism.* London, England: Friends of the Earth.

Iyengar, S., & Kinder, D. R. (1987). *News that matters: Television and American opinion.* Chicago, IL: University of Chicago Press.

Jacobs, J. P. (2014, June 5). Federal judges reject children's "public trust" suit against the EPA, other agencies. *E&E News.* Retrieved from http://www.eenews.net/greenwire/stories/1060000783/

Jacques, P. J., Dunlap, R. E., & Freeman, M. (2008). The organisation of denial: Conservative think tanks and environmental skepticism. *Environmental Politics, 17*(3), 349–385.

Jamieson, D. (2007). Justice: The heart of environmentalism. In P. C. Pezzullo & R. Sandler (Eds.), *Environmental justice and environmentalism: The social justice challenge to the environmental movement* (pp. 85–101). Cambridge, MA: MIT Press.

Jamieson, K. H., & Stromer-Galley, J. (2001). Hybrid genres. In T. O. Sloane (Ed.), *Encyclopedia of rhetoric* (pp. 361–363). Oxford, England: Oxford University Press.

Janis, I. L. (1977). *Victims of groupthink.* Boston, MA: Houghton Mifflin.

Jasinski, J. (2001). *Sourcebook on rhetoric: Key concepts in contemporary rhetorical studies.* Thousand Oaks, CA: Sage.

Jenner, E. (2012). News photographs and environmental agenda setting. *Policy Studies Journal, 40,* 274–301.

Johnson, L. (2008, November 28). Environmentalist turns to online campaign to protect B.C. forest. *CBC News.* Retrieved from http://www.cbc.ca/news

Johnston, K. (2008, October 29). Shut down for speaking up: SLAPP suits continue to chill free speech, despite legislated remedies. *New Times, 23*(13). Retrieved from http://www.newtimesslo.com

Johnstone, I. (2014, May 4). The age of Anthropocene: Was 1950 the year human activity began to leave an indelible mark on the geology of Earth? *The Indpendent* (UK). Retrieved from http://www.independent.co.uk/news/science/the-age-of-anthropocene-was-1950-the-year-human-activity-began-to-leave-an-indelible-mark-on-the-geology-of-earth-9321344.html

Jones, E. (2013, February 14). Green memes. Retrieved from http://www.greeniacs.com/GreeniacsArticles/Environmental-News/Green-Memes.html

Jones, L. (1996a, May 13). "Howdy, neighbor!" As a last resort, Westerners start talking to each other. *High Country News, 28,* pp. 1, 6, 8.

Jones, L. (1996b, May 13). Some not-so-easy steps to successful collaboration [Sidebar]. *High Country News, 28,* pp. 1, 6, 8.

Jonsson, P. (2010, May 29). BP "top kill" live feed makes stars out of disaster bots. *Christian Science Monitor.* Retrieved from http://www.csmonitor .com/

Jordan, C. (2009–present). Gyre. In *Running the numbers II: Portraits of global mass culture.* Retrieved from http://www .chrisjordan.com/gallery/rtn2/#gyre2

Karpf, D. (2010). Online political mobilization from the advocacy group's perspective: Looking beyond clicktivism. *Policy & Internet, 2*(4), Article 2.

Karpf, D. A. (2011). *The MoveOn effect: The unexpected transformation of American political advocacy.* New York, NY: Oxford University Press.

Kaufman, L. (2010, October 19). Kansans scoff at global warming but embrace cleaner energy. *New York Times,* pp. A1, 4.

Kazis, R., & Grossman, R. L. (1991). *Fear at work: Job blackmail, labor and the environment* (New ed.). Philadelphia, PA: New Society.

Ketler, A. (2013, September 17). India declares dolphins and whales as "non-human persons," dolphin shows banned. *Collective Evolution.* Retrieved from http://www .collective-evolution.com/2013/09/17/

india-declares-dolphins-whales-as-non-human-persons/

Killingsworth, M. J., & Palmer, J. S. (1996). Millennial ecology: The apocalyptic narrative from *Silent Spring* to global warming. In C. G. Herndl & S. C. Brown (Eds.), *Green culture: Environmental rhetoric in contemporary America* (pp. 21–45). Madison: University of Wisconsin Press.

Kinsella, W. J. (2004). Public expertise: A foundation for citizen participation in energy and environmental decisions. In S. P. Depoe, J. W. Delicath, & M-F. A. Elsenbeer (Eds.), *Communication and public participation in environmental decision making* (pp. 83–95). Albany: State University of New York Press.

Kinsella, W. J. (2008). Introduction: Narratives, rhetorical genres, and environmental conflict: Responses to Schwarze's "environmental melodrama." *Environmental Communication: A Journal of Nature and Culture, 2,* 78–79.

Kinsella, W. J., & J. Mullen. (2008). Becoming Hanford downwinders: Producing community and challenging discursive containment. In B. C. Taylor, W. J. Kinsella, S. P. Depoe, & M. S. Metzler (Eds.), *Nuclear legacies: Communication, controversy, and the U.S. nuclear weapons complex* (pp. 78-108). Lexington, KY: Lexington Books.

King, E. (2014, January 27). Tunisia embeds climate change in constitution. *RTCC [Responding to climate change]*. Retrieved from http://www.rtcc.org/2014/01/27/tunisia-embeds-climate-change-in-constititon

Klein, N. (2014, April 21) The change within: The obstacles we face are not just external. *The Nation*. Retrieved from http://www.thenation.com/article/179460/change-within-obstacles-we-face-are-not-just-external

Klinkenborg, V. (2013, September 22). Silencing scientists. *New York Times,* p. 10.

Kolbert, L (2014. *The sixth extinction: An unnatural history.* New York, NY: Henry Holt.

Kollmuss, A., & Agyeman, J. (2002). Mind the gap: Why do people act environmentally and what are the barriers to pro-environmental behavior? *Environmental Education Research, 8*(3), 96–119.

Krimsky, S. (2007). Risk communication in the Internet age: The rise of disorganized skepticism. *Environmental Hazards, 7,* 157–164.

Kristof, N. (2014, January 14). "Neglected topic" winner: Climate change. *New York Times*. p. SR11.

Krugman, P. (2009, September 27). Cassandras of climate. *New York Times,* p. A21.

Laclau, E., & Mouffe, C. (2001). *Hegemony and socialist strategy: Toward a radical democracy* (2nd ed.). London, England: Verso.

LaDuke, W. (2002, November/December). The salt woman and the coal mine. *Sierra,* pp. 44–47, 73.

Lakoff, G. (2010). Why it matters how we frame the environment. *Environmental Communication: A Journal of Nature and Culture, 4*(1), 70–81.

Latour, B. (2004). *Politics of nature: How to bring science into democracy* (C. Porter, Trans.). Cambridge, MA: Harvard University Press. (Originally published 1999, Paris: Editions la Découverte)

Lavelle, M., & Coyle, M. (1992, September 21). Unequal protection: The racial divide in environmental law. *National Law Journal,* S1, S2.

Layton, L. (2008, September 17). Study links chemical BPA to health problems. *Washington Post,* p. A3.

Leavenworth, S. (2014, April 26). New Chinese laws tackle worries over pollution. *News & Observer,* p. 8A.

Lee, C. (1996). Environment: Where we live, work, play, and learn. *Race, Poverty, and the Environment, 6,* 6.

Leopold, A. (1949). *A Sand County almanac*. New York: Ballantine Books. (Original work published 1966)

Lesly, P. (1992). Coping with opposition groups. *Public Relations Review, 18*(4), 325–334.

Lester, E. A., & Hutchins, B. (2012). The power of the unseen: Environmental conflict, the media and invisibility. *Media, Culture & Society, 34*(7), 847–863.

Lester, L. (2010). *Media and environment: Conflict, politics, and the news.* Cambridge, UK: Polity.

Lester, L. (2011, March). Species of the month: Anti-whaling, mediated visibility, and the news. *Environmental Communication: A Journal of Nature and Culture, 5*(1): 124–139.

Lester, L., & Hutchins, B. (2009). Power games: Environmental protest, news media and the Internet. *Media, Culture & Society, 31*(4), pp. 579–595.

Lindenfeld, L., Hall, D. M., McGreavy, B., Silka, L., & Hart, D. (2012). Creating a place for environmental communication research in sustainability science, *Environmental Communication: A Journal of Nature and Culture, 6*(1), 23–43.

Lindstrom, M. J., & Smith, Z. A. (2001). *The National Environmental Policy Act: Judicial misconstruction, legislative indifference, & executive neglect.* College Station: Texas A&M University Press.

Lindzen, R. S. (2009, July 26). Resisting climate hysteria. *Quadrant Online.* Retrieved from http://www.quadrant.org

Lippmann, W. (1922). *Public opinion.* New York, NY: Harcourt, Brace.

LoBianco, T. (2008). Groups spend millions in "clean coal" ad war. *Washington Times.* Retrieved from http://www.washingtontimes.com

Louv, R. (2008). *Last child in the woods: Saving our children from nature-deficit disorder.* New York, NY: Algonquin Books.

Lovelock, J. (2006, January 16). The earth is about to catch a morbid fever that may last as long as 100,000 years. *The Independent.* Retrieved from http://www.independent.co.uk/

Lujan v. Defenders of Wildlife, 504 U.S. 555 (1992).

Lundgren, R. E., & McMakin, A. H. (2009). *Risk communication: A handbook for communicating environmental, safety, and health risks* (4th ed.). Hoboken, NJ: John Wiley.

Luntz Research Companies. (2001). The environment: A cleaner, safer, healthier America. In *Straight Talk* (pp. 131–146). Retrieved from http://www.ewg.org

Maathai, W. (2008). *Unbowed: A memoir.* New York, NY: Vintage.

Mackintosh, B. (1999). *The National Park Service: A brief history.* ParkNet, National Park Service website. Retrieved from http://www.cr.nps.gov/history/hisnps/npshistory/npshisto.htm

Maibach, E., Wilson, K., & Witte, J. (2010). A national survey of news directors about climate change: Preliminary findings. Fairfax, VA: George Mason University. Retrieved from http://www.climatechangecommunication.org

Makower, J. (2010, July 19). Walmart and the sustainability index: One year later. Retrieved from http://www.greenbiz.com/blog

Mandel, J. (2010, October 5). Don't say "retrofit," say "upgrade"—study. *Greenwire.* Retrieved from http://www.eenews.net/gw

Mann, B. (2001, May 26). Bringing good things to life? *On the media.* New York: WNYC. Retrieved from http://www.onthemedia.org

Mann, M. E. (2014, January 19). If you see something, say something. *New York Times*, p. 8.

Marafiote, T. (2008). The American dream: Technology, tourism, and the transformation of wilderness. *Environmental Communication: A Journal of Nature and Culture, 2,* 154–172.

Marketing and communications. (2009, January 8). *GreenBiz.com*. Retrieved from http://www.greenbiz.com

Markets campaign. (n.d.). *The rainforest solutions project*. Retrieved from http://www.savethegreatbear.org/

Markowitz, G., & Rosner, D. (2002). *Deceit and denial: The deadly politics of industrial pollution*. Berkeley: University of California Press.

Marston, B. (2001, May 7). A modest chief moved the Forest Service miles down the road. *High Country News, 33*(9). Retrieved from http://www.hcn.org

Martell, A., & Patran, J. (2014, January 7). Social media empowers anti-mining activists. *Prosperity Saskatchewan*. Retrieved from http://prosperitysaskatchewan.wordpress.com/2014/01/09/social-media-empowers-anti-mining-activists/

Martin, A. (2006, October 24). Meat labels hope to lure the sensitive carnivore. *New York Times*. Retrieved from http://www.nytimes.com

Martin, T. (2007). Muting the voice of the local in the age of the global: How communication practices compromised public participation in India's Allain Dunhangan environmental impact assessment. *Environmental Communication: A Journal of Nature and Culture, 1*, 171–193.

Mary Robinson Foundation—Climate Justice. (2013). *Climate justice: An intergenerational approach*. Retrieved from http://www.mrfcj.org/media/pdf/Intergenerational-Equity-Position-Paper-2013–11–16.pdf

Matsa, K. E. & Mitchell, A. (2014). 8 key takeaways about social media and news. State of the news media 2014. Pew Research Center. Retrieved from http://www.journalism.org/2014/03/26/8-key-takeaways-about-social-media-and-news/

Massachusetts et al. v. Environmental Protection Agency et al. (2007). Supreme Court of the United States. Retrieved from http://www.supremecourtus.gov

Mayer, C. (2009). Precursors of rhetoric culture theory. In I. Strecker & S. Tyler (Eds.), *Culture and rhetoric* (pp. 31–48). New York, NY: Berghahn Books.

McCalmont, L. (2014, August 20). Leonardo DiCaprio voices climate change film. *Politico*. Retrieved from http://www.politico.com/story/2014/08/leonardo-dicaprio-climate-change-movie-110185.html

McCloskey, M. (1996, May 13). The skeptic: Collaboration has its limits. *High Country News, 28*, p. 7.

McCombs, M., & Shaw, D. (1972). The agenda setting function of the mass media. *Public Opinion Quarterly, 36*, 176–187.

McKibben, B. (2014, May, 21). A call to arms: An invitation to demand action on climate change. *Rolling Stone*. Retrieved from http://www.rollingstone.com/politics/news/a-call-to-arms-an-invitation-to-demand-action-on-climate-change-20140521

McLeod, K. (2014). *Pranksters: Making mischief in the modern world*. New York, NY: New York University Press.

McNair, B. (1994). *News and journalism in the UK*. London and New York: Routledge.

Merchant, B. (2009, July 14). Walmart's sustainability index: The greenest thing ever to happen to retail? Retrieved from http://www.treehugger.com

Merchant, C. (1995). *Earthcare: Women and the environment*. London, England: Routledge.

Merchant, C. (2005). *The Columbia guide to American environmental history*. New York: Columbia University Press.

Michaels, D., & Monforton, C. (2005). Manufacturing uncertainty: Contested science and the protection of the public's health and environment. *Public Health Matters, 95*(1), 39–48.

Mieszkowski, K. (2008, November/December). Big green brother. *Mother*

Jones. Retrieved from http://www.moth erjones.com/environment/2008/11/big -green-brother

Miller, M. M., & Riechert, B. P. (2000). Interest group strategies and journalistic norms: News media framing of environmental issues. In S. Allan, B. Adam, & C. Carter (Eds.), *Environmental risks and the media* (pp. 45–54). London, England: Routledge.

Milstein, T. (2011). Nature identification: The power of pointing and naming. *Environmental Communication: A Journal of Nature and Culture, 5*(1), 3–24.

Milstein, T. (2012). Greening communication. In S.D. Fassbinder, A. J. Nocella II, & R. Kahn (Eds.), *Greening the academy: Ecopedagogy through the liberal arts* (pp. 161–174). Rotterdam, Netherlands: Sense Publishers.

Mims, C. (2011). Solar-power oil field runs on sunshine, irony. *Grist*. Retrieved from http://grist.org/list/2011-08-04-solar -powered-oil-field-runs-on-sunshine -irony/

Mitchell, A. (2014, March 26). State of the news media: Overview. *Pew Research Journalism Project*. Retrieved from http:// www.journalism.org/2014/03/26/state -of-the-news-media- 2014-overview

Mitchell, S. (2007, March 28). The impossibility of a green Wal-Mart. Retrieved from http://www.grist.org

Montague, P. (1999, July 1). The uses of scientific uncertainty. *Rachel's Environment & Health News, 657*. Retrieved from http://www.rachel.org

Mooney, C. (2005, May-June). Some like it hot. *Mother Jones*. Retrieved from http://www.motherjones.com/environ ment/2005/05/some-it-hot

Mooney, C. (2006). *The Republican war on science.* New York, NY: Basic Books.

Mooney, C. (2013, July 8). Why Obama ditched green jobs from his climate change rhetoric. Retrieved from http://www.theguard ian.com/environment/2013/jul/08/ obama-green-jobs-climate-change

Morello, L. (2011, February 2). Award-winning scientists ask Congress to take a "fresh look" at climate change. *Climatewire*. Retrieved from http:// www.eenes.net/climatewire

Moser, S. C. (2010). Communicating climate change: History, challenges, process and future directions. *Wiley Interdisciplinary Reviews: Climate Change, 1*(1), 31–53.

Moyers, B. (Writer), & Jones, S. (Producer) (2001). *Trade secrets* [Documentary]. New York, NY: Public Affairs Television. Retrieved from http://www.pbs.org/ tradesecrets

MSNBC.com. (2010, June 8). I would have fired BP chief by now, Obama says. Retrieved from http://www.msnbc .msn.com

Mufson, S. (2008, January 18). Coal industry plugs into the campaign. *Washington Post*, p. D1. Retrieved from http://www .washingtonpost.com

Murray, D., Schwartz, J., & Lichter, S. (2001). *It ain't necessarily so, how media make and unmake the scientific picture of reality.* Lanham, MD: Rowman & Littlefield.

Myers, T. A., Nisbet, M. C., Maibach, E. W., & Leiserowitz, A. A. (2012). A public health frame arouses hopeful emotions about climate change. *Climate Change, 13*, 1105–1112.

Naess, A. (2000). Avalanches as social constructions. *Environmental Ethics, 22*(3), 335–336.

Namrouqa, H. (2012, December 21). More environmentalists using social media for activism. Retrieved from http://www .hispanicbusiness.com/2012/12/21/ more_environmentalists_using_social _media_for.htm

NASA. (n.d.). Global climate change. Retrieved from http://climate.nasa.gov/ effects/

Nash, R. (2001). *Wilderness and the American mind* (4th ed.). New Haven, CT: Yale University Press.

National Climate Assessment. (2014). U.S. Global Change Research Program.

Retrieved from http://nca2014.global-change.gov/highlights

National Environmental Justice Advisory Council Subcommittee on Waste and Facility Siting. (1996). *Environmental justice, urban revitalization, and brownfields: The search for authentic signs of hope* (Report Number EPA 500-R-96-002). Washington, DC: U.S. Environmental Protection Agency.

National Environmental Policy Act, 42 U.S.C.A. A7 4321 *et seq.* (1969).

National Institute of Environmental Health Sciences. (2003). *University of Wisconsin Milwaukee community outreach and education program.* Retrieved from http://www.apps.niehs.gov

National Research Council. (1996). *Understanding risk: Informing decisions in a democratic society.* Washington, DC: National Academy Press.

National Research Council Committee on Environmental Epidemiology. (1991). *Environmental epidemiology: Vol. 1. Public health and hazardous wastes.* Washington, DC: National Academy Press.

Nicodemus, R. (2012, April 2). Rwanda first country in the world to ban the plastic bag. *The Delicious Day: Live Happy. Explore Daily.* Retrieved from http://thedeliciousday.com/environment/rwanda-plastic-bag-ban/

Nils-Udo, B. (2002). *Towards nature.* (K. McVey, Trans.). Retrieved from http://greenmuseum.org/content/artist_content/ct_id-64__artist_id-36.html

Nocera, J. (2013, October 5). A fracking Rorschach test. *New York Times*, p. A19.

Nolo: Law for all. (2014). Public trust doctrine. Retrieved from http://www.nolo.com/dictionary/public-trust-doctrine-term.html

NRDC. (2014). Switchboard. "It's our air" ad. Retrieved from http://switchboard.nrdc.org/blogs/paltman/Ozone%20print%20ad.pdf

Obama revives panel on environmental justice. (2010, September 23). *USA Today*, p. 2A.

Office of Environmental Health Hazard Assessment. (2001). *A guide to health risk assessment.* California Environmental Protection Agency. Retrieved from http://oehha.ca.gov/pdf/HRSguide2001.pdf

Office of the Inspector General. (2004, March 1). *EPA needs to consistently implement the intent of the Executive Order on environmental justice.* Washington, DC: Environmental Protection Agency. Retrieved from http://www.epa.gov/oig/reports

OgilvyEarth. (2011). *Mainstream green: Moving sustainability from niche to normal.* Retrieved from http://www.ogilvyearth.com

Ohio: Geologists link earthquakes to gas drilling. (2014, April 12). *New York Times*, p. A14.

Olson, L., Finnegan, C., & Hope, D. S. (Eds.). (2008). *Visual rhetoric: A reader in communication and American culture.* Thousand Oaks, CA: Sage.

Only political activism and class struggle can save the planet. (2010). In I. Angus (Ed.), *The global fight for climate justice: Anticapitalist responses to global warming and environmental destruction* (pp. 182–186). Black Point, Nova Scotia, Canada: Fernwood Publishing.

O. R. (2013, January 21). Environmental activism and social media: Walking hand in hand. Retrieved from http://www.k2seo.com/environmental-activism-and-social-media-walking-hand-in-hand/

Oravec, C. (1981). John Muir, Yosemite, and the sublime response: A study in the rhetoric of preservationism. *Quarterly Journal of Speech, 67,* 245–258.

Oravec, C. L. (2004). Naming, interpretation, policy, and poetry. In S. L. Senecah (Ed.), *Environmental communication yearbook* (Vol. 1, pp. 1–14). Mahwah, NJ: Erlbaum.

Orr, D. W. (1992). *Ecological literacy: Education and the transition to a postmodern world.* Albany: State University of New York Press.

Orr, D. W. (1994). *Earth in mind: On education, environment, and the human prospect.* Washington, DC: Island Press.

Ostman, R. E., & Parker, J. L. (1987). Impacts of education, age, newspaper, and television on environmental knowledge, concerns, and behaviors. *Journal of Environmental Education, 19,* 3–9.

Ottman, J. A. (2003). *Hey, corporate America, it's time to think about products.* Retrieved from http://www.green marketing.com

Ottman, J. A. (2011). *The new rules of green marketing.* San Francisco, CA: Berrett-Koehler.

Pal, M., & Dutta, M. J. (2012). Organizing resistance on the Internet: The case of the international campaign for justice in Bhopal. *Communication, Culture & Critique, 5*(2), pp. 230–251.

Palen, J. (1998, August). *SEJ's origin.* Paper presented at the national convention of the Association for Education in Journalism and Mass Communications. Retrieved from http://www.sej.org/sejs-history

Palevsky, M. (2011, May 12). Yes Men hoax uses Twitter, Facebook to put Peabody Energy on the defensive. Retrieved from http://www.poynter.org

Park, C. C. (2001). *The environment: Principles and applications.* London, England: Routledge.

Parker, L., Johnson, K., & Locy, T. (2002, May 15). Post-9/11, government stingy with information. *USA Today,* p. 1A. Retrieved from http://www.usatoday.com/ news/nation

Parker-Pope, T. (2008, October 30). Panel faults F.D.A. on stance that chemical in plastic is safe. *New York Times,* p. A21.

Passary, A. (2014, January 30). Use of sonar in US naval exercise harming marine mammals, lawsuit claims. *Tech Times.* Retrieved from http://www.techtimes.com/articles/3047/20140130/use-of-sonar-in-us-naval-exercise-harming-marine-mammals-lawsuit-claims.htm

Peaceful Uprising. (n.d.). Tim's story. Retrieved from http://www.peacefuluprising.org/tim-dechristopher/tims-story

Peeples, J. (2013). Imaging toxins. *Environmental Communication: A Journal of Nature and Culture, 7*(2), 191–210.

Pender, G. (1993, June 1). Residents still not satisfied: Plant cleanup fails to ease Columbia fears. *Hattiesburg American,* 1.

People's Climate. (2014). http://peoples climate.org

People's Declaration for Climate Justice (Sumberklampok Declaration). (2007, December 7). Bali, Indonesia. Retrieved from http://peoples climatemovement .net

Pérez-Peña, R. (2014, April 1). College classes use art to brace for climate change. *New York Times,* pp. A12, A15.

Petermann, A., & Langelle, O. (2010). Crisis, challenge, and mass action. In I. Angus (Ed.), *The global fight for climate justice: Anticapitalist responses to global warming and environmental destruction* (pp. 186–195). Black Point, Nova Scotia, Canada: Fernwood Publishing.

Peterson, M. N., Peterson, M. J., & Peterson, T. R. (2007). Moving toward Sustainability: Integrating Social Practice and Material Processes. In R. Sandler & P. C. Pezzullo (Eds.), *Environmental justice and environmentalism: The social justice challenge to the environmental movement* (pp. 189–222). Cambridge, MA: MIT Press.

Peterson, T. R. (1997). *Sharing the earth: The rhetoric of sustainable development.* Columbia: University of South Carolina Press.

Pew Research Center. (2004). *The state of the news media 2004.* Retrieved from http://www.stateofthemedia.org/files/2011/01/execsum.pdf

Pezzullo, P. C. (2001). Performing critical interruptions: Rhetorical invention and narratives of the environmental justice movement. *Western Journal of Communication, 64,* 1–25.

Pezzullo, P. C. (2003). Touring "Cancer Alley," Louisiana: Performances of community and memory for environmental justice. *Text and Performance Quarterly, 23,* 226–252.

Pezzullo, P. C. (2004). Toxic tours: Communicating the "presence" of chemical contamination. In S. P. Depoe, J. W. Delicath, & M.-F. A. Elsenbeer (Eds.), *Communication and public participation in environmental decision making* (pp. 235–254). Albany: State University of New York Press.

Pezzullo, P. C. (2006). Articulating anti-toxic activism to "sexy" superstars: The cultural politics of *A Civil Action* and *Erin Brockovich. Environmental Communication Yearbook, 3,* 21-48.

Pezzullo, P. C. (2007). *Toxic tourism: Rhetorics of travel, pollution, and environmental justice.* Tuscaloosa: University of Alabama Press.

Pezzullo, P. C. (2011). Contextualizing boycotts and buycotts: The impure politics of consumer-based advocacy in an age of global ecological crises. *Communication and Critical/Cultural Studies, 8*(2), 124–145.

Pezzullo, P. C. (2014). Contaminated children: Debating the banality, precarity, and futurity of chemical safety. *Resilience: A Journal of the Environmental Humanities, 1*(2), 21 pp. http://www .jstor.org/table/10.5250/resilience.1.2.004

Pezzullo, P. C., & Sandler, R. (2007). Introduction: Revisiting the environmental justice challenge to environmentalism. In R. Sandler & P. C. Pezzullo (Eds.), *Environmental justice and environmentalism: The social justice challenge to the environmental movement* (pp. 1–24). Cambridge, MA: MIT Press.

Pirages, D. C., & Ehrlich, P. R. (1974). *Ark II: Social response to environmental imperatives.* San Francisco, CA: Freeman.

Plec, E., & Pettenger, M. (2012). Greenwashing consumption: The didactic framing of ExxonMobil's energy solutions.

Environmental Communication 6(4), 459–476.

Plough, A., & Krimsky, S. (1987). The emergence of risk communication studies: Social and political context. *Science, Technology, & Human Values, 12,* 4–10.

Policy Consensus Initiative. (2004). *State collaboration leads to successful wind farm siting.* Retrieved from http://www .policyconsensus.org

Pomerantz, D. (2012, August 1). Facebook means business in mission to "Unfriend Coal." *Greenpeace.* Retrieved from http://www.greenpeace.org/interna tional/en/news/Blogs/makingwaves/ facebook-means-business-in-mission -to-unfrien/blog/41619/

Pompper, D. (2004). At the 20th century's close: Framing the public policy issue of environmental risk. In S. L. Senecah (Ed.), *The environmental communication yearbook 1* (pp. 99–134). Mahwah, NJ: Erlbaum.

Pope, C. (2011, July/August). Bevies of black swans. *SIERRA,* p. 66.

Potomac Watershed Partnership. (2009). *Ecosystem management initiative.* University of Michigan. Retrieved from http://www.snre.umich.edu/ecomgt/ cases/potomac/index.htm

Price, J. (2014, April 4). Star of "Cosmos" faces the big questions. *News & Observer,* pp. 1A, 2A.

Priest, S. (2015). Mapping media's role in environmental thought and action. In A. Hanson & R. Cox (Eds.). *Routledge handbook of environment and communication.* London, England: Routledge.

Pring, G. W., & Canan, P. (1996). *SLAPPs: Getting sued for speaking out.* Philadelphia, PA: Temple University Press.

Proceedings: The first national people of color environmental leadership summit. (1991, October 24–27). Washington, DC: United Church of Christ Commission for Racial Justice.

Quiñones, M. (2014, April 25). Federal judge finds coal company liable for

W.V. selenium dischares. *E&E News.* Retrieved from http://www.eenews .net/greenwire/stories/1059998482/

Raffensperger, C. (1998). Editor's note: The precautionary principle—a fact sheet. *The Networker, 3*(1), para. 1. Retrieved from http://www.sehn.org

Rampton, S. (2002). Sludge, biosolids and the propaganda model of communication. *New Solutions, 12*(4), 347–353. Retrieved from http://www.sludgenews.org

Rampton, S., & Stauber, J. (2002). *Trust us, we're experts!* New York, NY: Jeremy P. Tarcher/Putnam.

Reckelhoff-Dangel, C., & Petersen, D. (2007, August). *Risk communication in action: The risk communication workbook.* Environmental Protection Agency. Cincinnati, OH: Office of Research and Development, National Risk Management Research Laboratory. Retrieved from http://www.epa.gov

Red Lodge Clearinghouse. (2008, April 11). *Quincy Library Group.* A project of the Natural Resources Law Center at the University of Colorado Law School. Retrieved from http://rlch.org

Reich, M. R. (1991). *Toxic politics: Responding to chemical disasters.* Ithaca, NY: Cornell University Press.

Report of the United Nations Secretary-General. (2013). *Intergenerational solidarity and the needs of future generations.* III.A.36. Retrieved from http://sustain abledevelopment.un.org/content/ documents/2006future.pdf

Review. (2003, June 5). Subhankar Banarjee, *Arctic National Wildlife Refuge: Seasons of life and land. Planet.* Retrieved July 17, 2004, from www.mountaineersbooks .org

Revkin, A. (2005, June 8). Bush aide softened greenhouse gas links to global warming. *New York Times.* Retrieved from http://www.nytimes.com

Revkin, A. C. (2006a, January 29). Climate expert says NASA tried to silence him. *New York Times,* p. A1.

Revkin, A. C. (2006b, February 8). A young Bush appointee resigns his post at NASA. *New York Times,* p. A11.

Revkin, A. C. (2007, April 3). The climate divide: Reports from four fronts in the war on warming. *New York Times.* Retrieved from http://www.nytimes.com

Rio declaration on environment and development. (1992). United Nations Conference on Environment and Development. Rio de Janeiro. Retrieved from http://www.unep.org

Rising Tide North America. (2008, December 1). *Climate activists invade DC offices of environmental defense.* Retrieved from http://www.risingtide northamerica.org

Robbins, D. (2014, April 14). *Years of living dangerously:* Is this the new trend? Retrieved from http://mediamatters .org/blog/2014/04/14/years-of-living -dangerously-is-this-the-new-tre/198878

Roberts, J. T. (2007). Globalizing environmental justice. In R. Sandler & P. C. Pezzullo (Eds.), *Environmental justice and environmentalism: The social justice challenge to the environmental movement* (pp. 285–307). Cambridge, MA: MIT Press.

Roberts, M. (2011, January 18). Wal-Mart environmentalism. Retrieved from http://pennpoliticalreview.org

Rogers, E. M., & Storey, J. D. (1987). Communication campaigns. In C. R. Berger & S. H. Chaffee (Eds.), *Handbook of communication science* (pp. 817–846). Newbury Park, CA: Sage.

Rogers, R. A. (1998). Overcoming the objectification of nature in constituive theories: Toward a transhuman, materialist theory of communication. *Western Journal of Communication, 62,* 244–272.

Roosevelt, M. (2011, June 10). Pressured by Greenpeace, Mattel cuts off sub-supplier APP. *Los Angeles Times.* Retrieved from http://articles.latimes.com

Ropeik, D. (2010). *How risky is it, really? Why our fears don't always match the facts.* Columbus, OH: McGraw-Hill.

Roser, H. (2006, February 10). Great Bear compromise set stage for wilderness future. *Newwest*. Retrieved from http://newwest.net/main/article/great_bear_compromise_sets_stage_for_wilderness_future/

Roser-Renouf, C., Stenhouse, N., Rolfe-Redding, J. Maibach, E., & Leiserowitz, A. (2014). *Engaging diverse audiences with climate change: Message strategies for global warming's six Americas*. Center for Climate Change Communication. Retrieved from http://www.climatechangecommunication.org/report/engaging-diverse-audiences-climate-change-message-strategies-global-warmings-six-americas

Ross, A. (1994). *The Chicago gangster theory of life: Nature's debt to society*. London, England: Verso.

Rudolf, J. C. (2010, May 3). Social media and the spill. *New York Times'* Green Blog. Retrieved from http://green.blogs.nytimes.com

Russill, C. (2008). Tipping point forewarnings in climate change communication: Some implications of an emerging trend. *Environmental Communication: A Journal of Nature and Culture, 2,* 133–153.

Rykiel, E. J., Jr. (2001). Scientific objectivity, value systems, and policymaking. *BioScience, 51,* 433–436.

Sachsman, D., & Valenti, J. M. (2015). Environmental reporters. In A. Hansen & R. Cox (Eds.). *Routledge handbook of environment and communication*. London, England: Routledge.

Sachsman, D. B., Simon, J., & Valenti, J. (2002, June). The environment reporters of New England. *Science Communication, 23,* 410–441.

Sacred Land Film Project. (2003). *Zuni Salt Lake*. Retrieved from http://www.sacredland.org/zuni_salt_lake

Sale, K. (1993). *The green revolution: The American environmental movement 1962–1992*. New York, NY: Hill & Wang.

Salmon, C.T., Post, L.A., & Christensen, R. E. (2003). *Mobilizing public will for social change*. Lansing: The Communications Consortium Media Center, Michigan State University.

Sandler, R., & Pezzullo, P. C. (Eds.). (2007). *Environmental justice and environmentalism: The social justice challenge to the environmental movement*. Cambridge, MA: MIT Press.

Sandman, P. (1987). Risk communication: Facing public outrage. *EPA Journal, 13*(9), 21–22.

Sandman, P. R. (2011). *Peter M. Sandman risk communication website*. Retrieved from http://www.psandman.com

Sandweiss, S. (1998). The social construction of environmental justice. In D. E. Camacho (Ed.), *Environmental injustices, political struggles* (pp. 31–58). Durham, NC: Duke University Press.

Saño, N. Y. (2013, November 13). It's time to stop this madness. Retrieved from http://www.rtcc.org/2013/11/11/its-time-to-stop-this-madness-philippines-plea-at-un-climate-talks/

Santos, S. L., & Chess, C. (2003). Evaluating citizen advisory boards: The importance of theory and participant-based criteria and practical implications. *Risk Analysis, 23,* 269–279.

Sarkar, B., & Walker, J. (2010). *Documentary testimonies: Global archives of suffering*. New York, NY: Routledge.

Schlechtweg, H. P. (1992). Framing Earth First! The *MacNeil-Lehrer NewsHour* and redwood summer. In C. L. Oravec & J. G. Cantrill (Eds.), *The conference on the discourse of environmental advocacy* (pp. 262–287). Salt Lake City: University of Utah Humanities Center.

Schmidt, A., Ivanova, A., & Schaefer, M. S. (2013). Media attention for climate change around the world: A comparative analysis of newspaper coverage in 27 countries. *Global Environmental Change-Human and Policy Dimensions, 23*(5), 1233–1248.

Schneider, S. H. (2009). *Science as a contact sport: Inside the battle to save earth's climate.* Washington, DC: National Geographic.

Schor, E. (2011, February 1). Enviro groups' public health pivot in support of EPA regs hitting red states too. *New York Times.* Retrieved from http://www.nytimes.com/

Schor, E. (2013, October 8). Celeb-studded documentary uses scientists, everyday Americans to tell less inconvenient truths. *Greenwire.* Retrieved from http://www.eenews.net/greenwire/stories/1059988528/

Schueler, D. (1992). Southern exposure. *Sierra, 77,* 45–47.

Schultz, P. W., & Zelezny, L. (2003). Reframing environmental messages to be congruent with American values. *Research in Human Ecology, 10,* 126–136.

Schwab, J. (1994). *Deeper shades of green: The rise of blue-collar and minority environmentalism in America.* San Francisco, CA: Sierra Club Books.

Schwarze, S. (2006). Environmental melodrama. *Quarterly Journal of Speech, 92*(3), 239–261.

Science and Environmental Health Network. (1998, January 26). *Wingspread conference on the precautionary principle.* Retrieved from http://www.sehn.org

Seciwa, C. (2003, August 5). *Zuni Salt Lake and sanctuary zone protected for future generations.* [News release]. Zuni Pueblo, NM: Zuni Salt Lake Coalition.

Segerberg, A., & Bennett, W. L. (2011). Social media and the organization of collective action: Using Twitter to explore the ecologies of two climate change protests. *Communication Review, 14*(3), 197–215.

Senecah, S. L. (2004). The trinity of voice: The role of practical theory in planning and evaluating the effectiveness of environmental participatory processes. In S. P. Depoe, J. W. Delicath, & M.-F. A. Elsenbeer (Eds.), *Communication and public participation in environmental decision making* (pp. 13–33). Albany: State University of New York Press.

Shabecoff, P. (2000). *Earth rising: American environmentalism in the 21st century.* Washington, DC: Island Press.

Shabecoff, P. (2003). *A fierce green fire: The American environmental movement* (Rev. ed.). Washington, DC: Island Press.

Shanahan, J. (1993). Television and the cultivation of environmental concern: 1988–92. In A. Hansen (Ed.), *The mass media and environmental issues* (pp. 181–197). Leicester, UK: Leicester University Press.

Shanahan, J., & McComas, K. (1999). *Nature stories: Depictions of the environment and their effects.* Cresskill, NJ: Hampton Press.

Shanahan, J., McComas, K, & Deline, M. B. (2015). Representations of the environment on television and their effects. In A. Hansen & R. Cox (Eds.). *Routledge handbook of environment and communication.* London, England: *Routledge.*

Shannon, C., & Weaver, W. (1949). *The mathematical theory of communication.* Urbana: University of Illinois Press.

Sheppard, K. (2013, November 13). Walmart's sustainability results don't match promises, report finds. *Huffington Post.* Retrieved from http://www.huffingtonpost.com/2013/11/13/walmart-sustainability_n_4263032.html

Shipp, B. (2014, June 12). Hazardous chemical lists no longer public record in Texas. *ABC News 8.* Retrieved from http://www.wfaa.com/news/texas-news/Exclusive-Hazardous-chemical-lists-no-longer-public-record-in-Texas-262943831.html

Shiva, V. (2008). From water crisis to water culture: Dr. Vandana Shiva, an interview by Andy Opel. *Cultural Studies 22*(3–4), 498–509.

Sierra Club. (2014). Beyond coal. Retrieved from http://content.sierraclub.org/coal/about-the-campaign

Sierra Club v. Morton, 405 U.S. 727. (1972). Justia US Supreme Court. Retrieved from http://supreme.justia.com/cases/federal/us/405/727/case.html

Singhal, A. (2001, September). *Facilitating community participation through communication.* Retrieved from http://utminers.utep.edu/asinghal/reports/Singhal-UNICEF-Participation-Report.pdf

Slobodkin, L. B. (2000). Proclaiming a new ecological discipline. *Bulletin of the Ecological Society of America, 81,* 223–226.

Slovic, P. (1987). Perceptions of risk. *Science, 236,* 280–285.

Smerecnik, K. R., & Renegar, V. R. (2010). Capitalistic agency: The rhetoric of BP's helios power campaign. *Environmental Communication 4*(2), 152–171.

Smith, M., Sterritt, A., & Armstrong, P. (2007, May 14). From conflict to collaboration: The story of the Great Bear Rainforest. Retrieved from http://www.forestethics.org/

Smith, T. M. (1998). *The myth of green marketing: Tending our goats at the edge of apocalypse.* Toronto, Ontario, Canada: University of Toronto Press.

Soraghan, M. (2014, April 14). Ohio makes strongest link yet between shaking and fracking. *E&E News.* Retrieved from http://www.eenews.net/energywire/2014/04/14/stories/1059997836

Soroka, S. N. (2002). Issue attributes and agenda-setting by media, the public, and policy-makers in Canada. *International Journal of Public Opinion Research, 14*(3), 264–285.

Sorry. (n.d.). 10:10 UK. Posted by franny. Available at http://www.1010global.org/uk/2010/10/sorry

Soulé, M. E. (Ed.). (1985). What is conservation biology? *BioScience, 35,* 727–734.

Soulé, M. E. (1986). *Conservation biology: The science of scarcity and diversity.* Sunderland, MA: Sinauer Associates.

Soulé, M. E. (1987). History of the Society for Conservation Biology: How and why we got here. *Conservation Biology, 1,* 4–5.

SourceWatch. (2009, January 5). *American Coalition for Clean Coal Electricity. (ACCCE).* Retrieved from http://www.sourcewatch.org.

SourceWatch. (2013, September 27). Retrieved from http://www.sourcewatch.org/index.php/Fracking_chemicals

SourceWatch. (2014, March 21). Fracking. Retrieved from http://www.sourcewatch.org/index.php?title=Fracking

SouthWest Organizing Project. (1990, March 16). Letter to the "Group of Ten" national environmental organizations. Albuquerque, NM. Retrieved from http://soa.utexas.edu

Spiess, B., & Ruskin, L. (2001, November 4). 2,000-acre query: ANWR bill provision caps development, but what does it mean? *Anchorage Daily News.* Retrieved from http://www.adn.com

Spread the word. (2014). *The Meatrix.* Retrieved from http://www.themeatrix.com/spread

Stearns, M. L. (2000, March 2–3). *From Lujan to Laidlaw: A preliminary model of environmental standing.* Paper presented at the Duke University Law and Policy Forum Symposium on Citizen Suits and the Future of Standing in the 21st Century. Retrieved from http://www.law.duke.edu/journals

Stein, R. (2010, October 28). Study: BPA has effect on sperm. *Washington Post,* p. A16.

Stephan, M. (2002). Environmental information disclosure programs: They work, but why? *Social Science Quarterly, 83*(1), 190–205.

Stephenson, W. (2012, December 13) "I'd rather fight like hell": Naomi Klein's fierce new resolve to fight for climate justice. *Commondreams.org.* Phoenix, AZ. Retrieved from http://www.commondreams.org/headline/2012/12/13

Stern, P., Dietz, T., Abel, T., Guagnano, G., & Kalof, L. (1999). A value-belief-norm theory of support for social movements: The case of environmentalism. *Human Ecology Review, 6*(2), 81–97.

Steward, R. B., & Krier, J. (Eds.). (1978). *Environmental law and public policy.* New York, NY: Bobbs-Merrill.

Stine, R. (2011, August 5). Social media and environmental campaigning: Brand lessons from Barbie. Retrieved from http://www.ethicalcorp.com

Stone, C. (1996). *Should trees have standing? And other essays on law, morals and the environment* (Rev. ed.). Dobbs Ferry, NY: Oceana.

Stone, D. (2002). *Policy paradox: The art of political decision making* (Rev. ed.). New York, NY: Norton.

Strom, S. (2014, September 4). Antibiotics eliminated in hatchery, Perdue says. *New York Times*, p. B3.

Stromberg, J. (2013, January). *What is the Anthropocene and are we in it? Smithsonian Magazine.* Retrieved from http://www .smithsonianmag.com/science-nature/ What-is-the-Anthropocene-and-Are-We -in-It-183828201.html#ixzz2mze5BiJx

Sullivan, M. (2013, November 24). After changes, how green is the *Times?* Sunday review. *New York Times*, p. 1.

Sunshine and shadows: The national security archive FOIA audit. (2010, March 15). Retrieved from http://www.gwu.edu

Sunstein, C. R. (2004). *Risk and reason: Safety, law, and the environment.* Cambridge, UK: Cambridge University Press.

Sustainable Life Media. (2008, June 26). Canada bans "green" and "eco-friendly" from product labels. Retrieved from http://www.sustainablelife media.com

Switzer, J. V. (1997). *Green backlash: The history and politics of environmental opposition in the U.S.* Boulder, CO: Lynne Rienner.

Synopsis. (2014). *Chasing ice.* Retrieved from http://www.chasingice.com/about-the -film/synopsis

Szabo, L. (2011, January 14). Pregnant women rife with chemicals. *USA Today*, p. 3A.

Sze, J. (2011) Asian American, immigrant and refugee environmental justice activism under neoliberal urbanism. *Asian American Law Journal, 18*, 5–23.

Tabuchi, H., Sanger, D. E., & Bradsher, K. (2011, March 14). Japan faces potential nuclear disaster as radiation levels rise. *New York Times,* p. A1.

Taleb, N. N. (2010). *The black swan.* (2nd ed.). London, England: Penguin.

Tatum, B. D. (2013). The journey to green: Becoming sustainable spelman. In P. F. Bartlett & G. W. Chase (Eds.), *Sustainability in higher education: Stories and strategies for transformation* (pp. 153-162). Cambridge, MA: MIT Press.

Tavernise, S. (2013, December 12). FDA restricts antibiotics use for livestock. *New York Times*, A1. Retrieved from http://www.nytimes.com/2013/12/12/ health/fda-to-phase-out-use-of-some -antibiotics-in-animals-raised-for-meat .html?pagewanted=all&_r=1&

Taylor, K., & Kaplan, T. (2014, July 1). Towns in New York can prohibit fracking, state's highest court rules. *New York Times*, p. A15.

Terhune, P., & Terhune, G. (1998, October 8-10). *QLG case study.* Prepared for engaging, empowering, and negotiating community: Strategies for conservation and development workshop. Sponsored by the Conservation and Development Forum, West Virginia University, and the Center for Economic Options. Retrieved from http://www.qlg.org/pub

TerraChoice Environmental Marketing Inc. (2010). The sins of greenwashing: Home and family edition. Retrieved from http://sinsofgreenwashing.org/ findings/the-seven-sins/

Thigpen, K. G., & Petering, D. (2004, September). Fish tales to ensure health. *Environmental Health Perspectives, 112*(13), p. A738.

Thomas, I. (2001, March 16). Web censorship. Retrieved from http://cartome.org

Thomashow, M. (2014). *The nine elements of a sustainable campus.* Cambridge, MA: MIT Press.

Thoreau, H. D. (1893/1932). Walking. In *Excursions: The writings of Henry David Thoreau* (Riverside ed., Vol. 9, pp. 251–304). Boston, MA: Houghton Mifflin.

Tindall, D. B. (1995). What is environmental sociology? An inquiry into the paradigmatic status of environmental sociology. In M. D. Mehta & E. Ouellet (Eds.), *Environmental sociology: Theory and practice* (pp. 33–59). North York, Ontario, Canada: Captus Press.

Todd, A. M. (2002). Prime-time subversion: The environmental rhetoric of *The Simpsons*. In M. Meister & P. Japp (Eds.), *EnviroPop: Studies in environmental rhetoric* (pp. 63–80). Westport, CT: Greenwood Press.

Todd, A. M. (2010). Anthropocentric distance in *National Geographic's* environmental aesethetic. *Environmental Communication: A Journal of Nature and Culture, 4*(2), 206–224.

Tokar, B. (2010). *Toward climate justice: Perspectives on the climate crisis and social change.* Porsgrunn, Norway: Communalism Press.

Torgerson, D. (1999). *The promise of green politics: Environmentalism and the public sphere.* Durham, NC: Duke University Press.

Top U.S. scientists warn Congress on the dangers of climate change. (2010, May 20). *PhysicsWorld.com.* Retrieved from http://physicsworld.com/

Topalian, N. (2013). *"Ana Ma Bkeb" campaign promotes street cleanliness in Lebanon.* Retrieved from http://al-shorfa.com/en_GB/articles/meii/features/2013/06/21/feature-02

Udell, C. (2007, May 19). Dirty air harms our most vulnerable population—children [Op-Ed]. *Salt Lake Tribune.* Reprinted by Utah Moms for Clean Air. Retrieved from http://blog.utahmomsforcleanair.org/press/

Union of Concerned Scientists. (2004, February). *Scientific integrity in policymaking: An investigation into the Bush administration's misuse of science.* Cambridge, MA: Author.

Union of Concerned Scientists. (2008). *Freedom to speak? A report card on federal agency media policies.* Retrieved from http://www.ucsusa.org

Union of Concerned Scientists. (2009). Environmental impacts of coal power: Air pollution. Retrieved from http://www.ucsusa.org/

Union of Concerned Scientists and Penguin Classic Books. Thoreau's legacy: American stories about global warming (Richard Hayes, Ed.). New York, NY: Penguin. 2009. Retrieved from http://www.ucsusa.org/americanstories/project-authors.html

United Nations. (2011). International year of forests—2011. Retrieved from http://www.un.org/en

United Nations Economic Commission for Europe. (2008, March 12). *Text of the [Aarhus] convention.* Retrieved from http://www.unece.org

United Nations Framework Convention on Climate Change (1992). Retrieved from http://unfccc.int/resource/docs/convkp/conveng.pdf

United Nations Framework Convention on Climate Change. (1994). Article 2 Objective. Retrieved from http://unfccc.int/essential_background/convention/background/items/1349.php

United States Environmental Protection Agency. (2007, August). *Risk communication in action: The tools of message mapping.* Retrieved from http://www.epa.gov/

United States Environmental Protection Agency. (2009). *Region 9 progress report: People.* Retrieved from http://www.epa.gov/region9/annualreport/09/communities.html

United States Environmental Protection Agency. (2010a, March). *Scoping materials for initial design of EPA research study on potential relationships between hydraulic fracturing and drinking water resources.* Retrieved from http://yosemite.epa.gov

United States Environmental Protection Agency. (2010b, August). *Risk assessment: Basic information.* Retrieved from http://epa.gov

United States Environmental Protection Agency. (2010c, October 1). *Mercury: Health effects.* Retrieved from http://www.epa.gov

United States Environmental Protection Agency. (2010d, October 7). *What is the Toxic Release Inventory (TRI) program?* Retrieved from http://www.epa.gov

United States Environmental Protection Agency. (2010e, November 11). *EPA releases reports on dioxin emitted during Deepwater Horizon BP spill.* Retrieved from http://www.epa.gov/

United States Food and Drug Administration. (2010, August 6). *Gulf of Mexico oil spill: Questions and answers.* Retrieved from http://www.fda.gov/

United States Food and Drug Administration & United States Environmental Protection Agency. (2004). *What you need to know about mercury in fish and shellfish.* Retrieved from http://www.epa.gov/

UPI. (2013, April 16). Wolf numbers down after U.S. federal species protections removed. Retrieved from http://www.upi.com/Science_News/2013/04/16/Wolf-numbers-down-after-US-federal-species-protections-removed/UPI-32131366156622/

U.S. Cancer Institute issues stark warning on environmental cancer risk. (2010, May 17). *Ecologist.* Retrieved from http://www.theecologist.org/

U.S. General Accounting Office. (1983). *Sitting of hazardous waste landfills and their correlation with racial and economic status of surrounding communities.* Washington, DC: U.S. General Accounting Office.

Valtin, T. (2003, November). Zuni Salt Lake saved. *Planet: Sierra Club Activist Resource, 1.*

Van Lopik, W. (2013). Learning sustainability in a tribal college context. In P. F. Bartlett & G. W. Chase (Eds.), *Sustainability in higher education: Stories and strategies for transformation* (pp. 105–114). Cambridge, MA: MIT Press.

Vanhemert, K. (2013, October 7). *13 of the year's best infographics.* Retrieved from http://www.wired.com/design/2013/10/13-sterling-pieces-of-data-viz-from-the-best-american-infographic-2013/#slideid-257861

Van Tuyn, P. (2000). "Who do you think you are?" Tales from the trenches of the environmental standing battle. *Environmental Law, 30*(1), 41–49.

Vatz, R. E. (1973). The myth of the rhetorical situation. *Philosophy and Rhetoric, 6,* 154–161.

Veil, R. S., Buehner, T., & Palenchar, J. M. (2011). A work-in-progress literature review: Incorporating social media in risk and crisis communication. *Journal of Contingencies and Crisis Management, 19*(2), 110–122.

Victory and new threats at Zuni Salt Lake, New Mexico. (2003, Winter). *The Citizen, 6.*

Vidal, J. (2013, November 21). Inaction on climate change will increase civil unrest, warn leading groups. *Guardian* [UK]. Retrieved from http://www.theguardian.com/global-development/2013/nov/21/climate-change-increase-civil-unrest-warning

Visser, N. (2014, September 21). Hundreds of thousands turn out for People's Climate March in New York City. *HuffPost Green.* Retrieved from http://www.huffingtonpost.com/2014/09/21/peoples-climate-march_n_5857902.html

Visual.ly. (2013). *Rhinos in danger: Just when you thought that things could not get worse.* Retrieved from http://visual.ly/rhinos-danger-20

Vitello, P. (2008, September 24). Gore's call to action. *New York Times.* Retrieved from http://thecaucus.blogs.nytimes.com/2008/09/24/gores-call-to-action/?_php=true&_type=blogs&_r=0

Vucetich, J. A., & Nelson, M. P. (2010, August 1). The moral obligations of scientists. *Chronicle of Higher Education.* Retrieved from http://chronicle.com

Walker, G. B. (2004). The roadless area initiative as national policy: Is public participation an oxymoron? In S. P. Depoe, J. W. Delicath, & M-F. A. Elsenbeer (Eds.), *Communication and public participation in environmental decision making* (pp. 113–135). Albany: State University of New York Press.

Walker, G. B. (2007). Communication in environmental policy decision-making: From concepts to structured conversations. *Environmental Communication: A Journal of Nature and Culture, 1*(1), 99–110.

Walker, G. B., Daniels, S. E., & Emborg, J. (2015). Public participation in environmental policy decision-making. In A. Hansen. & R. Cox (Eds.). *Routledge handbook of environment and communication.* London, England: Routledge.

Walmart. (2011, August). Sustainability facts. Retrieved from http://walmrt stores.com

Walsh, B. (2010, June 25). Assessing the health effects of the oil spill. *TIME.com.* Retrieved from http://www.time.com/

Ward, B. (2008, December 18). 2008's year-long fall-off in climate coverage. *The Yale Forum on Climate Change and the Media.* Retrieved from http://www .yaleclimatemediaforum.org

Weiss, E. B. (1989*). In fairness to future generations: International law, common patrimony, and intergenerational equity.* Ardsley, NY: Transnational.

Western Environmental Law Center. (2014, June 11). Youth win reversal in critical climate recovery case. [Press release]. Retrieved from http://www.westernlaw .org/article/youth-win-reversal-critical-cli mate-recovery-case-press-release-61114

We will make this right. (2010, June 1). [Advertisement]. *New York Times,* p. A11.

White, D. M. (1950). The "gatekeeper": A case study in the selection of news. *Journalism Quarterly, 27*(4), 383–390.

White, H. L. (1998). Race, class, and environmental hazards. In D. E. Camacho (Ed.), *Environmental injustices, political struggle: Race, class, and the environment.* Durham, NC: Duke University Press.

White House. (2009, January 21). Memorandum for the heads of executive departments and agencies. Press Office. Retrieved from http://www.whitehouse .gov/

White House (2014, September 23). Remarks by the President at U.N. Climate Change Summit, United Nations Headquarters, New York, New York. Retrieved from http://www.whitehouse.gov/the-press-office/2014/09/23/remarks-president-un-climate-change-summit

The Wilderness Act of 1964, 16 U.S.C. §§ 1131-1136 (1964).

Williams, B. A., & Matheny, A. R. (1995). *Democracy, dialogue, and environmental disputes.* New Haven, CT: Yale University Press.

Williams, C. (2012, October 5). Should nature have a legal standing? *Mother Nature Network.* Retrieved from http:// www.mnn.com/earth-matters/wilder ness-resources/stories/should-nature -have-a-legal-standing

Williams, D. E., & Olaniran, B. A. (1994). Exxon's decision-making flaws: The hypervigilant response to the *Valdez* grounding. *Public Relations Review, 20,* 5–18.

Williams, R. (1980). *Problems in materialism and culture.* London, England: Verso.

Wilmoth, A. (2014, April 12). Fracking and horizontal drilling work best in the U.S. *The Oklahoman.* Retrieved from http://newsok.com/fracking-and-hori

zontal-drilling-work-best-in-the-u.s./article/3954027

Wilson, E. O. (2002). *The future of life*. New York, NY: Knopf.

Wolfe, D. (2008). The ecological jeremiad, the American myth, and the vivid force of color in Dr. Seuss's *The Lorax*. *Environmental Communication: A Journal of Nature and Culture, 2*, 3–24.

Wolfe, D. (2009). The video rhizome: Taking technology seriously in *The Meatrix*. *Environmental Communication: A Journal of Nature and Culture, 3*(3), pp. 317–334.

Wondolleck, J. M., & Yaffee, S. L. (2000). *Making collaboration work: Lessons from innovation in natural resource management*. Washington, DC: Island Press.

Wood, J. T. (2009). *Communication in our lives* (5th ed.). Boston, MA: Wadsworth Cengage Learning.

World Resources Institute. (n.d.). Aqueduct: Measuring and mapping water risk. Retrieved from http://www.wri.org/our-work/project/aqueduct

WWF. (2011, February 3). The energy report—100% renewable energy by 2050. Retrieved from http://www.wwf.org.uk/wwf_articles.cfm?unewsid=4565

Wyss, B. (2008). *Covering the environment: How journalists work the green beat*. London, England, & New York, NY: Routledge.

Yaccino, S. (2012, November 1). As wolves' numbers rise, so does friction between guardians and hunters. *New York Times*. Retrieved from http://www.nytimes.com/2012/11/02/us/friction-between-wolf-hunters-and-protectors-rises.html?_r=0

Yale Forum on Climate Change and the Media. (2011, February). Science societies' annual meeting agendas focusing increasingly on communications issues. Retrieved from http://www.yaleclimatemediaforum.org/

Yale Project on Climate Change Communication and George Mason University Center for Climate Change Communication. (2013, April). *How Americans communicate about global warming*. Report. Retrieved from http://environment.yale.edu/climate-communication/files/Communication-April-2013.pdf

Yardley, W., & Pérez-Pena, R. (2009, March 16). Seattle paper shifts entirely to the web. *New York Times*. Retrieved from http://www.nytimes.com

Yopp, J. J., McAdams, K. C., & Thornburg, R. M. (2009). *Reaching audiences: A guide to media writing* (5th ed.). Boston, MA: Allyn & Bacon.

Zuni Salt Lake Coalition. (2001, October 6–7). [Zuni Salt Lake Coalition's campaign plan: Edward's kitchen. Notes from first meeting of coalition members]. Unpublished raw data.

Zuni Salt Lake Coalition. (2003). Background. Retrieved from http://www.zunisaltlakecoalition.org/

Index

Note: n in locator refers to endnote.